Universitext

Universitext is a series of textbooks that presents material from a wide variety of mathematical disciplines at master's level and beyond. The books, often well class-tested by their author, may have an informal, personal even experimental approach to their subject matter. Some of the most successful and established books in the series have evolved through several editions, always following the evolution of teaching curricula, to very polished texts.

Thus as research topics trickle down into graduate-level teaching, first textbooks written for new, cutting-edge courses may make their way into *Universitext*.

More information about this series at http://www.springer.com/series/223

Universitext

Universitext is a series of textbooks that presents material from a wide variety of mathematical disciplines at master's level and beyond. The books, often well class-tested by their author, may have an informal, personal even experimental approach to their subject matter. Some of the most successful and established books in the series have evolved through several editions, always following the evolution of teaching curricula, to very polished texts.

Thus as research topics trickle down into graduate-level teaching, first textbooks written for new, cutting-edge courses may make their way into *Universitext.*

More information about this series at http://www.springer.com/series/223

Norbert Steinmetz

Nevanlinna Theory, Normal Families, and Algebraic Differential Equations

Springer

Norbert Steinmetz
Fakultät für Mathematik
TU Dortmund
Dortmund
Germany

ISSN 0172-5939 ISSN 2191-6675 (electronic)
Universitext
ISBN 978-3-319-59799-7 ISBN 978-3-319-59800-0 (eBook)
DOI 10.1007/978-3-319-59800-0

Library of Congress Control Number: 2017942798

Printed on acid-free paper

This Springer imprint is published by Springer Nature
The registered company is Springer International Publishing AG
The registered company address is: Gewerbestrasse 11, 6330 Cham, Switzerland

Rolf Nevanlinna (1895–1980) Hans Wittich (1911–1984)

Photo: Privately owned

Introduction and Preface

Nevanlinna Theory, Normal Families, and Algebraic Differential Equations—how are these topics related to each other?

Zalcman's Re-scaling method set up a way to combine Nevanlinna Theory and Normal Families in both directions—to prove qualitative ('soft') results in Nevanlinna Theory by using Normal Family methods, and to prove normality criteria using results from Nevanlinna Theory as a pattern. Part of Schiff's *Normal Families* is dedicated to this interesting and fruitful connection. In the present Chap. 4, some old and new results in this direction are presented. Moreover, the connection of Normal Families with Algebraic Differential Equations is discussed on an elementary level.

In his seminal paper on the value distribution of meromorphic functions and their derivatives, Hayman opened a new field of application of Nevanlinna Theory. Some of his results on differential polynomials are outlined in his indispensable *Meromorphic functions*. He initiated a vast field of research in the 1970s and 80s, which to the author's knowledge has never appeared in book form. Nevanlinna himself was the first to apply his theory to problems of uniqueness of meromorphic functions, known as the Five- and Four-Value Theorems. The enormous progress initiated by Gundersen in the 1980s and early 90s has also never been presented in book form. In combination with introductory applications to Algebraic Differential Equations, these topics constitute Chap. 3.

The benefits of applying Nevanlinna Theory to the field of Algebraic Differential Equations were first recognised in the early 1930s, and then systematically since the 50s due to the pioneering work of Wittich. Apart from single chapters and a few remarks in the books [15, 84, 85, 96, 202], just Laine's monograph *Nevanlinna Theory and Complex Differential Equations* is dedicated to this field. Since the beginning of the new century much progress has been made in the context of Painlevé Differential Equations.[1] For the first time, Re-scaling and Normal Family

[1] We quote from the introduction to L.A. Rubel's *Entire and meromorphic functions* [146], where the author expressed the need to have more examples of interesting meromorphic functions. *One*

arguments were used to gain new insight into the nature of the so-called Painlevé transcendents and solutions to other algebraic differential equations. Inspired by Zalcman's Re-scaling Lemma and older ideas due to Yosida, the so-called Yosida classes entered the stage quite naturally. This material is presented in Chaps. 5 and 6, and in parts also in Chap. 4. The monograph [60] is concerned with Painlevé's Equations from the complex analytic point of view, but was published 'too early' to incorporate the newest developments, methods, and results. This, of course, may happen to every book. Regrettably, the theory of Differential Equations in the Complex Domain is not commonly accepted as a genuine part of Complex Analysis, though it is completely based on complex analytic techniques. Chapters 2, 3, and 4 urgently demonstrate that the theory of Algebraic Differential Equations provides indispensable tools, and their solutions often mark the range of validity of important results in Nevanlinna Theory.

Chapter 1 is included for the convenience of the reader. It provides material from classical Complex Analysis, which—apparently or seemingly—does not belong to the generally accepted background. This is particularly true for the topics of Ordinary Differential Equations and Asymptotic Expansions.

In Chap. 2 not only the classical Nevanlinna Theory is outlined, but also Cartan's Theory of Entire Curves and the Selberg–Valiron Theory of Algebroid Functions is briefly presented. This includes generalisations of the Second Main Theorem and various applications to problems in Complex Analysis which then become 'elementary',

The text is not written redundancy-free. Some of the problems are dealt with at an early stage using the methods at hand, and are picked up later on when new tools are available. Several examples and exercises require extensive computations, which in principle can be realised by hand. It is, however, much more convenient to use some computer algebra system like MAPLE. Non-experts like the author can use MAPLE like a separate sheet of paper to carry out auxiliary computations.

The present text was written within the first 2 years of my ultimate sabbatical, but has a long history. It developed from research in Nevanlinna Theory, Analytic Differential Equations, and related subfields of Complex Analysis starting in the late 1970s, with a break in the 90s, which were occupied by other activities. The choice of the material naturally depends on personal preferences, experiences, and skills. Rather than to aim at completeness of the presentation, I intended to explain the main ideas and results exemplarily.

I appreciate the support and criticism by friends and colleagues. Of course, the responsibility for mistakes and misinterpretations remains with me. I also profited very much from the survey by A. Eremenko and J. Langley within the English translation of the monograph *Value Distribution of Meromorphic Functions* by A.A. Gol'dberg and I.V. Ostrovskii, and, in particular, from helpful comments by the

promising source of such examples is the Painlevé transcendents. However, in spite of a growing literature on these functions, the unfortunate fact is that the "proofs" are incomplete and not rigorous [. . .]—and also in earlier and the earliest papers, one could add.

referees. It is thanks to them if the present version has become more reader-friendly than the version they commented on.

Finally, I gratefully acknowledge the seminal impact the wonderful books *Aufgaben und Lehrsätze aus der Analysis I und II* (Problems and Theorems in Analysis I and II) by G. Pólya and G. Szegö had on me since my time as a student. They accompanied me for life. I would be pleased if some of the exercises I posed in place of proofs and worked out examples came close to the spirit of Pólya and Szegö. Definitions and theorems form the skeleton of a theory, while it is brought to life by applications, examples, and exercises only.

Dortmund, Germany Norbert Steinmetz

Contents

Notation

$\mathbb{N}, \mathbb{N}_0, \mathbb{Z}$	Set of positive, non-negative, all integers
$\mathbb{Q}, \mathbb{R}$	Set of rational, real numbers
$\mathbb{C}, \widehat{\mathbb{C}}$	Complex plane, Riemann sphere $\mathbb{C} \cup \{\infty\}$
Re, Im	Real- and imaginary part
$\mathbb{D}, \mathbb{H}$	Unit disc, upper or right half-plane
$\Delta_\delta(p)$	Local disc $\|z - p\| < \delta\|p\|^{-\beta}$ of 'radius' δ
$\mathfrak{P}, \mathfrak{Q}$	Set of poles and zeros (of the function in question)
$\mathfrak{C}, \mathscr{L}$	Cluster set, lattice
$\mathfrak{Y}^0_{\alpha,\beta}, \mathfrak{Y}_{\alpha,\beta}$	Yosida classes
$(\frac{s}{t}, \varpi)$-string	Sequence satisfying $p_{k+1} = p_k + (\varpi + o(1))p_k^{-\frac{s}{t}}$
$\|\cdot\|, \chi(\cdot,\cdot)$	Absolute value, chordal metric on $\widehat{\mathbb{C}}$
$\\|\cdot\\|, \\|\cdot\\|_\infty$	Euclidean, maximum norm on $\mathbb{R}^n$ and $\mathbb{C}^n$
dist (z, A)	Euclidean distance of z to the set A
$f^\sharp, f^{\sharp_\alpha}$	Spherical derivative, modified spherical derivative
$S_f, \{f, z\}$	Schwarzian derivative
$W(w_1, \ldots, w_n)$	Wronskian determinant of $w_1, \ldots, w_n$
$\wp$	Weierstraß P-function
(a, b, c, d)	Cross-ratio of $a, b, c, d \in \widehat{\mathbb{C}}$
$M(r,f), m(r,f)$	Maximum modulus, Nevanlinna proximity function
$N(r,f), n(r,f)$	Integrated, non-integrated counting function of poles
$T(r,f), S(r,f)$	Nevanlinna characteristic, remainder term
$T_C(r, \mathfrak{g}), T_S(r, \mathfrak{f}), T_V(r, \mathfrak{f})$	Cartan, Selberg, Valiron characteristic
$\delta(a,f), \vartheta(a,f)$	Deficiency, ramification index of a
$\varrho(f), \varrho_a(f)$	Order of growth, exponent of convergence of a-points
ord	Order of Airy and Weber–Hermite solution
$\deg R, \deg_x R$	Degree, degree w.r.t. x of the rational function R
$d_\Omega, \mathsf{d}_\Omega$	Degree, weight of the differential polynomial Ω
$\text{res}_p f$	Residue of f at p
h_f	Phragmén–Lindelöf indicator function

$\asymp$	$\phi(r) \asymp \psi(r) \Leftrightarrow 1/C \leq \|\phi(r)/\psi(r)\| \leq C \ (r \geq r_0)$, also $a_n \asymp b_n \Leftrightarrow 1/C \leq \|a_n/b_n\| \leq C \ (n \geq n_0)$
$\sim$	$\phi(r) \sim \psi(r) \Leftrightarrow \phi(r)/\psi(r) \to 1$ as $r \to \infty$, also $f(z) \sim \sum_{k=0}^{\infty} c_k z^{-k}$, asymptotic series representing f
☕	End of proof, example, exercise, remark

Chapter 1
Selected Topics in Complex Analysis

In this chapter we will discuss several topics in Complex Analysis which usually are not or only incomprehensively considered in lectures and textbooks, but are of particular interest in the field of Analytic and Algebraic Differential Equations. Our standard reference is Ahlfors' forever young *Complex Analysis* [2].

1.1 Algebraic Functions

The quadratic equation $P(x, y) = xy^2 - 2x^2y + x + 1 = 0$ has two solutions $y = \phi_{1,2}(x) = x \pm \sqrt{\triangle_P(x)/x}$ $(\triangle_P = x^3 - x - 1)$; they are holomorphic except at $x = 0$, a zero of the leading coefficient, $x = \infty$, which always has to be considered separately, and the zeros x_1, x_2, x_3 of the *discriminant* $\triangle_P$ of P with respect to y. At $x = 0$ the solutions have 'algebraic poles', $\phi_{1,2}(x) = \pm ix^{-\frac{1}{2}} \pm \frac{i}{2}x^{\frac{1}{2}} + \cdots$, while the singularities at x_ν are 'algebraic': $\phi_{1,2} = x_\nu \pm c_1\sqrt{x - x_\nu} \pm c_2(x - x_\nu) + \cdots$. At $x = \infty$, $\phi_1(x) = 2x - \frac{1}{2}x^{-2} + \cdots$, say, has a pole and $\phi_2(x) = \frac{1}{2}x^{-1} + \frac{1}{2}x^{-2} + \cdots$ has a zero. Analytic continuation along closed curves avoiding the points x_ν and 0 leaves the branches ϕ_ν invariant or permutes them. The aim of this section is to extend these simple results to general algebraic equations[1]

$$P(x, y) = a_m(x)y^m + a_{m-1}(x)y^{m-1} + \cdots + a_0(x) = 0 \quad (a_m(x) \not\equiv 0), \qquad (1.1)$$

where P is an *irreducible* polynomial; irreducibility just means that P cannot be written non-trivially as P_1P_2. By $\triangle_P$ we denote its *discriminant* with respect to y.

---

[1]Though the variables are complex, we prefer to write x and y to emphasise their equivalence.

© Springer International Publishing AG 2017
N. Steinmetz, *Nevanlinna Theory, Normal Families, and Algebraic Differential Equations*, Universitext, DOI 10.1007/978-3-319-59800-0_1

1.1.1 Local Branches

For x fixed, Eq. (1.1) has m mutually distinct finite solutions $y_\mu = \phi_\mu(x)$ provided $a_m(x)\triangle_P(x) \neq 0$. Locally defined holomorphic functions solving some Eq. (1.1) are called *algebraic functions*, and, more precisely, *branches* of the *Algebraic Function* defined by (1.1). Local existence is guaranteed by

Theorem 1.1 *Suppose $P(x_0, y_0) = 0$ and $P_y(x_0, y_0) \neq 0$. Then in a neighbourhood of $x = x_0$ there exists a unique holomorphic solution $y = \phi(x)$ to Eq. (1.1) with $\phi(x_0) = y_0$.*

Proof For $\epsilon > 0$ sufficiently small, $P(x_0, y) \neq 0$ holds on $0 < |y - y_0| \leq \epsilon$, hence there exists a $\delta > 0$ such that $P(x, y) \neq 0$ holds on $|x - x_0| < \delta$, $|y - y_0| = \epsilon$. For x fixed, Eq. (1.1) has

$$n(x) = \frac{1}{2\pi i} \int_{|y-y_0|=\epsilon} \frac{P_y(x, y)}{P(x, y)}\, dy$$

solutions on $|y - y_0| < \epsilon$. Since $n(x)$ is integer-valued and depends continuously on x, $n(x) = n(x_0) = 1$ follows. Thus ϕ is defined on $|x - x_0| < \delta$. The proof that ϕ is holomorphic is part of the following exercise.   ☕

Exercise 1.1 Prove that $\phi(x) = \dfrac{1}{2\pi i} \int_{|y-y_0|=\epsilon} \dfrac{yP_y(x, y)}{P(x, y)}\, dy.$[2]

(**Hint.** Compute the residue of $yP_y(x, y)/P(x, y)$ at $y = \phi(x)$.)   ☕

1.1.2 Regular and Singular Points

Finite points x_0 such that $P(x_0, y) = 0$ has m finite and mutually distinct solutions are called *regular*, and *singular* otherwise. At regular points there exist m holomorphic branches. The set S of finite singular points consists of the zeros of $a_m(x)\triangle_P(x)$.

Theorem 1.2 *Any branch admits unrestricted analytic continuation in $\mathbb{C} \setminus S$. Analytic continuation along any closed curve avoiding S permutes the branches.*

Proof Given $x_0 \notin S$, let $r(x_0)$ denote the largest radius such that all branches are holomorphic on $|x - x_0| < r(x_0)$. Then a simple geometric argument shows that $|r(x_1) - r(x_2)| \leq |x_1 - x_2|$ holds, and this guarantees that every branch admits unrestricted analytic continuation in $\mathbb{C} \setminus S$. Actually, it turns out that $r(x) = \text{dist}(x, S)$. The second assertion is obvious.   ☕

---

[2]Note the similarity to the *Bürmann–Lagrange inversion formula*.

Let x_0 be any finite singularity. Analytic continuation along the circle $|x-x_0| = r$ with sufficiently small radius permutes the branches ϕ_μ, and repeating this q times (q suitably chosen) leaves them invariant. Thus analytic continuation along $|t| = r^{1/q}$ leaves the branches $s = \Phi_\mu(t) = \phi_\mu(x_0+t^q)$ of $P(x_0+t^q, s) = 0$ invariant. They are holomorphic on $0 < |t| < r^{1/q}$ with an isolated singularity at $t = 0$; the singularity is not essential, this following from $P(x_0+t^q, \Phi_\mu(t)) = 0$ and the Casorati–Weierstraß Theorem. This proves

Theorem 1.3 *Finite singularities other than ordinary poles are algebraic singularities or algebraic poles:*

$$\phi(x) = \sum_{n=k}^{\infty} c_n (x - x_0)^{n/q} \quad (c_k \neq 0)$$

holds on $|x-x_0| < \delta$, with $c_n \neq 0$ for at least one $n \not\equiv 0 \bmod q$. The series represents q distinct branches according to the chosen branch of $(x - x_0)^{1/q}$.

Remark 1.1 Series of this kind are also called *Puiseux series*. Algebraic and ordinary poles occur if and only if $a_m(x_0) = 0$. In this case one has to consider

$$\eta^m P(x, \alpha + 1/\eta) = P(x, \alpha)\eta^m + \cdots + a_m(x) = 0 \quad (m = \deg_y P,\ P(x_0, \alpha) \neq 0)$$

at $(x, \eta) = (x_0, 0)$. To examine the point $x = \infty$ one has just to discuss the equation

$$\xi^n P(1/\xi, y) = 0 \quad (n = \deg_x P)$$

at $\xi = 0$. Again $x = \infty$ may be regular, an algebraic singularity or an algebraic or ordinary pole for some or all branches. The reader is encouraged to investigate all cases in detail.

Exercise 1.2 Compute (using MAPLE, say) the discriminant w.r.t. y of

$$P(x, y) = y^3 - 3((\bar{a} - 1)x - 2)xy^2 - 3(2x - a + 1)xy - x^3.$$

Discuss, in particular, the singularities if $a = -\frac{1}{2} \pm \frac{i}{2}\sqrt{3}$.

Solution for $a = -\frac{1}{2} \pm \frac{i}{2}\sqrt{3}$: $\Delta_P(x) = x^3(x - a)^3(x + 1)^3$.

To sum up we can state the Main Theorem on algebraic functions:

Theorem 1.4 *The branches defined by Eq.* (1.1) *admit unrestricted analytic continuation in $\widehat{\mathbb{C}} \setminus S$, where now S may include $x = \infty$. The singularities are algebraic and algebraic poles.*[3] *Analytic continuation along closed curves in $\widehat{\mathbb{C}} \setminus S$ results in a permutation of the branches.*

[3] *Ordinary poles are not viewed as singularities.*

Exercise 1.3 The statement of Theorem 1.4 may be reversed. Prove that if $\phi_1, \ldots, \phi_m$ are holomorphic functions on some domain D, such that

- each ϕ_μ admits unrestricted analytic continuation in $\widehat{\mathbb{C}} \setminus S$, where S is a finite set;
- analytic continuation along any closed curve in $\widehat{\mathbb{C}} \setminus S$ results in a permutation of the branches ϕ_μ;
- the points in S are algebraic singularities or algebraic poles for some or all ϕ_μ.

Then $\phi_1, \ldots, \phi_m$ are the branches of some algebraic equation $P(x, y) = 0$.
(**Hint.** Consider $\prod_{\mu=1}^{m}(y - \phi_\mu(x)) = y^m + \sum_{\mu=0}^{m-1} b_\mu(x) y^\mu$ and show that the b_μ are well-defined holomorphic functions on $\widehat{\mathbb{C}} \setminus S$ without essential singularities.)

We note that irreducibility of P means that *every* permutation may be achieved by choosing the closed curve appropriately. Otherwise the branches decompose in several independent subsets.

Exercise 1.4 Prove that if ϕ and ψ are local branches of Algebraic Functions, then so are $\phi + \psi$, $\phi\psi$, ϕ/ψ, ϕ', and ϕ^{-1}.
(**Hint.** For $\phi\psi$, say, consider all products $\phi_\nu \psi_\mu$; note, however, that the new equation might be reducible. The case of ϕ^{-1} is easier than expected.)

Example The equation for $\chi = \phi\psi$, where ϕ and ψ satisfy $y^2 - x = 0$ and $y^4 - x = 0$, respectively, obtained in this way is $(y^4 - x^3)^2 = 0$.

1.1.3 *The Newton Polygon*

In a neighbourhood of $x = \infty$ there are m solutions $\phi_\mu(x) = c_\mu x^{\rho_\mu} + \cdots$ ($c_\mu \neq 0$) to Eq. (1.1). To determine the principal terms cx^ρ write $a_\kappa(x) = A_\kappa x^{\alpha_\kappa} + \cdots$ ($A_\kappa = \alpha_\kappa = 0$ if $a_\kappa \equiv 0$) and observe that there must be a k and an ℓ such that $k\rho + \alpha_k = \ell\rho + \alpha_\ell \geq \nu\rho + \alpha_\nu$ holds for each ν, since the sum $\sum_{\alpha_\kappa + \kappa\rho = \max} A_\kappa c^\kappa$ must vanish. This implies $-\rho \leq \frac{\alpha_k - \alpha_\mu}{k - \mu}$ ($k > \mu$) and $-\rho \geq \frac{\alpha_\mu - \alpha_k}{\mu - k}$ ($k < \mu$), with equality if $\mu = \ell$, and has a simple geometric interpretation: let C denote the convex hull of the points $(0, 0), (0, \alpha_0), (1, \alpha_1), \ldots, (m, \alpha_m), (m, 0)$. The upper boundary of C is the graph of a concave polygon, called the *Newton–Puiseux polygon* (due to Newton, rediscovered by Puiseux), and $-\rho$ is one of its slopes.

Exercise 1.5 Suppose that the Newton–Puiseux polygon has vertices (k_j, α_{k_j}) ($0 = k_0 < k_1 < \cdots < k_s < k_{s+1} = m$), and slope $-\rho_j$ in $[k_j, k_{j+1}]$. Set $P_j(c) = \sum_\nu' A_\nu c^{\nu - k_j}$, where the sum runs over those indices $k_j \leq \nu \leq k_{j+1}$ such that the points (ν, α_ν) belong to the Newton–Puiseux polygon, hence $\alpha_\nu + \nu\rho_j = \max = \alpha_{k_j} + k_j\rho_j = \alpha_{k_{j+1}} + k_{j+1}\rho_j$ holds. Prove that there are *exactly* $k_{j+1} - k_j$ branches $\sim cx^{\rho_j}$ as $x \to \infty$, one for each zero c of P_j (counting multiplicities).

Exercise 1.6 Describe the analogue of the Newton–Puiseux polygon at $x = 0$.
(**Hint.** Consider $Q(\xi, y) = \xi^n P(1/\xi, y) = 0$ at $\xi = \infty$.)

Exercise 1.7 Prove that the algebraic equation $P(x, y) = 0$ in Exercise 1.2 has one regular and two singular solutions at $x = \infty$: $\phi_1(x) = 3(\bar{a} - 1)x^2 + \cdots$ and $\phi_{2,3}(x) = \pm\sqrt{\frac{x}{3(1-\bar{a})}} + \cdots$ Determine the principal terms of the solutions at $x = 0$ and, if $a = -\frac{1}{2} \pm \frac{i}{2}\sqrt{3}$, also at $x = a$ and $x = -1$.

1.1.4 Algebraic Curves

Any irreducible equation $P(x, y) = \sum_{\mu+\nu\le d} a_{\mu\nu} x^\nu y^\mu = 0$ of algebraic degree d defines a so-called *algebraic curve* and also a *compact Riemann surface*, which usually are identified with P. Algebraic curves have a *genus* $g = g(P)$. The most descriptive definition is

$$g = \tfrac{1}{2}(d - 1)(d - 2) - c(P),$$

where $c(P)$ denotes the number of 'double points' of P; the term $\frac{1}{2}(d - 1)(d - 2)$ is the greatest number of double points an algebraic curve of degree d can have. Two cases are of particular interest. Algebraic curves of

- *genus zero* have a rational parametrisation $x = r(t)$, $y = s(t)$, that is, $P(r(t), s(t)) \equiv 0$ holds, and are also called *rational curves*.
- *genus one* are parametrised by elliptic functions $x = r(t)$, $y = s(t)$, and are also called *elliptic curves*.

Remark 1.2 Parametrisations are, of course, not unique. The circle $x^2 + y^2 = 1$ has the rational parametrisation $x = \frac{1-t^2}{1+t^2}$, $y = \frac{2t}{1+t^2}$ and the entire parametrisation $x = \cos t$, $y = \sin t$. The Weierstraß functions $x = \wp(t; g_2, g_3)$, $y = \wp'(t; g_2, g_3)$ $(g_2^3 - 27g_3^2 \neq 0)$ parametrise the elliptic curve $y^2 = 4x^3 - g_2 x - g_3$ (more on elliptic functions in Sect. 1.5.1). In both cases there exists a parametrisation (r, s) of smallest algebraic degree resp. elliptic order equal to $(\deg_y P, \deg_x P)$. Any pair (f, g) of non-constant meromorphic functions satisfying $P(f(z), g(z)) \equiv 0$ may be written as $(f, g) = (r \circ h, s \circ h)$, where h is a non-constant meromorphic resp. entire function (see [10]).

Example 1.1 The algebraic curve

- in Exercise 1.2 with $a = -\frac{1}{2} \pm \frac{i}{2}\sqrt{3}$ has genus zero with parametrisation $x = r(t) = \frac{1}{9}(1 - a)\frac{t(t-3)^2}{(t-1)^2}$, $y = s(t) = -\frac{1}{9}(1 - a)\frac{t^2(t-3)}{(t-1)}$.
- $x^3 + y^3 + 3xy + \kappa = 0$ is reducible if $\kappa = -1$, has genus zero if $\kappa = 0$ (and is known as the *folium of Descartes*) with parametrisation $x = \frac{-3t^2}{t^3+1}$, $y = \frac{-3t}{t^3+1}$, and has genus one otherwise.

1.2 Normal Families

1.2.1 *Sequences of Holomorphic Functions*

Convergence of sequences of *holomorphic* functions (f_n) on some domain D is always measured with respect to the euclidean metric, and is always assumed to be locally uniform. The most important results in this context are the Theorems of Weierstraß and Hurwitz.

Weierstraß' Theorem *Suppose f_n tends to f, locally uniformly on D. Then f is holomorphic on D, and the sequence (f_n') tends to f', again locally uniformly on D.*

Hurwitz' Theorem *Suppose f_n tends to $f \not\equiv 0$, locally uniformly on D. Then any zero z_0 of f is the limit of some sequence (z_n) of zeros of f_n $(n \geq n_0)$. Any m-fold zero z_0 is 'accompanied' by zeros of f_n on $|z - z_0| < \rho_n \to 0$ of 'total' multiplicity m, that is, f_n has zeros $z_{n,j}$ with multiplicities $m_{n,j}$ such that $z_{n,j} \to z_0$ and $\sum_j m_{n,j} = m$, and* vice versa.

A family $\mathscr{F}$ of *holomorphic* functions on some common domain of definition D is called *normal* if every sequence in $\mathscr{F}$ contains a subsequence that converges either to some holomorphic function f or else to infinity, locally uniformly on D. A sequence (f_n) is called normal if the countable family $\mathscr{F} = \{f_n : n \in \mathbb{N}\}$ is normal, that is, if every subsequence (f_{n_k}) itself contains a convergent subsequence.

Normality is a Local Property *The family $\mathscr{F}$ is normal on D if and only if it is normal at every $z_0 \in D$*, that is, if to every $z_0 \in D$ there exists an $r(z_0) > 0$ such that the family $\mathscr{F}_{z_0}$ of functions $f \in \mathscr{F}$ restricted to $|z - z_0| < r(z_0)$ is normal on this disc. This is a corollary of Cantor's well-known diagonal argument.

Vitali's Theorem *If (f_n) is normal on D and $f_n(z)$ converges pointwise on some non-discrete set $E \subset D$, then the sequence (f_n) converges locally uniformly on D.*

Apparently, the best-known normality criterion is

Montel's Criterion *A family $\mathscr{F}$ of holomorphic functions on some domain D is normal if $\mathscr{F}$ is locally bounded, in which case all limit functions are holomorphic.*

Remark 1.3 Montel's Criterion is a corollary of the *Arzelà–Ascoli Theorem* and Cauchy's Integral Formula; the latter implies that the family $\mathscr{F}'$ of derivatives is locally bounded if $\mathscr{F}$ is, hence $\mathscr{F}$ is locally equi-continuous.

1.2.2 *Sequences of Meromorphic Functions*

In the context of nonlinear algebraic differential equations it is unavoidable to also consider sequences of *meromorphic* functions. Here convergence has to be measured with respect to the *chordal metric* $\chi(a, b) = \frac{1}{2}\|\sigma^{-1}(a) - \sigma^{-1}(b)\|_{\mathbb{R}^3}$

on the Riemann sphere; σ denotes the stereographic projection $S^2 \longrightarrow \widehat{\mathbb{C}}$. In terms of $a, b \in \widehat{\mathbb{C}}$

$$\chi(a,\infty) = \frac{1}{\sqrt{1+|a|^2}} \quad \text{and} \quad \chi(a,b) = \frac{|a-b|}{\sqrt{1+|a|^2}\sqrt{1+|b|^2}} \quad (a, b \in \mathbb{C})$$

holds. By definition, χ is invariant under rotations of the sphere $S(z) = \frac{\alpha z+\beta}{-\bar{\beta}z+\bar{\alpha}}$. In the complex plane $\mathbb{C}$, the euclidean and the chordal metric are locally equivalent. Weierstraß' Theorem holds in parts.

Theorem 1.5 *Suppose that the sequence of meromorphic functions (f_n) tends to $f \not\equiv \infty$ with respect to the chordal metric, locally uniformly on some domain D. Then f is meromorphic on D.*

Proof The limit function is continuous (as a map $D \longrightarrow \widehat{\mathbb{C}}$). If $f(z_0)$ is finite, then f is bounded on some disc $\Delta : |z - z_0| < \delta$, and f_n tends to f with respect to the euclidean metric. Thus f is holomorphic on this disc. If, however, $f(z_0) = \infty$, the same argument applied to $(1/f_n)$ shows that $1/f$ is holomorphic on Δ, and either z_0 is a pole of f or else $f \equiv \infty$ holds on Δ. The assertion now follows from the classical connectivity argument: the set where f is meromorphic as well as its complement with respect to D (where $f \equiv \infty$) is open.

Chordal convergence is delicate and differs in many respect from euclidean convergence. The reason for this is that poles and zeros of f_n may 'collide', and also that the chordal metric is not compatible with the arithmetic structure of the complex numbers. A simple necessary condition for $f_n \to f \not\equiv \infty$ is that the poles of f_n are uniformly separated from the zeros: to every $z_0 \in D$ there exists an $r(z_0) > 0$ such that $|z - z_0| < r(z_0)$ does not simultaneously contain poles and zeros of any f_n.

Exercise 1.8 The sequences $f_n(z) = (z - 1/n)^{-1}$ and $g_n(z) = (z + 1/n)^{-1}$ converge to z^{-1}, uniformly on the unit disc $\mathbb{D}$, say. Prove, however, that $f_n + g_n$ does not converge to $2z^{-1}$ in any neighbourhood of the origin.

Exercise 1.9 Suppose that f_n is meromorphic on D and tends to $f \not\equiv \infty$, and ϕ_n is holomorphic and tends to $\phi \not\equiv \infty$, locally uniformly on D. Prove that $f_n + \phi_n$ tends to $f + \phi$, and $\phi_n f_n$ tends to ϕf, if in the latter case the poles of f are separated from the zeros of ϕ.
(**Hint.** It suffices to consider neighbourhoods of poles of f, where $|\phi_n(z)| \le \frac{1}{2}|f_n(z)|$ may be assumed in the first case, and $0 < \delta < |\phi_n(z)| \le K$ in the second.)

It is obvious that Hurwitz' Theorem also remains valid for sequences of meromorphic functions. If f_n tends to f on $|z - z_0| < r$ and f has an m-fold pole at z_0, hence is zero-free on $|z - z_0| < \rho$, then $1/f_n$ ($n \ge n_0$) is holomorphic on $|z - z_0| < \rho$ and tends to $1/f$ also in the euclidean metric. By Hurwitz' Theorem for holomorphic sequences, $1/f_n$ has m zeros on $|z - z_0| < \delta$ counted with multiplicities, and f_n has m poles there ($n \ge n_\delta$).

Exercise 1.10 Prove that the sequence $f_n(z) = (z^2 - n^{-2})^{-1}$ tends to $f(z) = z^{-2}$, uniformly on $\mathbb{D}$, say, while $f_n' \to f'$ fails. Note that the poles of f_n are neither uniformly separated from each other nor from the zero of f_n' at $z = 0$.

The second part of Weierstraß' Theorem fails to hold for sequences of meromorphic functions. Suppose f_n converges to f, locally uniformly on $|z| < r$, and f has a pole of order m at $z = 0$, and no other. For $f_n' \to f'$ it is necessary that there exists a $\rho > 0$ such that f_n' $(n \ge n_\rho)$ has no zeros on $|z| < \rho$, and f_n has only one pole (of order m) on $|z| < \rho$, since each pole of f_n of order ℓ is a pole of f_n' of order $\ell + 1$.

Exercise 1.11 Prove that the just mentioned first condition implies the second. (**Hint.** Apply the Argument Principle to f_n' and f', noting that f_n''/f_n' tends to f''/f', uniformly on $|z| = \rho$ w.r.t. the euclidean metric.)

The second condition is decisive for the validity of Weierstraß' Theorem.

Theorem 1.6 *Suppose that f_n converges to f, locally uniformly on $|z| < r$, and f has a pole of order m at $z = 0$, and no other. Then the above conditions are also sufficient in order that $f_n' \to f'$, locally uniformly on some neighbourhood of $z = 0$.*

Proof We have to show that the second condition is sufficient. Assume that f_n has an m-fold pole at $b_n \to 0$ and no other on $|z| < \rho$, that is, $f_n(z) = \psi_n(z)(z - b_n)^{-m}$ holds, where ψ_n is holomorphic and tends to $\psi = z^m f(z)$ as $n \to \infty$ w.r.t. the euclidean metric, uniformly on $|z| = \rho$. Then $\psi_n \to \psi$ on $|z| < \rho$ holds by the Maximum Principle. From $\psi(0) \ne 0$ and $\psi_n' \to \psi'$ it then follows that

$$f_n'(z) = \frac{(z - b_n)\psi_n'(z) - m\psi_n(z)}{(z - b_n)^{m+1}} \to \frac{z\psi'(z) - m\psi(z)}{z^{m+1}} = f'(z),$$

uniformly on some neighbourhood of $z = 0$.

Remark 1.4 In any case the sequence f_n' tends to f', locally uniformly on the punctured disc $0 < |z| < \rho$. If the above conditions are violated, then some of the poles of f_n' collide with zeros of f_n', and in the limit multiplicities disappear as do zeros of f_n'. If $f_n = 1/P_n$, P_n a polynomial of degree m, the equivalence of both conditions follows from the Gauß–Lucas Theorem (the zeros of P' are contained in the convex hull of the zeros of P).

Example 1.2 In the context of algebraic differential equations it is not always necessary to know that $f_n \to f$ implies $f_n' \to f'$, $f_n'' \to f''$, etc. Suppose $\Omega_n[w]$ is a sequence of polynomials in $w, w', \ldots, w^{(m)}$ with coefficients that are holomorphic on some domain D, and such that $\Omega_n[w] \to \Omega[w]$ as $n \to \infty$, locally uniformly with respect to the euclidean metric on $D \times \mathbb{C}^{m+1}$. Suppose also that $\Omega_n[f_n](z) \equiv 0$ and $f_n \to f \not\equiv \infty$, locally uniformly on D with respect to the chordal metric. Then $f_n^{(k)}$ tends to $f^{(k)}$, locally uniformly with respect to the euclidean metric on the domain $D \setminus \{\text{poles of } f\}$, and $\Omega[f](z) \equiv 0$ holds on D.

1.2.3 Normal Families of Meromorphic Functions

The definition of *normality* applies to families $\mathscr{F}$ of meromorphic functions word-by-word. In the context of algebraic differential equations the limit function ∞ is often excluded for intrinsic reasons. For meromorphic functions the distortion is measured by the *spherical derivative*

$$f^\sharp(z) = \lim_{h\to 0} \frac{\chi(f(z+h), f(z))}{|h|} = \frac{|f'(z)|}{1+|f(z)|^2}$$

rather than $|f'|$, and the part of Montel's Criterion is played by

Marty's Criterion *A family $\mathscr{F}$ of meromorphic functions is normal if and only if the family $\mathscr{F}^\sharp$ of spherical derivatives $f^\sharp$ is locally bounded.*

Exercise 1.12 The family $\mathfrak{P}_d$ of all polynomials $P(z;\mathbf{a}) = a_d z^d + \cdots + a_0$ of fixed degree $d > 2$ and $\|\mathbf{a}\| = \max_{0\le\nu\le d} |a_\nu| \le 1$ is not normal on $1 < |z| \le \infty$ (∞ is included). Thus convergence $\mathbf{a}^{(n)} \to \mathbf{a}$ is not sufficient for $P(z;\mathbf{a}^{(n)}) \to P(z;\mathbf{a})$. For example, $P_n(z) = \frac{1}{n}z^d - z^{d-1}$ satisfies $P_n^\sharp(n) = n^{d-2}$. Prove, however, that the subfamily of polynomials with $|a_d| = 1$ is normal.
(**Hint.** Prove that $|P(z)| \ge \frac{1}{2}|z|^d$ and $|P'(z)| \le 2d|z|^{d-1}$ holds on $|z| > 2d$.)

1.3 Ordinary Differential Equations

As in the theory of ordinary differential equations in the real domain we will consider initial value problems for single differential equations

$$w' = f(z, w), \; w(z_0) = w_0,$$

systems

$$\mathfrak{w}' = \mathfrak{f}(z, \mathfrak{w}), \quad \mathfrak{w}(z_0) = \mathfrak{w}_0, \tag{1.2}$$

with $\mathfrak{w} = (w_1, \ldots, w_n)$ and $\mathfrak{f} = (f_1, \ldots, f_n)$, and higher-order differential equations

$$w^{(n)} = f(z, w, w', \cdots, w^{(n-1)}), \quad w^{(\nu)}(z_0) = w_\nu^0 \; (0 \le \nu < n) \tag{1.3}$$

in the complex domain. Since every n-th order equation may be transformed into a system $\mathfrak{y}' = \mathfrak{f}(z, \mathfrak{y})$, for example into $y_\nu' = y_{\nu+1}$ $(1 \le \nu < n)$ and $y_n' = f(z, y_1, \ldots, y_n)$ via $y_\nu = w^{(\nu-1)}$ $(1 \le \nu \le n)$, all results obtained for systems (1.2) also hold *mutatis mutandis* for higher-order problems (1.3). Standard references in the complex domain are the books by Bieberbach [15], Golubew [58], Hille [84, 85], Ince [94], and Laine [102]. Since we expect holomorphic solutions, the components of the right-hand side $\mathfrak{f}$ have to be holomorphic on some domain $G \subset \mathbb{C} \times \mathbb{C}^n$. It is thus necessary to give a short introduction to the field of holomorphic functions of several variables.

1.3.1 Holomorphic Functions of Several Variables

Let $G \subset \mathbb{C}^n$ be any domain. A function $f : G \longrightarrow \mathbb{C}$ is called *holomorphic* if the partial derivatives

$$f_{z_\nu}(\mathfrak{z}) = \lim_{h\to 0} \frac{f(\mathfrak{z} + h\mathfrak{e}_\nu) - f(\mathfrak{z})}{h} \quad (1 \leq \nu \leq n,\ \mathfrak{z} = (z_1, \ldots, z_n) \in G)$$

exist; $\mathfrak{e}_\nu$ denotes the ν-th unit vector and h is complex. The one-dimensional theory then yields ($z_1, \ldots, z_{n-1}$ fixed)

$$f(z_1, \ldots, z_{n-1}, z_n) = \frac{1}{2\pi i} \int_{\mathfrak{C}_n} \frac{f(z_1, \ldots, z_{n-1}, \zeta_n)}{\zeta_n - z_n}\, d\zeta_n \quad (|z_n - z_\nu^\circ| < r_n),$$

and iteration finally gives Cauchy's Integral Formula

$$f(\mathfrak{z}) = \frac{1}{(2\pi i)^n} \int_{\mathfrak{C}_n} \cdots \int_{\mathfrak{C}_1} \frac{f(\zeta_1, \ldots, \zeta_n)}{(\zeta_1 - z_1) \cdots (\zeta_n - z_n)}\, d\zeta_1 \cdots d\zeta_n \quad (|z_\nu - z_\nu^\circ| < r_\nu);$$

$\mathfrak{C}_\nu$ denotes the positively oriented circle $|z - z_\nu^\circ| = r_\nu$, and f is assumed to be holomorphic on the poly-cylinder $\{\mathfrak{z} : |z_\nu - z_\nu^\circ| < R_\nu,\ 1 \leq \nu \leq n\}$ with $R_\nu > r_\nu$. By a famous Theorem of Hartogs (see [101], for example) the existence of the partial derivatives implies the continuity of f, in contrast to the real case. It is then easy to derive

$$f(\mathfrak{z}) = \sum_{\mathfrak{p}\in\mathbb{N}_0^n} c_{\mathfrak{p}} (\mathfrak{z} - \mathfrak{z}^\circ)^{\mathfrak{p}} \quad (|z_\nu - z_\nu^\circ| < r_\nu),$$

$$c_{\mathfrak{p}} = \frac{1}{\mathfrak{p}!} (\nabla^{\mathfrak{p}} f)(\mathfrak{z}_0) = \frac{1}{(2\pi i)^n} \int_{\mathfrak{C}_n} \cdots \int_{\mathfrak{C}_1} \frac{f(\zeta_1, \ldots, \zeta_n)}{(\zeta_1 - z_1^\circ)^{p_1+1} \cdots (\zeta_n - z_n^\circ)^{p_n+1}}\, d\zeta_1 \cdots d\zeta_n,$$

with the usual and useful abbreviations $\mathfrak{p} = (p_1, \ldots, p_n) \in \mathbb{N}_0^n$ (multi-index), $\mathfrak{p}! = p_1! p_2! \cdots p_n!$, and $\mathfrak{z}^{\mathfrak{p}} = z_1^{p_1} z_2^{p_2} \cdots z_n^{p_n}$. In particular, it follows that the partial derivatives are also holomorphic. We note some useful corollaries, which may be proved just as in one complex dimension.

– $|c_{\mathfrak{p}}| \leq \dfrac{M}{r_1^{p_1} \cdots r_n^{p_n}}$ with $M = \max |f(\mathfrak{z})|$ on $\{\mathfrak{z} : |z_\nu - z_\nu^\circ| = r_\nu,\ 1 \leq \nu \leq n\}$.

- The chain rule, in particular $\frac{d}{dz}f(\mathfrak{g}(z)) = \sum_{\nu=1}^{n} f_{w_\nu}(\mathfrak{g}(z))g'_\nu(z)$ ($z \in D \subset \mathbb{C}, f : G \subset \mathbb{C}^n \longrightarrow \mathbb{C}$, and $\mathfrak{g} = (g_1, \ldots, g_n) : D \longrightarrow G$).
- Locally f satisfies some Lipschitz condition.
- Functions $f, g : G \longrightarrow \mathbb{C}$ already agree on G if they agree on $E_1 \times \cdots \times E_n$, where each set E_ν has an accumulation point z_ν° with $(z_1^\circ, \ldots, z_n^\circ) \in G$.

1.3.2 Cauchy's Existence Theorem

Existence and uniqueness of solutions to the initial value problem (1.2) can be proved just as in the real (Picard–Lindelöf) case. For $z_0 = 0$, $\mathfrak{w}_0 = \mathfrak{o}$, say, the initial value problem is equivalent to the integral equation

$$\mathfrak{w}(z) = \int_0^z \mathfrak{f}(\zeta, \mathfrak{w}(\zeta))\,d\zeta. \tag{1.4}$$

Assuming $Z(r, R) = \{z : |z| \le r\} \times \{\mathfrak{w} : \|\mathfrak{w}\|_\infty \le R\} \subset G$ and $\|\mathfrak{f}(z, \mathfrak{w})\|_\infty \le M$ on $Z(r, R)$ we obtain a Lipschitz condition

$$\|\mathfrak{f}(z, \mathfrak{u}) - \mathfrak{f}(z, \mathfrak{v})\|_\infty \le L\|\mathfrak{u} - \mathfrak{v}\|_\infty$$

on $Z(\rho, \mathrm{P})$, where $\rho < r$ and $\mathrm{P} < R$ are chosen in such a way that $\rho M \le \mathrm{P}$ holds. The space $\mathfrak{B}$ of bounded holomorphic functions

$$\mathfrak{u} : \{z : |z| < \rho\} \longrightarrow \mathbb{C}^n, \quad \mathfrak{u}(0) = \mathfrak{o}$$

will be endowed with the norm

$$\|\mathfrak{u}\| = \sup\{\|\mathfrak{u}(z)\|_\infty\, e^{-2L|z|} : |z| < \rho\}.$$

Then $\mathfrak{B}$ is a Banach space, since norm convergence and uniform convergence agree. The operator T defined by the right-hand side of (1.4) then turns out to be a contracting self map of the closed ball $\{\mathfrak{u} \in \mathfrak{B} : \|\mathfrak{u}\| \le \mathrm{P}\}$. The inequality $\|(T\mathfrak{u})(z)\|_\infty e^{-2L|z|} \le M\rho \le \mathrm{P}$ is obvious, and $\|T\mathfrak{u} - T\mathfrak{v}\| \le \frac{1}{2}\|\mathfrak{u} - \mathfrak{v}\|$ follows from

$$\|(T\mathfrak{u})(z) - (T\mathfrak{v})(z)\|_\infty \le L \int_0^z \|\mathfrak{u}(t) - \mathfrak{v}(t)\|_\infty e^{-2L|t|} e^{2L|t|}\,|dt| \le L\|\mathfrak{u} - \mathfrak{v}\| \frac{e^{2L|z|} - 1}{2L}.$$

The unique fixed point then yields the desired unique solution to the integral equation and also to the initial value problem.

1.3.3 Linear Differential Equations and Systems

Linear systems may be compactly written as

$$\mathfrak{w}' = A(z)\mathfrak{w} + \mathfrak{b}(z),$$

where the entries of the $n \times n$-matrix A and the n-vector $\mathfrak{b}$ are holomorphic on some domain D. Since this time we may work on the cylinder $G = D \times \mathbb{C}^n$, there is no restriction like $\rho M \leq$P, hence every solution exists on $|z - z_0| < \operatorname{dist}(z_0, \partial D)$. In particular, for $D = \mathbb{C}$ the components of the solutions are entire functions.

Theorem 1.7 *Suppose that the entries of A and $\mathfrak{b}$ are polynomials. Then the components of every solution are entire functions of finite order, that is,* $\log|w_\nu(z)| = O(|z|^\lambda)$ *holds for some $\lambda > 0$ that is independent of the solution $\mathfrak{w}$.*

Proof We fix θ and $r > 0$, and set $u(\rho) = \|\mathfrak{w}(\rho e^{i\theta})\|_\infty$, $b = \max\limits_{|z|\leq r} \|\mathfrak{b}(z)\|_\infty$, and $a = \max\limits_{|z|\leq r} \|A(z)\|$ (operator norm to the maximum norm on $\mathbb{C}^n$). Then $b \leq \beta(1 + r^k)$ and $a \leq \alpha(1 + r^\ell)$ hold, and u satisfies

$$u(\rho) \leq u(0) + br + a\int_0^\rho u(t)\,dt \quad (0 \leq \rho \leq r).$$

Gronwall's Lemma[4] yields $u(\rho) \leq (u(0) + br)e^{a\rho}$. For $|z| = \rho = r$ this gives

$$|w_\nu(z)| \leq (\|\mathfrak{w}(0)\|_\infty + \beta(|z| + |z|^{k+1}))e^{\alpha(|z|+|z|^{\ell+1})}.$$

The solutions to the homogenous problem $\mathfrak{w}' = A(z)\mathfrak{w}$ form a vector space of dimension n. The Wronskian determinant

$$W = W(\mathfrak{w}_1, \mathfrak{w}_2, \ldots, \mathfrak{w}_n) = \begin{vmatrix} w_{11} & w_{21} & \cdots & w_{n1} \\ w_{12} & w_{22} & \cdots & w_{n2} \\ \vdots & \vdots & & \vdots \\ w_{1n} & w_{2n} & \cdots & w_{nn} \end{vmatrix}$$

of any n-tuple of solutions $\mathfrak{w}_\nu$ satisfies $W' = \operatorname{trace} A(z)\, W$, hence $W(z) \neq 0$ if and only if the $\mathfrak{w}_\nu$ form a *basis*, also called a *fundamental system* or *fundamental set*.

---

[4]**Gronwall's Lemma.** *Suppose u is continuous and satisfies $u(t) \leq \alpha + \beta\int_0^t u(s)\,ds$ on $[0, c)$ for some $\beta \geq 0$. Then $u(t) \leq \alpha e^{\beta t}$ holds on $[0, c)$.* To prove this set $v(t) = \alpha + \beta\int_0^t u(s)\,ds$. Then $u(t) \leq v(t)$ implies $\frac{d}{dt}v(t)e^{-\beta t} = (v'(t) - \beta v(t))e^{-\beta t} = \beta(u(t) - v(t))e^{-\beta t} \leq 0$, hence $v(t)e^{-\beta t}$ decreases and is $\leq v(0) = \alpha$, that is, $u(t) \leq v(t) \leq \alpha e^{\beta t}$ holds.

Linear equations will be written as

$$w^{(n)} + a_{n-1}(z)w^{(n-1)} + \cdots + a_1(z)w' + a_0(z)w = b(z)$$

(a_ν, b holomorphic on some domain $D \subset \mathbb{C}$); each linear equation is equivalent to some linear system $y_\nu' = y_{\nu+1}$ $(1 \le \nu < n)$ and $y_n' = b(z) - a_0(z)y_1 - \cdots - a_{n-1}(z)y_n$ via $y_\nu = w^{(\nu-1)}$. For $D = \mathbb{C}$ the solutions are entire functions; they have finite order of growth if the coefficients are polynomials. The Wronskian determinant

$$W = W(w_1, w_2, \ldots, w_n) = \begin{vmatrix} w_1 & w_2 & \cdots & w_n \\ w_1' & w_2' & \cdots & w_n' \\ \vdots & \vdots & & \vdots \\ w_1^{(n-1)} & w_2^{(n-1)} & \cdots & w_n^{(n-1)} \end{vmatrix}$$

of any n-tuple of solutions to the homogeneous equation satisfies $W' = -a_{n-1}(z)W$.

Exercise 1.13 Prove the following rules for Wronskian determinants ($f_1, \ldots, f_n, f$ meromorphic, $w_1, \ldots w_m$ meromorphic and linearly independent).

1. $W(ff_1, \ldots, ff_n) = f^n W(f_1, \ldots, f_n)$.
2. $w_1, \ldots, w_m$ form a fundamental system to $L[w] = \dfrac{W(w_1, \ldots, w_m, w)}{W(w_1, \ldots, w_m)} = 0$.
   In other words, L can be rediscovered from $w_1, \ldots, w_m$.
3. $\dfrac{W(w_1, \ldots, w_m, f_1, \ldots, f_n)}{W(w_1, \ldots, w_m)} = W(L[f_1], \ldots, L[f_n])$.
4. $W(f_1, \ldots, f_n, f) = (-1)^{n+1} f^{n+1} W((f_1/f)', \ldots, (f_n/f)')$.

For more useful results on Wronskian and other determinants, see Muir [123].

1.3.4 Some Algebraic Aspects

We consider linear differential operators K, L, M, etc., with coefficients in some field $\mathfrak{F}$ that is closed with respect to differentiation, for example, the field of rational functions. By $u_1, \ldots, u_k$, $v_1, \ldots v_\ell$, etc., we denote (local) fundamental sets to $K[u] = 0$, $L[v] = 0$, etc. Without proof we quote an amalgam of results from the book of Ince [94] and the papers by Frank and Wittich [50] and Spigler [159]. The latter contains quite general and useful results in the real and complex domain.

- There exists some linear operator M over the same field $\mathfrak{F}$ such that the products $w = u_\kappa v_\lambda$ satisfy $M[w] = 0$. The order of M may be chosen as small as possible, namely equal to the rank of the Wronskian matrix of the functions $u_\kappa v_\lambda$. *Mutatis mutandis* the same is true for the sums $w = u_\kappa + v_\lambda$.
- Suppose K has rational coefficients. Then there exists some operator L with *polynomial* coefficients that annihilates the functions u_κ.

– Suppose the solutions to $K[u] = 0$ also satisfy $L_1[u] = \cdots = L_n[u] = 0$. Then there exist linear operators M_ν such that $L_\nu[u] = M_\nu[K[u]]$ holds.

1.3.5 Cauchy's 'calcul des limites'

Cauchy's method to prove existence was quite different from the 'modern' Picard–Lindelöf method. For simplicity we will describe it for a single differential equation

$$w' = f(z, w) = \sum_{j,k=0}^{\infty} a_{jk} z^j w^k, \; w(0) = 0. \tag{1.5}$$

The power series $\sum_{p=0}^{\infty} C_p z^p$ is said to *dominate* $\sum_{p=0}^{\infty} c_p z^p$ if $|c_p| \le C_p$ holds for every p. This will be expressed by writing $\sum_{p=0}^{\infty} c_p z^p \ll \sum_{p=0}^{\infty} C_p z^p$; the relation '$\ll$' may easily be extended to power series in several variables. The radius of convergence of the first power series is not less than the radius of convergence of the second. If f in (1.5) is bounded by M on $|z| < r$, $|w| < R$, then the power series for f is dominated by the geometric series

$$\sum_{j,k=0}^{\infty} A_{jk} z^j w^k = \frac{rRM}{(r-z)(R-w)} \quad (|a_{jk}| \le A_{jk} = Mr^{-j}R^{-k}).$$

Our problem (1.5) has a *formal* solution $\sum_{\ell=1}^{\infty} c_\ell z^\ell$. The coefficients have the form

$$c_\ell = P_\ell(a_{jk} : j + k < \ell),$$

where P_ℓ is a *universal* polynomial in the variables a_{jk} with *non-negative* coefficients (for example, $c_1 = a_{00}$ and $2!c_2 = a_{10} + a_{01}a_{00}$). To prove convergence of the formal power series solution we consider the comparison problem

$$W' = \frac{rRM}{(r-z)(R-W)}, \; W(0) = 0$$

(or any other problem $W' = F(z, W)$, $W(0) = 0$ with $f \ll F$). The solution

$$W(z) = R\Big(1 - \sqrt{1 + \frac{2Mr}{R} \log\Big(1 - \frac{z}{r}\Big)}\Big) \quad (\sqrt{1} = 1, \; \log 1 = 0)$$

exists on $|z| < \rho = r(1 - \exp(-\frac{R}{2Mr}))$, and the coefficients $C_\ell = P_\ell(A_{jk} : j + k < \ell)$ of its power series expansion $W = \sum_{\ell=1}^{\infty} C_\ell z^\ell$ are *positive*. From

$$|c_\ell| = |P_\ell(a_{jk} : j + k < \ell)| \le P_\ell(|a_{jk}| : j + k < \ell) \le P_\ell(A_{jk} : j + k < \ell) = C_\ell$$

it then follows that the solution to (1.5) also exists at least on $|z| < \rho$.

Exercise 1.14 The solution to $w' = z + w^2$, $w(0) = 0$ may be expressed in terms of so-called Airy functions and is holomorphic on $|z| < r \approx 1.986$ (the first pole). Prove that the power series ansatz $w = \sum_{n=1}^{\infty} c_n z^n$ leads to $c_1 = 0$, $c_2 = 1/2$, and $(n+1)c_{n+1} = \sum_{k=2}^{n-2} c_k c_{n-k}$ $(n \geq 2)$, and deduce $c_n = |c_n| \leq (1/2)^{(1+2n)/5}$. (**Hint.** Choose $B > 0$ and $A > 0$ such that $|c_n| \leq AB^n$ holds for every $n \geq 2$.) Thus the solution exists on $|z| < \sqrt[5]{4} \approx 1.32$. The Picard–Lindelöf method yields existence on $|z| < \rho_1 \approx 0.63$; ρ_1 is the largest number subject to $\rho_1(\rho_1 + \mathsf{P}^2) \leq \mathsf{P}$ for some $\mathsf{P} > 0$; the inequality $|w'| \leq \rho_2 + |w|^2$ on $|z| < \rho_2$ yields existence on $|z| < (\pi/2)^{2/3} \approx 1.35$, and (iii) Cauchy's 'calcul des limites' yields existence on $|z| < \rho \approx 0.25$.

1.3.6 The Complex Implicit Function Theorem

This theorem, which allows us to 'solve' equations

$$\mathfrak{f}(\mathfrak{z}, \mathfrak{w}) = \mathfrak{o}, \ \mathfrak{w}(\mathfrak{z}_0) = \mathfrak{w}_0 \quad (\mathfrak{z} \in \mathbb{C}^n, \ \mathfrak{w} \in \mathbb{C}^m)$$

with regular Jacobian matrix $A = \mathfrak{f}_{\mathfrak{w}}(\mathfrak{z}_0, \mathfrak{w}_0)$ for $\mathfrak{w} = \mathfrak{w}(\mathfrak{z})$, may be proved in the 'modern' way and also by using Cauchy's 'calcul des limites'. Again it suffices to describe the method for $m = n = 1$, that is, to consider $f(z, w) = 0$ with $f(0, 0) = 0$ and $A = f_w(0, 0) \neq 0$. It is convenient to rewrite the given problem as a fixed-point equation

$$w = F(z, w) = w - A^{-1} f(z, w) = \sum_{s,q=0}^{\infty} a_{sq} z^s w^q$$

with $a_{00} = F(0, 0) = 0$ and $a_{01} = F_w(0, 0) = 0$. It is clear that this problem has a unique *formal* solution $\sum_{p=1}^{\infty} c_p z^p$, and again $c_p = Q_p(a_{sq} : s + q \leq p)$ holds, where Q_p is a universal polynomial with non-negative coefficient. To prove convergence on some disc $|z| < \rho$ and to simplify the argument we introduce new variables Rw and rz, if necessary, to achieve that F is holomorphic on $|z| \leq 1$, $|w| \leq 1$, and satisfies $|F(z, w)| \leq 1$. Then F is dominated by

$$\frac{1}{(1-z)(1-w)} - 1 - w = \frac{z + (1-z)w^2}{(1-z)(1-w)} = \Phi(z, w),$$

and the new problem $W = \Phi(z, W)$, $W(0) = 0$, has the unique holomorphic solution

$$W(z) = \frac{1 - z - \sqrt{1 - 10z + 9z^2}}{4(1-z)} = \sum_{p=1}^{\infty} C_p z^p \quad (|z| < \tfrac{1}{9}).$$

The solution to the original problem then also exists on $|z| < \frac{1}{9}$ since $|c_p| \le C_p$ and $|w(z)| \le W(|z|) < \frac{1}{4}$ hold.

1.3.7 Dependence on Parameters and Initial Values

We will now consider initial value problems

$$\mathfrak{w}' = \mathfrak{f}(z, \mathfrak{w}, \lambda), \; \mathfrak{w}(z_0) = \mathfrak{w}_0,$$

where the right-hand side $\mathfrak{f} : G \times \Lambda \longrightarrow \mathbb{C}^n$ is holomorphic on some subdomain of $\mathbb{C}^{1+n+p}$. For the moment the initial values are assumed to be independent of λ. Since the numbers r, R, M, hence also ρ and P, in the proof of Cauchy s Theorem can be determined uniformly with respect to λ on every compact subset of Λ, the local solution, denoted $\mathfrak{w}(z; \lambda)$, exists on $|z - z_0| < \rho$ and is a *holomorphic* function of λ. This follows from the fact that the local solution is obtained by successive approximation: $\mathfrak{w}_0(z; \lambda) \equiv \mathfrak{w}_0$ and $\mathfrak{w}_{k+1}(z; \lambda) = \mathfrak{w}_0 + \int_{z_0}^{z} \mathfrak{f}(t, \mathfrak{w}_k(t, \lambda), \lambda)\, dt$, hence $\mathfrak{w}_k$ is also holomorphic in λ. In contrast to the real case the differentiability with respect to λ is a by-product of uniform convergence $\mathfrak{w}_k(z; \lambda) \to \mathfrak{w}(z; \lambda)$.

The case of varying initial values may be reduced to the case just handled by considering the initial value problem

$$\tilde{\mathfrak{w}}' = \mathfrak{f}(z + z_0, \tilde{\mathfrak{w}} + \mathfrak{w}_0) = \tilde{\mathfrak{f}}(z, \tilde{\mathfrak{w}}, z_0, \mathfrak{w}_0), \; \tilde{\mathfrak{w}}(0) = \mathfrak{o}$$

for $\tilde{\mathfrak{w}}(z) = \mathfrak{w}(z + z_0) - \mathfrak{w}_0$. Obviously the right-hand side $\tilde{\mathfrak{f}}$ is holomorphic in all its variables. We also note that 'analytic dependence' may be replaced with 'continuous dependence'.

Remark 1.5 Suppose $\|\mathfrak{f}(z, \mathfrak{w}, \lambda)\| \le M$ holds on $|z| \le \rho$, $\|\mathfrak{w}\| \le \mathsf{P}$, $\|\lambda\| \le L$ with $\rho M \le \mathsf{P}$. Then the family $\big(\mathfrak{w}(z; \lambda)\big)_{\|\lambda\| \le L}$ of solutions to $\mathfrak{w}' = \mathfrak{f}(z, \mathfrak{w})$, $\mathfrak{w}(0) = \mathfrak{o}$, is normal on $|z| < \rho$ since the family $\big(\|\mathfrak{w}'(z; \lambda)\|_\infty\big)_{\|\lambda\| \le L}$ is bounded. Thus the Theorem on analytic resp. continuous dependence on parameters may be viewed as a corollary of Montel's Criterion and uniqueness of the solutions.

1.3.8 Analytic Continuation

Analytic continuation of solutions to some system (1.2) always means analytic continuation *as solutions*.

Example 1.3 Let f be holomorphic on some domain $D \subset \mathbb{C}$ having natural boundary ∂D, that is, f cannot be continued beyond any boundary point. Then the problem $w' = 1 + (w - z)f(w)$, $w(z_0) = z_0$, has the unique solution $w = z$, which, however, exists as a solution only on D.

If analytic continuation is possible, then it is possible by solving initial value problems. Denote the solution to the initial value problem (1.2) by $\mathfrak{w}(z; z_0, \mathfrak{w}_0)$; it exists on $|z - z_0| < \rho$, where $\rho = \rho(z_0, \mathfrak{w}_0)$ is chosen as large as possible. For $0 < |z_1 - z_0| < \rho(z_0, \mathfrak{w}_0)$ the solution to $\mathfrak{w}' = \mathfrak{f}(z, \mathfrak{w})$, $\mathfrak{w}(z_1) = \mathfrak{w}(z_1; z_0, \mathfrak{w}_0) = \mathfrak{w}_1$ exists at least on $|z - z_1| < \rho(z_0, \mathfrak{w}_0) - |z_1 - z_0|$, hence $\rho(z_1, \mathfrak{w}_1) \ge \rho(z_0, \mathfrak{w}_0) - |z_1 - z_0|$ holds. Analytic continuation beyond $|z - z_0| = r(z_0, \mathfrak{w}_0)$ is obtained if and only if $\rho(z_1, \mathfrak{w}_1) > \rho(z_0, \mathfrak{w}_0) - |z_1 - z_0|$.

The solutions to the linear problem $\mathfrak{w}' = A(z)\mathfrak{w} + \mathfrak{b}(z)$, with A, $\mathfrak{b}$ holomorphic on some domain D, exist on $|z - z_0| < \operatorname{dist}(z_0, \partial D)$. Hence we have

Theorem 1.8 *The solutions to* $\mathfrak{w}' = A(z)\mathfrak{w} + \mathfrak{b}(z)$ *admit unrestricted analytic continuation on D. If D is simply connected, every solution is holomorphic on D.*

1.3.9 Painlevé's Theorem

Any solution to $\mathfrak{x}' = \mathfrak{f}(t, \mathfrak{x})$ in the real domain has a maximal interval (α, β) of existence. As $t \to \alpha+$ and $t \to \beta-$, the graph $\{(t, \mathfrak{x}(t)) : \alpha < t < \beta\}$ leaves every compact subset of the domain of definition of the right-hand side $\mathfrak{f}$, and never comes back. In the complex case a maximal domain of existence does not necessarily exist. The analogous result is known as

Painlevé's Theorem *Suppose that the solution* $\mathfrak{w}$ *to* (1.2) *admits analytic continuation along every sub-arc* $\gamma|_{[0,t]}$, $0 < t < 1$, *of the arc* $\gamma : [0, 1] \longrightarrow \mathbb{C}$, *and let the continuation along* $\gamma|_{[0,t]}$ *at the end point* $z = \gamma(t)$ *be denoted by* $\omega(t)$. *Then* $\mathfrak{w}$ *admits analytic continuation along* γ *if and only if there exists some sequence* $t_k \to 1$ *such that* $\omega(t_k) \to \omega^*$ *with* $(z^*, \omega^*) \in G$ $(z^* = \gamma(1))$. *In other words, if* $\mathfrak{w}$ *cannot be continued along* γ, *then as* $t \to 1$, *the curve* $(\gamma(t), \omega(t))|_{[0,1)}$ *leaves every compact subset of G and never comes back – it accumulates on* ∂G.

Proof For $(z^*, \omega^*) \in G$, the solution $\mathfrak{w}(z; z^*, \omega^*)$ exists on $|z - z^*| < \rho^*$. For $\epsilon > 0$ sufficiently small, the solution $\mathfrak{w}(z; \tilde{z}_0, \tilde{\mathfrak{w}}_0)$ with $|\tilde{z}_0 - z^*| < \epsilon$ and $\|\tilde{\mathfrak{w}}_0 - \mathfrak{w}^*\| < \epsilon$ exists on $|z - \tilde{z}_0| < \rho^*/2$, say. But this is true for $\tilde{z}_0 = \gamma(t_k)$ and $\tilde{\mathfrak{w}}_0 = \omega(t_k)$ if k is sufficiently large, and $\mathfrak{w}(z; z^*, \omega^*)$ provides the analytic continuation of $\mathfrak{w}(z; \gamma(t_k), \omega(t_k))$ into the end-point of γ.

Remark 1.6 It is obvious that analytic continuation respects holomorphic and continuous dependence on parameters. If, for $\lambda = \lambda_0$, the solution $\mathfrak{w}(z; \lambda_0)$ to the initial value problem $\mathfrak{w}' = \mathfrak{f}(z, \mathfrak{w}, \lambda_0)$, $\mathfrak{w}(z_0) = \mathfrak{w}_0$, admits analytic continuation along γ, then this is also true for $\mathfrak{w}(z; \lambda)$ if λ is sufficiently close to λ_0.

Example $w' = \lambda + w^2$, $w(0) = 0$, with $w(z; \lambda) = \sqrt{\lambda}\tan(\sqrt{\lambda}z)$.

We will give two applications of Painlevé's Theorem.

Example 1.4 The first application concerns Riccati equations

$$w' = a(z) + b(z)w + c(z)w^2, \tag{1.6}$$

where a, b, c are holomorphic on some domain $D \subset \mathbb{C}$ and $c(z) \not\equiv 0$. Then *every solution to Eq.* (1.6) *admits unrestricted analytic continuation in D, the only singularities are poles.* The case when (with the notation used in the proof of Painlevé's Theorem) $\omega(t_k)$ tends to some finite limit ω^* is settled by Painlevé's Theorem. If, however, $\omega(t)$ tends to ∞ as $t \to 1$, Painlevé's Theorem applies to the equation $y' = -c(z) - b(z)y - a(z)y^2$ for $y = 1/w$, and z^* is a zero of y and a pole of w; $y \not\equiv 0$ follows from $c(z) \not\equiv 0$.

Remark 1.7 We note that even for differential equations $w' = P(w)$, P any polynomial of degree $\deg P > 2$, singularities other than poles may occur. A simple example is $w' = w^3$; the non-trivial solutions $w = (-2z + c)^{-1/2}$ have an algebraic pole at $z = \frac{1}{2}c = z_0 + \frac{1}{2}w_0^{-2}$. In the present case the term 'moving singularity' is self-explanatory.

Exercise 1.15 A different proof for Riccati equations is available. Prove that there are (infinitely many) linear systems $u' = a_{11}(z)u + a_{12}(z)v$, $v' = a_{21}(z)u + a_{22}(z)v$ with coefficients related to a, b, c, such that $w = u/v$ solves Eq. (1.6), and every solution is obtained this way.

Example 1.5 The next application concerns the implicit initial value problem

$$w'^2 = P(w), \ w(z_0) = w_0 \quad (P(w_0) \neq 0), \tag{1.7}$$

where P is any polynomial of degree at most 4. The problem is equivalent to $w' = v$, $v' = \frac{1}{2}P'(w)$, $w(z_0) = w_0$, $v(z_0) = w_0'$, if only initial values with $w_0'^2 = P(w_0)$ are admitted. We claim that *every solution to Eq.* (1.7) *is meromorphic in the plane.* To prove this we again have to discuss two possibilities: if $\omega(t_k)$ (notation as above) is bounded, then so is $P(\omega(t_k))$. Assuming $\omega(t_k) \to \omega^*$, Painlevé's Theorem immediately applies to some subsequence such that $(P(\omega(t_{k_j})))^{1/2}$ converges. If, however, $\omega(t) \to \infty$, the transformation $w = 1/y$ leads to $y'^2 = Q(z, y) = y^4 P(z, 1/y)$ ($\deg_y Q \le 4$), and Painlevé's Theorem yields a zero of y and a pole of the original solution. The theory of elliptic functions may be built upon Eq. (1.7).

For further applications of Painlevé's Theorem, in particular to implicit first-order equations $P(z, w, w') = 0$ and higher-order equations and systems, the reader is referred to the textbooks quoted in the beginning of this section, and, of course, to Chaps. 5 and 6.

1.4 Asymptotic Expansions

In this section we will consider infinite series $\sum_{\nu=m}^{\infty} a_\nu z^{-\nu/q}$ which, whether convergent or not, in some sense represent holomorphic functions on open sectors $|\arg z - \theta_0| < \eta$; q and m are integers.

1.4.1 Asymptotic Series

Although $q > 1$ and $m < 0$ are unavoidable in the asymptotic theory of algebraic differential equations, it suffices to consider holomorphic functions f and series $\sum_{\nu=0}^{\infty} a_\nu z^{-\nu}$ on sectors $S : |\arg z| < \eta$. Then f is said to have an *asymptotic expansion*

$$f(z) \sim \sum_{\nu=0}^{\infty} a_\nu z^{-\nu} \text{ on } S$$

if for every $n \in \mathbb{N}_0$ and every proper sub-sector $S_\delta : |\arg z| < \eta - \delta$,

$$f(z) = \sum_{\nu=0}^{n} a_\nu z^{-\nu} + O(|z|^{-n-1}) \text{ as } z \to \infty \text{ on } S_\delta$$

holds. We note some elementary facts:

- the coefficients a_ν are uniquely determined: $a_0 = \lim_{z\to\infty} f(z)$, $a_1 = \lim_{z\to\infty} z(f(z) - a_0)$, $a_2 = \lim_{z\to\infty} z^2(f(z) - a_0 - a_1 z^{-1})$, etc.;
- the derivative of f has the asymptotic expansion $f'(z) \sim \sum_{\nu=0}^{\infty} -\nu a_\nu z^{-\nu-1}$;
- $f(z) \sim \sum_{\nu=2}^{\infty} a_\nu z^{-\nu}$ has a primitive $F(z) \sim \sum_{\nu=2}^{\infty} \frac{a_\nu}{-\nu+1} z^{-\nu+1}$;
- if g also has an asymptotic expansion on S, then so have $f + g$, fg, and g/f (provided $a_0 \neq 0$). The coefficients are obtained in an obvious way; for example, the product fg has the asymptotic expansion $\sum_{n=0}^{\infty} c_n z^{-n}$ with $c_n = \sum_{\nu=0}^{n} a_\nu b_{n-\nu}$.

The second assertion requires a *proof.* Given $\delta > 0$, Cauchy's Theorem applies to $f_n(\zeta) = f(\zeta) - \sum_{\nu=0}^{n} a_\nu \zeta^{-\nu}$ on the disc $|\zeta - z| \le |z| \sin\delta$ ($z \in S_{2\delta}$) and yields

$$f'(z) = \sum_{\nu=0}^{n} -\nu a_\nu z^{-\nu-1} + \frac{1}{2\pi i} \int_{|\zeta - z| = |z|\sin\delta} \frac{f_n(\zeta)}{(\zeta - z)^2}\, d\zeta.$$

Since $f_n(\zeta) = O(|\zeta|^{-n-1}) = O(|z|^{-n-1})$ as $z \to \infty$, uniformly on $|\zeta - z| = |z| \sin\delta$, we obtain $f'(z) = \sum_{\nu=0}^{n} -\nu a_\nu z^{-\nu-1} + O(|z|^{-n-2})$ as $z \to \infty$ on $S_{2\delta}$. ☕

Example 1.6 Asymptotic series representing holomorphic functions need not be convergent. In fact, convergence at z_0 implies convergence on $|z| > |z_0|$. On the other hand, *every* asymptotic series represents some holomorphic function on some sector. One of the most famous asymptotic expansions is known as *Stirling's formula,*

$$\log \varGamma(z) - \tfrac{1}{2}\log 2\pi - (z - \tfrac{1}{2})\log z + z \sim \sum_{n=1}^{\infty} \frac{B_{2n}}{2n(2n-1)} z^{-2n+1};$$

B_n is the n-th Bernoulli number. The series converges nowhere.

Exercise 1.16 Suppose $f(z) \sim \sum_{k=0}^{\infty} c_k z^{-k}$ holds on $|\arg z| < \eta$, and let $\varPhi$ be holomorphic on some neighbourhood of c_0. Prove that $\varPhi \circ f$ has an asymptotic expansion on $|\arg z| < \eta - \delta$, $|z| > r_\delta$, $\delta > 0$ arbitrary.

1.4.2 Asymptotic Integration of Differential Equations

We quote from Wasow [197] two jewels in the asymptotic theory of differential equations, and start with linear systems

$$z^{-q}\mathfrak{y}' = A(z)\mathfrak{y} \tag{1.8}$$

of *rank* $q+1 \geq 1$; A is a holomorphic $n \times n$-matrix that has an asymptotic expansion

$$A(z) \sim \sum_{r=0}^{\infty} A_r z^{-r} \quad (A_0 \neq \mathcal{O})$$

on some sector S with central angle $\frac{\pi}{q+1}$.

Theorem 1.9 (Wasow [197], Theorem 12.3 and 19.1) *There exists a* holomorphic *fundamental matrix*

$$Y(z) = \hat{Y}(z) z^G \operatorname{diag}\,(e^{Q_1(z)}, \ldots, e^{Q_n(z)})$$

such that

- $\hat{Y}(z) \sim \sum_{k=0}^{\infty} \hat{Y}_k z^{-k/p}$ *holds on S with* $\hat{Y}_0 \neq \mathcal{O}$; $p \in \{1, \ldots, n\}$ *is some integer;*
- *G is a constant matrix;*
- $Q_1(z), \ldots, Q_n(z)$ *are polynomials in* $z^{1/p}$.

If, in addition, the eigenvalues $\lambda_1, \ldots, \lambda_n$ of A_0 are mutually distinct, then $p = 1$, $\det \hat{Y}_0 \neq 0$, *G is a diagonal matrix, and Q_ν has leading term* $\frac{\lambda_\nu}{q+1} z^{q+1}$.

Remark 1.8 Any linear differential equation

$$w^{(n)} + a_{n-1}(z)w^{(n-1)} + \cdots + a_1(z)w' + a_0(z)w = b(z) \tag{1.9}$$

with analytic coefficients having asymptotic expansions on some sector may be transformed into some linear system (1.8). In particular, this holds if the coefficients a_ν and b are polynomials. The procedure, however, is not unique and does not always reveal the rank. In most applications it suffices to have the analogue of Theorem 1.9 for linear equations in the much older form going back to Sternberg [187], see Theorem 5.3.

The next theorem deals with nonlinear systems of differential equations

$$z^{-q}\mathfrak{y}' = A\mathfrak{y} + z^{-1}\mathfrak{f}(z, \mathfrak{y}), \tag{1.10}$$

and will be stated in a form that is adapted to our purposes. Again q is some non-negative integer, S is any sector with central angle $\le \frac{\pi}{q+1}$, A is a regular $n\times n$-matrix, and $\mathfrak{f}$ is holomorphic on $S \times \{\mathfrak{y} : \|\mathfrak{y}\| < \delta\}$ with asymptotic expansion

$$\mathfrak{f}(z, \mathfrak{y}) \sim \sum_{r=0}^{\infty} \mathfrak{f}_r(\mathfrak{y})z^{-r} \quad (\mathfrak{f}_r \text{ holomorphic on } \|\mathfrak{y}\| < \delta),$$

uniformly with respect to $\mathfrak{y}$ on every closed sub-sector of S.

Theorem 1.10 (Wasow [197], Theorem 12.1 and 14.1) *Under the assumptions stated above, the system* (1.10) *has a solution with asymptotic expansion*

$$\mathfrak{y} \sim \sum_{k=1}^{\infty} \mathfrak{c}_k z^{-k} \quad \text{on } S.$$

Remark 1.9 The asymptotic series may represent several solutions. The coefficients $\mathfrak{c}_k$ are uniquely determined: from $A\mathfrak{y} = z^{-q}\mathfrak{y}' - z^{-1}\mathfrak{f}(z, \mathfrak{y})$ it follows that

$$A\mathfrak{c}_1 = -\lim_{z\to\infty} \mathfrak{f}(z, \mathfrak{o}) \quad \text{and} \quad A\mathfrak{c}_{k+1} = \mathfrak{P}_k(\mathfrak{c}_j : j \le k),$$

where $\mathfrak{P}_k$ is a vector-valued polynomial in the variables $\mathfrak{c}_j$, with coefficients arising from the coefficients in the asymptotic expansion of $\mathfrak{f}$. In particular, the asymptotic series coincides with the unique formal series solution.

Example 1.7 Theorem 1.10 does not immediately apply to the differential equation $w' = z^2 + w^2$. It is, however, obvious that the principal terms of asymptotic solutions, if any, are $\pm iz$. We set $w = y \pm iz$ to obtain the differential equation

$$z^{-1}y' = \pm 2iy + z^{-1}(y^2 \mp i),$$

to which Theorem 1.10 applies. Given any quadrant S there exists a solution $w_\pm$ having an asymptotic expansion on S with principal term $\pm iz$. It is not hard to show that $w_\pm(z) \sim \pm iz + \frac{1}{2}z^{-1} \pm \frac{3}{8}z^{-3} - \frac{3}{4}z^{-5} \pm \cdots$ holds.

Exercise 1.17 Theorem 1.11 also does not apply to $w' = z - w^2$. It is plausible that the first approximation is $w \sim \pm\sqrt{z}$. The change of variables $z = t^2$, $w(z) = tv(t)$ leads to $t^{-2}\dot{v} = 2-2v^2-t^{-3}v = 2(1-v)(1+v)-t^{-3}v$. Theorem 1.11 applies to the differential equation for $y = v \mp 1$. Compute the first few terms of the asymptotic expansion for y to obtain the asymptotic series $\pm z^{\frac{1}{2}} - \frac{1}{4}z^{-1} \mp \frac{5}{32}z^{-5/2} - \frac{15}{64}z^{-4} + \cdots$ in the z-plane. To every sector with central angle $2\pi/3$ there exist some solution $w_\pm$ having this asymptotic expansion.

1.4.3 Asymptotic Integration of Algebraic Differential Equations

In contrast to Theorems 1.9 and 1.10, the next theorem on the existence of asymptotic expansions applies to *specific* solutions. The equation we will consider is

$$Q[w] = P(z, w); \tag{1.11}$$

- $P(z, w)$ is a polynomial in w and rational in z;
- $Q[w]$ is a finite sum of terms $a_M(z)w^{\ell_M}M[w]$ with
  $a_M(z) = A_M z^{\alpha_M}(1 + o(1))$ as $z \to \infty$, $A_M \neq 0$, $\alpha_M \in \mathbb{Z}$,
  $M[w] = w'^{\ell_1} \cdots w^{(m)\ell_m}$, $\ell_\nu = \ell_\nu(M)$, and
  $\mathsf{d}_M = 2\ell_1 + \cdots + (m+1)\ell_m \geq \alpha_M + 2$ for every M.

Theorem 1.11 *Let w be any solution to the algebraic differential equation* (1.11) *satisfying $w(z) \to c_0$ as $z \to \infty$ on some sector S, and assume that $P(z, c_0) \to 0$ and $P_w(z, c_0) \to c \neq 0$ as $z \to \infty$ on S. Then w has an asymptotic expansion*

$$w \sim \sum_{k=0}^{\infty} c_k z^{-k} \quad \text{on } S.$$

Proof We will start with

$$w(z) = \sum_{\nu=0}^{n} c_\nu z^{-\nu} + o(|z|^{-n}) = \psi_n(z) + o(|z|^{-n}), \tag{1.12}$$

which is true for $n = 0$. To proceed we need

$$w^{\ell_M}M[w] = \psi_n^{\ell_M}M[\psi_n] + o(|z|^{-\mathsf{d}_M-n+1}) \quad (z \to \infty); \tag{1.13}$$

the proof will be given below. Using (1.13) we obtain

$$Q[w] = Q[\psi_n] + o(|z|^{\max_M(\alpha_M - \mathrm{d}_M) - n + 1}) = Q[\psi_n] + o(|z|^{-n-1}),$$

hence w satisfies

$$P(z, w) = Q[\psi_n](z) + o(|z|^{-n-1}). \tag{1.14}$$

It follows from $P(z, c_0) \to 0$ and $P_w(z, c_0) \to c \neq 0$ that for every rational function R that tends to zero as $z \to \infty$, the algebraic equation $P(z, y) = R(z)$ has a unique solution that tends to c_0 as $z \to \infty$. In particular, the equation

$$P(z, y) = Q[\psi_n](z) \tag{1.15}$$

has a unique solution $y_n(z) = \sum_{\nu=0}^{\infty} a_\nu^{[n]} z^{-\nu}$ about $z = \infty$ with $a_0^{[n]} = c_0$, and from (1.14), (1.15), and

$$P(z, w(z)) - P(z, y_n(z)) = \int_{y_n(z)}^{w(z)} P_\zeta(z, \zeta)\, d\zeta = (c + o(1))(w(z) - y_n(z))$$

as $z \to \infty$ (we integrate along the line segment in $|\zeta - c_0| < \delta$ from $y_n(z)$ to $w(z)$) it follows that $w(z) - y_n(z) = o(|z|^{-n-1})$, hence (1.12) holds with $n+1$ and $\psi_{n+1}(z) = \sum_{\nu=0}^{n+1} a_\nu^{[n]} z^{-\nu}$ in place of n and ψ_n, respectively; we note that $a_\nu^{[k]} = c_\nu$ holds for $0 \le \nu \le k \le n$, while the new coefficient is $c_{n+1} = a_{n+1}^{[n]}$. To prove (1.13) we first consider the case of $\psi_n(z) \equiv c_0$, hence $M[\psi_n] = 0$. Then

$$\begin{aligned} w^{(k)}(z) &= \psi_n^{(k)}(z) + o(|z|^{-n-k}) = o(|z|^{-n-k}), \\ w^{\ell_M} M[w] &= o(|z|^{\ell_1(-n-1) + \cdots + l_m(-n-m)}) = o(|z|^{-\mathrm{d}_M - n + 1}) \end{aligned}$$

is true since $-\sum_{k=1}^{m}(n+k)\ell_k = -\mathrm{d}_M - (n-1)\sum_{k=1}^{m} \ell_k \le -\mathrm{d}_M - n + 1$. Thus (1.13) holds with $M[\psi_n] \equiv 0$. Now let c_ν $(\nu \ge 1)$ be the first non-zero coefficient. Then

$$\begin{aligned} \psi_n^{(k)}(z) &= O(|z|^{-\nu-k}), \\ w^{(k)}(z) &= \psi_n^{(k)}(z) + o(|z|^{-n-k}) = O(|z|^{-\nu-k}), \\ (w^{(k)}(z))^{\ell_k} &= (\psi_n^{(k)}(z))^{\ell_k} + O(|z|^{-(\nu+k)(\ell_k-1)}) o(|z|^{-n-k}) \\ &= (\psi_n^{(k)}(z))^{\ell_k} + o(|z|^{-k\ell_k - n - \nu(\ell_k-1)}), \end{aligned}$$

and $M[w] = M[\psi_n] + R_n$ holds, with remainder term

$$\begin{aligned} R_n &= \sum_{\ell_j > 0} \prod_{k \neq j} (\psi_n^{(k)}(z))^{\ell_k} o(|z|^{-j\ell_j - n - \nu(\ell_j - 1)}) \\ &= \sum_{\ell_j > 0} o\Big(|z|^{-\sum_{k \neq j} \ell_k(k+\nu) - j\ell_j - n - \nu(\ell_j - 1)}\Big) = o(|z|^{-\mathrm{d}_M - n + 1}). \end{aligned}$$

This is true since now $\nu - \sum_{k=1}^m \ell_k(\nu-1) \le 1$, and thus it follows that

$$\begin{aligned} w^{\ell_M} M[w] &= \psi_n^{\ell_M} M[\psi_n] + o(|z|^{-n}) M[\psi_n] + o(|z|^{-\mathrm{d}_M - n + 1}) \\ &= \psi_n^{\ell_M} M[\psi_n] + o\Big(|z|^{-n-\sum_{k=1}^m (k+\nu)\ell_k}\Big) + o(|z|^{-\mathrm{d}_M - n+1}) \\ &= \psi_n^{\ell_M} M[\psi_n] + o(|z|^{-\mathrm{d}_M - n+1}), \end{aligned}$$

this finishing the proof of Theorem 1.11.

1.5 Miscellanea

1.5.1 Elliptic Functions

Non-constant meromorphic functions with $\mathbb{R}$-linearly independent primitive periods ϑ_1 and ϑ_2 ($\mathrm{Im}\,\frac{\vartheta_2}{\vartheta_1} \ne 0$) are called *elliptic* (also doubly periodic). The periods span the lattice $\mathscr{L} = \vartheta_1\mathbb{Z} \oplus \vartheta_2\mathbb{Z}$, and the basis may be chosen to satisfy $|\vartheta_1| \le |\vartheta_2| \le |\vartheta_1 \pm \vartheta_2|$. For every basis the parallelogram $P = \{s\vartheta_1 + t\vartheta_2 : 0 \le s, t < 1\}$ has the same area $|\mathrm{Im}\,(\overline{\vartheta}_1\vartheta_2)|$. The basic properties of elliptic functions are known as

Liouville's Theorems *Let f be an elliptic function and $(\vartheta_1, \vartheta_2)$ any basis of the corresponding lattice. Then*

- *f has (infinitely many) poles;*
- $\sum_{z\in P} \mathrm{res}_z f = 0$;
- *in every parallelogram $c + P$, f assumes every value in $\widehat{\mathbb{C}}$ equally often.*

This number is called the *elliptic order*; it is also the mapping degree of $f : P \longrightarrow \widehat{\mathbb{C}}$. Every elliptic function f may be written as $f = R(\wp) + S(\wp)\wp'$, where R and S are uniquely determined rational functions, and $\wp$ denotes the fundamental

Weierstraß P-function Given any lattice $\mathscr{L}$, the Weierstraß P-function is defined by the Mittag-Leffler series

$$\wp(z) = \frac{1}{z^2} + \sum_{\vartheta\in\mathscr{L}\setminus\{0\}} \frac{1}{(z-\vartheta)^2} - \frac{1}{\vartheta^2}; \tag{1.16}$$

its periodicity is proved via the periodicity of $\wp'(z) = -\sum_{\vartheta\in\mathscr{L}} \dfrac{2}{(z-\vartheta)^3}$ and $\wp(-z) = \wp(z)$. The P-function has elliptic order two and is the negative derivative of the

- Weierstraß *Zeta-function* $\zeta(z) = \dfrac{1}{z} + \sum_{\vartheta\in\mathscr{L}\setminus\{0\}} \dfrac{1}{z-\vartheta} + \dfrac{1}{\vartheta} + \dfrac{z}{\vartheta^2}$, which itself is the logarithmic derivative of the

- Weierstraß *Sigma-function* $\sigma(z) = z \prod_{\vartheta \in \mathscr{L} \setminus \{0\}} \left(1 - \frac{z}{\vartheta}\right) e^{\frac{z}{\vartheta} + \frac{z^2}{2\vartheta^2}}$.

The series (on $\mathbb{C} \setminus \mathscr{L}$) and the product converge absolutely and locally uniformly since $\sum_{\vartheta \in \mathscr{L} \setminus \{0\}} |\vartheta|^{-s}$ converges for $s > 2$. The P-function satisfies

$$\wp'^2 = 4\wp^3 - g_2\wp - g_3 = 4(\wp - e_1)(\wp - e_2)(\wp - e_3)$$

with $g_2 = 60 \sum_{\vartheta \in \mathscr{L} \setminus \{0\}} \vartheta^{-4}$, $g_3 = 140 \sum_{\vartheta \in \mathscr{L} \setminus \{0\}} \vartheta^{-6}$ and $g_2^3 - 27g_3^2 \neq 0$, and

$$e_1 = \wp(\tfrac{1}{2}\vartheta_1),\ e_2 = \wp(\tfrac{1}{2}\vartheta_2),\ e_3 = \wp(\tfrac{1}{2}(\vartheta_1 + \vartheta_2)) \quad (e_1 + e_2 + e_3 = 0).$$

The critical values e_1, e_2, e_3, and $e_4 = \infty$ are always assumed with multiplicity two.

Jacobi's Elliptic Functions The theory of elliptic functions can also be built on Jacobi's functions sn ('sinus amplitudinis'), cn ('cosinus amplitudinis'), and dn ('delta amplitudinis'), which may be defined by the initial value problem

$$\begin{aligned} x_1' &= x_2x_3, \ x_1(0) = 0, \\ x_2' &= -x_1x_3, \ x_2(0) = 1, \\ x_3' &= -\kappa^2 x_1x_2, \ x_3(0) = 1 \qquad (\kappa^2 \neq 0, 1). \end{aligned}$$

Exercise 1.18 Deduce $x_1^2 + x_2^2 = \kappa^2 x_1^2 + x_3^2 = 1$ and

$$\begin{aligned} x_1'^2 &= (1 - x_1^2)(1 - \kappa^2 x_1^2), \\ x_2'^2 &= (1 - x_2^2)(1 - \kappa^2 + \kappa^2 x_2^2), \\ x_3'^2 &= (1 - x_3^2)(\kappa^2 - 1 + x_3^2). \end{aligned}$$

(**Hint.** Compute $x_1x_1' + x_2x_2'$ etc.)

For $\kappa^2 \neq 0, 1$ the solutions are elliptic functions. Note, however, that they have different period lattices, in classical notation spanned by $(4K, 2iK')$, $(4K, 2K + 2iK')$, and $(2K, 4iK')$, respectively (Fig. 1.1). For $\kappa^2 = 0$, sn and cn degenerate to the ordinary sine and cosine. What happens if $\kappa^2 = 1$?

1.5.2 *The Phragmén–Lindelöf Principle*

Let D be any domain and assume that the boundary of D is divided into two disjoint parts α and β. The unique bounded harmonic function on D with boundary values 1 on $\alpha \setminus \bar{\beta}$ and 0 on $\beta \setminus \bar{\alpha}$ (one cannot expect boundary values on $\bar{\alpha} \cap \bar{\beta}$) is called the *harmonic measure* of α and denoted $\omega(z, \alpha, D)$. We will not discuss conditions

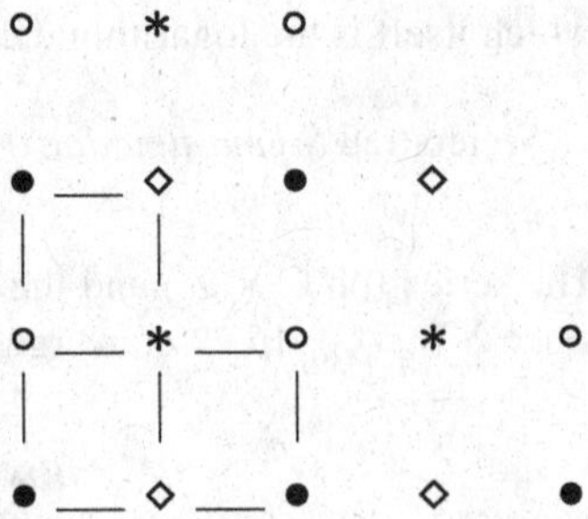

Fig. 1.1 The distribution of zeros of sn $(K, K' > 0)$; $\frac{d}{dz}\text{sn}\,\bullet = \text{cn}\,\bullet = \text{dn}\,\bullet = 1$; $\frac{d}{dz}\text{sn}\,\circ = 1$, $\text{cn}\,\circ = \text{dn}\,\circ = -1$; $\frac{d}{dz}\text{sn}\,* = \text{dn}\,* = -1$, $\text{cn}\,* = 1$; $\frac{d}{dz}\text{sn}\,\diamond = \text{cn}\,\diamond = -1$, $\text{dn}\,\diamond = 1$

that ensure the existence of ω, but work only in cases where ω is explicitly known or may be estimated appropriately.

Exercise 1.19 Prove that

- $\omega(z, \alpha, D) \le \omega(\phi(z), \phi(\alpha), D')$ holds on D, where $\phi : D \longrightarrow D'$ is holomorphic and has a continuous extension to ∂D such that $\phi(\alpha) \subset \partial D'$.
- $\omega(z, \alpha, D) \le \omega(z, \alpha, D')$ holds on D if $D \subset D'$ and $\alpha \subset \partial D \cap \partial D'$.

Determine the harmonic measure in the following cases:

- $D = \{z : 0 < \arg z < \theta\}$, $\alpha = (0, \infty)$.
- $D = \{z \in \mathbb{D} : \operatorname{Im} z > 0\}$, $\alpha = (-1, 1)$. (**Hint.** Thales' Theorem.)

Exercise 1.20 Let f be holomorphic on some domain D with boundary values $\limsup_{z\to\alpha} |f(z)| \le M$ and $\limsup_{z\to\beta} |f(z)| \le 1$. Prove the

Two-Constants Theorem $\quad |f(z)| \le M^{\omega(z,\alpha,D)} \quad (z \in D)$.

(**Hint.** $\log|f|$ may be viewed as a subharmonic function on D or a harmonic function on $D \setminus \{\text{zeros of } f\}$.)

The Phragmén–Lindelöf Principle is an immediate corollary of the Two-Constants Theorem when applied to holomorphic functions f on sectors Σ. We assume that

$$M_\Sigma(r,f) = \sup\{|f(z)| : |z| = r,\ z \in \Sigma\} \quad (r > 0)$$

is finite. In its simplest (and classical) form it says the following:

Theorem 1.12 *Let f be holomorphic on the upper half-plane $\mathbb{H}$ with boundary values* $\limsup_{z\to x} |f(z)| \le 1$ $(x \in \mathbb{R})$. *Then either* $M_{\mathbb{H}}(r,f) \le 1$ $(r > 0)$ *or else* $\liminf_{r\to\infty} \frac{1}{r} \log M_{\mathbb{H}}(r,f) > 0$ *holds.*

Proof The harmonic measure of the semi-circle $\alpha : |z| = r,\ \operatorname{Im} z > 0$ measured in the semi-disc $D : |z| < r,\ \operatorname{Im} z > 0$ is

$$\omega(z, \alpha, D) = \frac{2}{\pi} \arctan \frac{2r \operatorname{Im} z}{r^2 - |z|^2} \le \frac{4r|z|}{\pi(r^2 - |z|^2)}$$

(see Exercise 1.19), and the Two-Constants yields

$$\log|f(z)| \le \frac{4r|z|}{\pi(r^2-|z|^2)}\log M_{\mathbb{H}}(r,f) = \frac{4r^2|z|}{\pi(r^2-|z|^2)}\frac{\log M_{\mathbb{H}}(r,f)}{r}.$$ ☕

Exercise 1.21 Prove that

- the alternative in Theorem 1.12 remains true if f is holomorphic only on the domain $\mathbb{H}\setminus\{z:|z|\le r_0\}$ and satisfies $|f(z)|\le 1$ on $|z|=r_0$, $\operatorname{Im} z>0$. (**Hint.** $\omega(z,\alpha_r,D\cap D_r)\le\omega(z,\alpha_r,D_r)$.)
- replacing the upper half-plane $\mathbb{H}$ in Theorem 1.12 with any sector Σ with central angle δ turns the alternative into $M_\Sigma(r,f)\le 1$ or $\liminf_{r\to\infty}\frac{1}{r^{\pi/\delta}}\log M_\Sigma(r,f)>0$. (**Hint.** Consider $f(z^{\delta/\pi})$ on some appropriate half-plane.) ☕

The version of the Phragmén–Lindelöf Principle we will use here is

Theorem 1.13 *Let f be holomorphic on the upper half-plane $\mathbb{H}$ and continuous on $\mathbb{H}\cup\mathbb{R}$, and assume $\log^+|f(z)|=o(|z|)$ as $z\to\infty$ on $\mathbb{H}$. Then f is bounded on $\mathbb{H}$ if it is bounded on $\mathbb{R}$, in which case the following is true:*

- *If f tends to a as $z=x\to+\infty$ along the positive real axis, then f tends to a as $z\to\infty$ on every sector $0\le\arg z\le\pi-\epsilon<\pi$.*
- *If, in addition, f has a limit as $z=x\to-\infty$, then f tends to a as $z\to\infty$ on $\mathbb{H}$.*

Proof We assume that f is bounded on $\mathbb{R}$, hence also on $\mathbb{H}$ by Theorem 1.12. To prove the first assertion we may assume that $a=0$ and $|f(z)|\le 1$ on $\mathbb{H}$ (otherwise consider $(f(z)-a)/M$ in place of f, M sufficiently large). Given $\epsilon>0$ there exists some $x_0>0$ such that $|f(x)|<\epsilon$ holds on (x_0,∞), hence

$$|f(z)|<\epsilon^{\omega(z,(x_0,\infty),\mathbb{H})}=\epsilon^{(1-\frac{1}{\pi}\arg(z-x_0))}\le\epsilon^{\delta/2\pi}$$

holds on $0\le\arg(z-x_0)\le\pi-\delta/2$ by the Two-Constants Theorem, and also on $0\le\arg z\le\pi-\delta$, $|z|>x_0$ (by the 'inscribed angle Theorem', the rays $\arg z=\pi-\delta$ and $\arg(z-x_0)=\pi-\delta/2$ intersect at $x_0e^{i(\pi-\delta)}$). To prove the second assertion one has just to combine the first one with its counterpart on the negative real axis, noting that $\lim_{y\to+\infty}f(iy)=a$. ☕

Exercise 1.22 Prove that Theorem 1.13 *mutatis mutandis* remains true for functions f satisfying $\log^+|f(z)|=o(|z|^{\pi/\delta})$ as $z\to\infty$ on $\Sigma:|\arg z|<\delta/2$, $|z|>r_0$, and having finite limits as $z\to\infty$ on the rays $\sigma_\pm:\arg z=\pm\delta/2$. ☕

As a corollary we obtain:

Corollary 1.1 *Let f be holomorphic on $|\arg z|<\eta$, $|z|\ge r_0$ of finite order of growth ($\log^+|f(z)|=O(|z|^k)$ as $z\to\infty$), and assume that*

$$f(z)=\begin{cases}Ae^{\alpha z}z^{\lambda}(1+o(1)) & (z=re^{i\delta}\to\infty)\\ Be^{\beta z}z^{\mu}(1+o(1)) & (z=re^{-i\delta}\to\infty)\end{cases}$$

$(A, B, \alpha, \beta, \lambda, \mu \in \mathbb{C},\ AB \neq 0)$ *holds for every* $0 < \delta < \eta$. *Then*

- $\operatorname{Re}\alpha = \operatorname{Re}\beta$ *and* $\operatorname{Im}\alpha \le \operatorname{Im}\beta$, *and*
- $\alpha = \beta$ *implies* $f(z) = Ae^{\alpha z}z^{\lambda}(1 + o(1))$ *as* $z \to \infty$ *on* $|\arg z| \le \eta/2$, *say; in particular,* f *has only finitely many zeros on* $|\arg z| \le \eta/2$.

Proof To prove the first assertion we assume $\operatorname{Re}\alpha \neq \operatorname{Re}\beta$ and even $\operatorname{Re}\alpha > \operatorname{Re}\beta$ (otherwise consider $\overline{f(\bar z)}$ instead of f to replace α, β with $\bar\beta, \bar\alpha$). Then

$$h(z) = e^{-\alpha z}z^{-\lambda}f(z) \tag{1.17}$$

has finite order of growth on $|\arg z| < \eta$ and satisfies $h(re^{i\delta}) \to A \neq 0$ and

$$|f(re^{-i\delta})| = O(r^{\operatorname{Re}(\mu-\lambda)})\, e^{\operatorname{Re}(\beta-\alpha)r\cos\delta + \operatorname{Im}(\beta-\alpha)r\sin\delta} \tag{1.18}$$

as $r \to \infty$. The right-hand side of (1.18) tends to zero if $\delta > 0$ is chosen sufficiently small. This, however, contradicts Theorem 1.13 in the form given in Exercise 1.22. In the same way we obtain a contradiction if we assume $\operatorname{Re}\alpha = \operatorname{Re}\beta$ and $\operatorname{Im}\alpha > \operatorname{Im}\beta$; then h tends to A as $z = re^{i\delta} \to \infty$, while (1.18) for $\operatorname{Re}\alpha = \operatorname{Re}\beta$ again implies $f(re^{-i\delta}) \to 0$ as $r \to \infty$. This proves $\operatorname{Im}\alpha \le \operatorname{Im}\beta$. To prove the second assertion we assume ($\alpha = \beta$ and) $\operatorname{Re}\lambda > \operatorname{Re}\mu$. Then h, again defined by (1.17), tends to $A \neq 0$ as $z = re^{i\delta} \to \infty$, while $|h(re^{-i\delta})| = O\big(r^{\operatorname{Re}(\mu-\lambda)}\big)$ tends to zero as $r \to \infty$. This proves $\operatorname{Re}\lambda \le \operatorname{Re}\mu$, and in the same way we obtain $\operatorname{Re}\lambda \ge \operatorname{Re}\mu$ (replace f with $\overline{f(\bar z)}$), hence $\operatorname{Re}\lambda = \operatorname{Re}\mu = \kappa$, and h is bounded and tends to A as $z = re^{i\delta} \to \infty$. By the Phragmén–Lindelöf Principle, $h(z) \to A$ as $z \to \infty$ is true on $0 \le \arg z \le \delta$; in particular, $f(x) = Ae^{\alpha x}x^{\kappa + i\operatorname{Im}\lambda}(1 + o(1))$ holds as $x \to +\infty$ (x real). In the same way we obtain $f(x) = Be^{\alpha x}x^{\kappa + i\operatorname{Im}\mu}(1 + o(1))$, hence $\lambda = \mu$, $A = B$, and $f(z) = Ae^{\alpha z}z^{\lambda}(1 + o(1))$ holds on $|\arg z| < \eta$. ☕

Exercise 1.23 Let f be an entire function of finite order ($\log^+|f(z)| = O(|z|^k)$), and assume that $f(z) = e^{\alpha z^2 + a_1 z + a_0}z^{\lambda}(1 + o(1))$ and $f(z) = e^{\beta z^2 + b_1 z + b_0}z^{\mu}(1 + o(1))$ holds as $z \to \infty$ on $\delta < \arg z < \pi - \delta$ and $-\pi + \delta < \arg z < -\delta$, respectively ($0 < \delta < \pi/2$ arbitrary). Prove $\alpha = \beta$ and $\operatorname{Im} a_1 \le \operatorname{Im} b_1$. Prove also that $f(z) = e^{\alpha z^2 + a_1 z}P(z)$ holds with some polynomial P if $a_1 = b_1$.
(**Hint.** Consider $f(\sqrt{z})$ with $\operatorname{Re}\sqrt{z} > 0$ and $f(\sqrt{-z})$ with $\operatorname{Re}\sqrt{-z} > 0$ on $|\arg z| < \frac{\pi}{2}$ and $|\arg(-z)| < \frac{\pi}{2}$, respectively.) ☕

1.5.3 Wiman–Valiron Theory

Let $g(z) = \sum_{n=0}^{\infty} a_n z^n$ be a transcendental entire function. Since $|a_n|r^n$ tends to zero as $n \to \infty$, the sequence $(|a_n|r^n)$ has a maximum called the *maximum term*,

$$\mu(r) = \max_{0 \le n < \infty} |a_n|r^n. \tag{1.19}$$

Exercise 1.24 The inequality $\mu(r) \le M(r,g) = \max_{|z|=r} |g(z)|$ (maximum modulus) is obvious. Prove $M(r,g) \le 2\mu(2r)$.

The greatest index such that $|a_n|r^n = \mu(r)$ is called the *central index*,

$$\nu(r) = \max\{n : |a_n|r^n = \mu(r)\}. \tag{1.20}$$

The maximum term is (strictly) increasing since

$$\mu(r) = |a_{\nu(r)}|r^{\nu(r)} < |a_{\nu(r)}|\rho^{\nu(r)} \le \mu(\rho) \quad (r < \rho,\ \nu(r) > 0).$$

By the same reasoning (and its definition), the central index is also increasing, hence is piecewise constant with discontinuities (jumps) at $0 \le r_0 < r_1 < r_2 < \cdots$. From $\mu'(r)/\mu(r) = \nu(r)/r$ on each interval $r_k < r < r_{k+1}$ it follows that

$$\log \mu(r) - \log \mu(r') = \int_{r'}^{r} \frac{\nu(t)}{t}\,dt \quad (0 < r' < r),$$

hence the *order of growth* may be defined threefold:

$$\varrho(g) = \limsup_{r\to\infty} \frac{\log^+ \log^+ M(r,g)}{\log r} = \limsup_{r\to\infty} \frac{\log^+ \log^+ \mu(r)}{\log r} = \limsup_{r\to\infty} \frac{\log^+ \nu(r)}{\log r}.$$

Exercise 1.25 To determine $\nu = \nu(r)$ asymptotically it often suffices to use the inequalities $|a_n|r^n \le |a_\nu|r^\nu$ and $|a_m|r^m < |a_\nu|r^\nu$ for the largest index $n < \nu$ with $a_n \ne 0$ and the smallest index $m > \nu$ with $a_m \ne 0$, respectively. Prove in this manner $\nu(r) = [r]$ and $\nu(r) \sim \sqrt{r}/2$ for $g(z) = e^z$ and $g(z) = \cos\sqrt{z}$, respectively.

The central result of Wiman–Valiron theory is the relation between $\nu(r)$ and $g'(z)/g(z)$ at points $z = z_r$ on $|z| = r$ such that $|g(z_r)| = M(r,g)$.

Theorem 1.14 *For every transcendental entire function and every n there exists some set $F_n \subset (0,\infty)$ of finite* logarithmic measure $\int_{F_n} \frac{dt}{t}$ *such that*

$$\frac{g^{(n)}(z_r)}{g(z_r)} \sim \left(\frac{\nu(r)}{z_r}\right)^n \quad (r \to \infty,\ r \notin F_n)$$

holds at points z_r with $|z_r| = r$ and $|g(z_r)| = M(r,g)$.

An equivalent statement is that $\dfrac{g(z_r + \frac{r}{\nu(r)}\zeta)}{g(z_r)} \to e^\zeta$ as $r \to \infty$ outside F, locally uniformly with respect to ζ. The reader is referred to [12, 79, 193, 200]. It is almost obvious how this applies to determine the possible orders of growth of the transcendental solutions to linear differential equations

$$w^{(n)} + a_{n-1}(z)w^{(n-1)} + \cdots + a_1(z)w' + a_0(z)w = b(z)$$

with polynomial coefficients. Dividing by $w(z_r)$ yields

$$\big(\nu(r)/z_r\big)^n + a_{n-1}(z_r)(1+o(1))\big(\nu(r)/z_r\big)^{n-1} + \cdots + a_0(z_r)(1+o(1)) = o(1)$$

as $r \to \infty$ outside F (of finite logarithmic measure). Although this is not exactly an algebraic equation, it suffices to consider the simplified algebraic equation

$$y^n + \sum_{j=0}^{n-1} A_j x^{\alpha_j+n-j} y^j = 0 \quad (x = z_r,\ y = \nu(r))$$

(assuming $a_j(z) = A_j z^{\alpha_j} + \cdots$) to determine $y = \nu(r)$ approximately. In combination with the Newton–Puiseux method it turns out that the possible orders of growth coincide up to sign with the negative slopes of the Newton–Puiseux polygon. The exceptional set F can be neglected since $\nu(r)$ is increasing.

Exercise 1.26 Compute the order of growth of the transcendental solutions to the linear differential equation $w''' - 4z^2w' - 12zw = 0$. **Solution.** $\nu(r) \sim 2r^2$ and z_r is approximately real. Why?

Example 1.8 The results are not always satisfactorily. In case of the nonlinear differential equation $ww'' - w'^2 - ww' = 0$ we obtain

$$(1+\epsilon_2(r) - (1+\epsilon_1(r))^2)\nu(r)^2 - (1+\epsilon_1(r))z_r\nu(r) = 0.$$

Since $1+\epsilon_2(r)-(1+\epsilon_1(r))^2 = \epsilon_2(r)-2\epsilon_1(r)-\epsilon_1(r)^2$ cannot be controlled, nothing can be said about $\nu(r)$. The solution e^{e^z} has infinite order.

1.5.4 The Schwarzian Derivative

The expression

$$S_f = \left(\frac{f''}{f'}\right)' - \frac{1}{2}\left(\frac{f''}{f'}\right)^2 = \frac{f'''}{f'} - \frac{3}{2}\left(\frac{f''}{f'}\right)^2 \tag{1.21}$$

is called the *Schwarzian derivative* of f, classically also written $\{f, z\}$.

Exercise 1.27 Prove the chain-rule $S_{f\circ g} = (S_f \circ g)g'^2 + S_g$.

Exercise 1.28 Suppose that f is regular at z_0 and $f'(z_0) \neq 0$. Prove that $w_2 = (f')^{-\frac{1}{2}}$ (any branch on $|z-z_0| < \delta$), and also $w_1 = fw_2$ satisfies the linear differential equation

$$w'' + \tfrac{1}{2}S_f(z)w = 0, \tag{1.22}$$

hence $f = w_1/w_2$ holds. (**Hint.** Start with $w_2'/w_2 = -\frac{1}{2}f''/f'$.)

Exercise 1.29 Deduce that

- $S_f \equiv 0$ if and only if f is a Möbius transformation;
- $S_{M \circ f} = S_f$ for every Möbius transformation M;
- S_f is regular except at the critical points of f (zeros of f', multiple poles of f);
- $S_f(z) = \dfrac{\frac{1}{2}(1-p)^2}{(z-z_0)^2} + \dfrac{c_1}{(z-z_0)} + \cdots$ if z_0 is a $(p-1)$-fold critical point.

Remark 1.10 The Schwarzian derivative occurs in a natural way in several subfields of Complex Analysis. We mention

- conformal mapping of circular domains, one of its origins;
- univalence criteria of Nehari type;
- Nevanlinna Theory (equality in the Second Main Theorem);
- Teichmüller space;
- quasiconformal mappings;
- iteration of transcendental functions (and of real functions on real intervals);
- normal family criteria.

Exercise 1.30 Suppose f is holomorphic at $z = 0$ with $f'(0) \neq 0$. Prove that for mutually distinct constants a, b, c, d,

$$\frac{(f(az), f(bz), f(cz), f(dz))}{(a, b, c, d)} = 1 + \tfrac{1}{6}(a-b)(c-d)S_f(0)z^2 + O(|z|^3) \quad (z \to 0)$$

holds, where $(a, b, c, d) = \frac{(a-c)(b-d)}{(a-d)(b-c)}$ denotes the cross-ratio.

Exercise 1.31 Prove that the cross-ratio (w_1, w_2, w_3, w_4) of any four distinct solutions w_ν to the Riccati equation $w' = a(z) + b(z)w + c(z)w^2$ is constant.
(**Hint.** Compute the logarithmic derivative of the cross-ratio and make use of $\frac{w'_\mu - w'_\nu}{w_\mu - w_\nu} = b(z) + c(z)(w_\mu + w_\nu)$.)

Chapter 2
Nevanlinna Theory

In this chapter Nevanlinna Theory, Cartan's Theory of Entire Curves, and the Selberg–Valiron Theory of Algebroid Functions will be outlined, particularly with regard to the applications in the subsequent chapters and including recent developments in the context of the Second Main Theorem. Nevanlinna Theory provides the most effective tools in the modern theory of meromorphic functions, and even simplifies the older theory of entire functions considerably.[1]

2.1 The First Main Theorem

2.1.1 The Poisson–Jensen Formula

Nevanlinna Theory ([124, 126, 128]) is based on the *Poisson–Jensen formula* for meromorphic functions; 'meromorphic' always means meromorphic in the plane. By

$$P(z, re^{i\theta}) = \operatorname{Re}\frac{re^{i\theta}+z}{re^{i\theta}-z} = \frac{r^2-|z|^2}{|re^{i\theta}-z|^2} \quad \text{and} \quad g(z,c) = -\log\left|\frac{r(z-c)}{r^2-\bar{c}z}\right|$$

we denote the Poisson kernel and Green's function with pole at c of the disc $|z| < r$, respectively.

[1]H. Weyl (Meromorphic functions and analytic curves. Princeton University Press 1943) called it "one of the few great mathematical events of the [XX] century". W.K. Hayman dedicated his monograph [78] to Rolf Nevanlinna, "the creator of the modern theory of meromorphic functions", and A.A. Gol'dberg and I.V. Ostrovskii wrote in the preface to [57]: "After his [Nevanlinna's] work, the value distribution theory acquired, in some way, a complete form. The main classical results of the theory of entire functions have been included in Nevanlinna's Theory in a natural way."

N. Steinmetz, *Nevanlinna Theory, Normal Families, and Algebraic Differential Equations*, Universitext, DOI 10.1007/978-3-319-59800-0_2

Poisson–Jensen Formula *Let f be meromorphic with zeros and poles a_μ and b_ν, written down according to multiplicities. Then*

$$\begin{aligned}\log|f(z)| = \frac{1}{2\pi}\int_0^{2\pi} & P(z, re^{i\theta}) \log|f(re^{i\theta})|\,d\theta \\ & - \sum_{|a_\mu|<r} g(z, a_\mu) + \sum_{|b_\nu|<r} g(z, b_\nu)\end{aligned} \tag{2.1}$$

holds on $|z| < r$, $z \neq a_\mu, b_\nu$.

Proof The function $u(z) = \log|f(z)| + \sum_{|a_\mu|<r} g(z, a_\mu) - \sum_{|b_\nu|<r} g(z, b_\nu)$ is harmonic on the disc $|z| < r$ since the singularities a_μ and b_ν are removable, and has boundary values $\log|f(z)|$ on $|z| = r$. Thus (2.1) reduces to the Poisson integral formula for u if f has no zeros and poles on $|z| = r$. If, however, f has zeros $re^{i\alpha_j}$ and poles $re^{i\beta_k}$ on $|z| = r$, then (2.1) holds with r replaced with s ($r - \delta < s < r$) on $|z| < s$. Since $|f(z)| \prod_k |z - re^{i\beta_k}| \prod_j |z - re^{i\alpha_j}|^{-1}$ is bounded and bounded away from zero on $r - \delta < |z| \le r$, and every term $|se^{i\theta} - re^{i\gamma}|$ is bounded by $2r$ from above, and from below by r on $\pi/2 \le |\theta - \gamma| \le \pi$ and $r|\sin(\theta - \gamma)|$ on $|\theta - \gamma| \le \pi/2$, it follows that $|\log|f(se^{i\theta})||$ is dominated by

$$\textstyle\sum_j |\log|\sin(\theta - \alpha_j)|| + \sum_k |\log|\sin(\theta - \beta_k)|| + \text{const.}$$

The assertion then follows from Lebesgue's Theorem on dominated convergence by taking the limit $s \nearrow r$.

2.1.2 The Nevanlinna Functions

The particular case $z = 0$ in (2.1) with $f(0) \neq 0, \infty$ is known as *Jensen's formula*[2]

$$\log|f(0)| = \frac{1}{2\pi}\int_0^{2\pi} \log|f(re^{i\theta})|\,d\theta - \sum_{|a_\mu|<r} \log\frac{r}{|a_\mu|} + \sum_{|b_\nu|<r} \log\frac{r}{|b_\nu|}. \tag{2.2}$$

Exercise 2.1 Suppose that $f(z) = cz^p + \cdots$ ($c \neq 0$) holds at $z = 0$. Prove

$$\log|c| = \frac{1}{2\pi}\int_0^{2\pi} \log|f(re^{i\theta})|\,d\theta - p\log r - \sum_{|a_\mu|<r} \log\frac{r}{|a_\mu|} + \sum_{|b_\nu|<r} \log\frac{r}{|b_\nu|} \tag{2.3}$$

by applying Jensen's formula to $f(z)z^{-p}$.

[2] Jensen's paper Acta. Math. **22** (1899), 359–364, has the ambitious title 'Sur un nouvel et important théorème de la théorie des fonctions'.

The integral in the Poisson–Jensen formula will be re-written as follows. Set $\log^+ x = (\log x)^+$ and define the *proximity function* (*Schmiegungsfunktion*) of f by

$$m(r,f) = \frac{1}{2\pi}\int_0^{2\pi} \log^+ |f(re^{i\theta})|\, d\theta$$

to obtain

$$\frac{1}{2\pi}\int_0^{2\pi} \log|f(re^{i\theta})|\, d\theta = m(r,f) - m(r, 1/f).$$

The sums on the right-hand side of (2.2) and (2.3) may be converted into integrals. Since this procedure will also turn out to be extremely useful in different settings, we shall describe it in greater generality.

Exercise 2.2 Let $\phi : (0,\infty) \longrightarrow \mathbb{R}$ be continuously differentiable and (ξ_k) be any sequence in $(0,\infty)$ tending to infinity, with counting function $\nu(t) = \sum_{\xi_k<t} 1$. Prove by Abel's method of partial summation or else integration by parts that

$$\sum_{\xi_k\le r} \phi(\xi_k) = \int_0^r \phi(t)\, d\nu(t) = \phi(r)\nu(r) - \int_0^r \phi'(t)\nu(t)\, dt. \tag{2.4}$$

If $\nu(r)\phi(r) \to 0$ as $r \to \infty$ this implies

$$\sum_{k=1}^{\infty} \phi(\xi_k) = -\int_0^{\infty} \phi'(t)\nu(t)\, dt, \tag{2.5}$$

provided the series resp. the integral converges. Prove that $\nu(r)\phi(r) \to 0$ follows from the convergence of either side if ϕ is decreasing.

Returning to (2.3) we set $\nu(t) = n(t,f) - n(0,f)$, where $n(t,f)$ denotes the number of poles of f on $|z| \le t$, and $\phi(t) = \log(r/t)$ to obtain

$$\sum_{0<|b_\nu|<r} \log\frac{r}{|b_\nu|} = \int_0^r \frac{n(t,f) - n(0,f)}{t}\, dt.$$

The function

$$N(r,f) = \int_0^r \frac{n(t,f) - n(0,f)}{t}\, dt + n(0,f)\log r$$

is called the (integrated) *counting function* of poles of f. Since in (2.3) the number p equals $n(0,1/f) - n(0,f)$, the right-hand side of (2.2) and (2.3) is given by

$$m(r,f) - m(r,1/f) + N(r,f) - N(r,1/f).$$

2.1.3 The Nevanlinna Characteristic

The sum

$$T(r,f) = m(r,f) + N(r,f)$$

is called the Nevanlinna *characteristic* of f. From (2.3) we obtain a preliminary form of the First Main Theorem of Nevanlinna, namely

$$T(r, 1/f) = T(r,f) + \log|c|,$$

and $T(r, 1/(f-a)) = m(r,f-a)+N(r,f)+C(a)$ for some constant $C(a)$ depending on the given complex value $a \neq 0$. From $|\log^+|f-a|-\log^+|f|| \leq \log^+|a|+\log 2$ we eventually obtain the

First Main Theorem *For every non-constant meromorphic function f and complex number a we have*

$$T\Big(r, \frac{1}{f-a}\Big) = T(r,f) + O(1) \quad (r \to \infty).$$

It is obvious that the term $O(1)$ is unavoidable, but plays a marginal role only since $T(r,f) \geq \log r + O(1)$ holds anyway. We note some 'arithmetic' rules, which may easily be extended to more than two functions

Exercise 2.3 For non-constant meromorphic functions f_1 and f_2 and $r > 1$ prove

$$\begin{aligned} m(r,f_1+f_2) &\leq m(r,f_1) + m(r,f_2) + \log 2, \\ m(r,f_1f_2) &\leq m(r,f_1) + m(r,f_2), \\ N(r,f_1+f_2) &\leq N(r,f_1) + N(r,f_2), \\ N(r,f_1f_2) &\leq N(r,f_1) + N(r,f_2), \\ T(r,f_1+f_2) &\leq T(r,f_1) + T(r,f_2) + \log 2, \\ T(r,f_1f_2) &\leq T(r,f_1) + T(r,f_2). \end{aligned}$$

2.1.4 Valiron's Lemma

The First Main Theorem is equivalent to $T(r, S\circ f) = T(r,f) + O(1)$, where S is any Möbius transformation, and this again is a special case of Valiron's Lemma.

Exercise 2.4 Let $P(w) = a_d w^d + \cdots + a_0$ be any polynomial of degree $d \geq 2$ and f any non-constant meromorphic function, and set $F = P \circ f$. Use $|P(w)| \asymp |w|^d$ as $w \to \infty$ to prove $m(r,F) = dm(r,f) + O(1)$, hence deduce $T(r,F) = dT(r,f) + O(1)$ from the obvious equality $N(r,F) = dN(r,f)$.

Valiron's Lemma (**[192]**) *Let f be any non-constant meromorphic function and R any rational function of degree $d \geq 1$. Then $F = R \circ f$ has characteristic*

$$T(r, F) = dT(r, f) + O(1) \quad (r \to \infty).$$

Proof Write $R = P/Q$, where P and Q are mutually prime polynomials with $\deg P \leq \deg Q = d$ (otherwise consider $1/R = Q/P$ and use the First Main Theorem), and set $p = P \circ f$ and $q = Q \circ f$. For $\delta > 0$ sufficiently small, $1/C < |p(z)| < C$ holds on $E_0 = \{z : |q(z)| < \delta\}$, while on the complement of E_0, $|F(z)| + |1/q(z)|$ is bounded. Thus $\log^+ |F(z)| = \log^+ |q(z)|^{-1} + O(1)$ ($|z| \to \infty$) holds without restriction, this implying $m(r, F) = m(r, 1/q) + O(1)$. Since $N(r, F) = N(r, 1/q)$ is obviously true, the assertion follows from the First Main Theorem and Exercise 2.4.

Exercise 2.5 Valiron's Lemma may be completed as follows: prove that for every non-rational meromorphic function f and non-constant entire function g, the ratio $T(r, f \circ g)/T(r, g)$ tends to infinity as $r \to \infty$. (**Hint.** Assume that f has infinitely many simple poles $b_1, b_2, \ldots$, and prove $m(r, f \circ g) \geq \sum_{\nu=1}^{p} m(r, 1/(g - b_\nu)) + O(1)$.)
More on the growth of composite functions can be found in Clunie [29].

Exercise 2.6 Verify

$$\begin{aligned}
f \text{ rational of degree } d:&\quad T(r,f) = d \log r + O(1).\\
f \text{ non-rational}:&\quad \log r = o(T(r,f)) \ (r \to \infty).\\
f(z) = e^z:&\quad m(r,f) = T(r,f) = r/\pi.\\
f(z) = g(cz)\ (c \neq 0):&\quad T(r,f) = T(|c|r, g), \text{ and the same for } m \text{ and } N.\\
f(z) = g(z^p):&\quad T(r,f) = T(r^p, g), \text{ and the same for } m \text{ and } N.\\
f(z) = \cos z = \tfrac{1}{2}(e^{iz} + e^{-iz}):&\quad T(r,f) = m(r,f) = 2r/\pi + O(1).\\
f(z) = \tan z = -i\frac{e^{2iz} - 1}{e^{2iz} + 1}:&\quad T(r,f) = 2r/\pi + O(1), m(r,f) = O(1).\\
P(z) = a_d z^d + \cdots + a_0:&\quad T(r, e^P) = \frac{|a_d|}{\pi} r^d + O(r^{d-1}).
\end{aligned}$$

Exercise 2.7 The Gamma function has counting function of poles $r + O(\log r)$. To prove $m(r, \Gamma) = \dfrac{r}{\pi} \log r + O(r)$ use the Mittag-Leffler representation

$$\Gamma(z) = \sum_{\nu=0}^{\infty} \frac{(-1)^\nu}{\nu!(z + \nu)} + \int_1^\infty e^{-t} t^{z-1}\, dt = \sigma(z) + G(z)$$

on the half-plane $\operatorname{Re} z \leq 1$ to obtain $\log^+ |\Gamma(z)| = \log^+ \frac{1}{|z+\nu|} + O(1)$ for $\operatorname{Re} z \leq 1$, $-\frac{1}{2} < |z| - \nu \leq \frac{1}{2}$, and Stirling's formula $\log \Gamma(z) = z \log z + O(|z|)$ as $z \to \infty$ on $\operatorname{Re} z > 0$ to obtain $\log |\Gamma(re^{i\theta})| = r \log r \cos\theta + O(r)$ on $|\theta| < \pi/2$.

2.1.5 Cartan's Identity

There is yet another representation of the Nevanlinna characteristic, known as

Cartan's Identity ([21]) *For every non-constant meromorphic function f,*

$$T(r,f) = \frac{1}{2\pi}\int_0^{2\pi} N\Big(r, \frac{1}{f - e^{i\theta}}\Big)\, d\theta + C \tag{2.6}$$

holds; C is some constant ($C = \log^+ |f(0)|$ *if* $f(0) \neq \infty$).

Proof The identity $\frac{1}{2\pi}\int_0^{2\pi} \log|e^{i\theta} - a|\, d\theta = \log^+ |a|$ is nothing else than Jensen's formula for the function $z - a$ and radius $r = 1$. We set $a = f(re^{i\phi})$ and integrate with respect to ϕ to obtain

$$m(r,f) = \frac{1}{2\pi}\int_0^{2\pi} \frac{1}{2\pi}\int_0^{2\pi} \log|e^{i\theta} - f(re^{i\phi})|\, d\phi\, d\theta$$

by Fubini's Theorem.[3] By Jensen's formula the inner integral is (set $f_\theta = f - e^{i\theta}$ and assume $f(0) \neq \infty$ for the moment)

$$m(r,f_\theta) - m(r, 1/f_\theta) = N(r, 1/f_\theta) - N(r,f_\theta) + \log^+ |f(0) - e^{i\theta}|$$

with at most one exceptional θ, hence (2.6) with $C = \log^+ |f(0)|$ follows from $N(r,f_\theta) = N(r,f)$. If, however, f has a pole at $z = 0$ we set $F = 1/f$. Then the zeros of $F - e^{i\theta}$ coincide with the zeros of $f - e^{-i\theta}$, and Cartan's Identity for F and the fact that $T(r,f) = T(r,F) + C$ yields the assertion also in that case. ☕

Obviously $N(r,f)$ is continuous and increasing with respect to r, and

$$\frac{dN(r,f)}{d\log r} = r\frac{d}{dr}N(r,f) = n(r,f) \geq 0$$

holds except at the discontinuities of $n(r,f)$. Thus we have the following

Corollary 2.1 *$N(r,f)$, $T(r,f)$, and $m(r,f)$ are continuous functions of r, and $N(r,f)$ and $T(r,f)$ are increasing and convex functions of* $\log r$.

Exercise 2.8 Let $P(z) = z^d + a_{d-1}z^{d-1} + \cdots + a_0$ be any normalised polynomial. Then for $a > |a_{d-1}|$ and $r > r_c$, $P(\{z : |z| < r\})$ contains $\{w : |w| < r^d - ar^{d-1}\}$ and is contained in $\{w : |w| < r^d + ar^{d-1}\}$, hence

$$n(r^d - ar^{d-1}, 1/(f - c)) \leq n(r, 1/(f \circ P - c)) \leq n(r^d + ar^{d-1}, 1/(f - c))$$

[3] The integrability of $|\log|e^{i\theta} - f(re^{i\phi})||$ over $[0, 2\pi] \times [0, 2\pi]$ is obvious.

holds for every $c \in \mathbb{C}$. Use Cartan's Identity to prove

$$T(r, f \circ P) = T(r^d + \theta(r) r^{d-1}, f) \quad (|\theta(r)| \text{ bounded}).$$

Exercise 2.9 Prove that an elliptic function of elliptic order g has counting function of poles $N(r,f) = \frac{\pi g}{2A} r^2 + O(r)$, where A is the area of any primitive period parallelogram, and deduce $T(r,f) = \frac{\pi g}{2A} r^2 + O(r)$ using Cartan's Identity. In particular, $m(r,f) = O(r)$ holds; a more delicate analysis even yields $m(r,f) = O(1)$. (**Hint.** (Packing and covering) Let $p(r)$ be the largest number of parallelograms $P + \varpi$, $\varpi \in \mathscr{L}$, that are contained in $|z| < r$, and $q(r)$ the smallest number of parallelograms that are needed to cover this disc. Prove $q(r) \leq p(r + d)$ with $d = \operatorname{diam} P$. Deduce $q(r) = p(r) + O(r) = \frac{\pi}{A} r^2 + O(r)$ and $n(r,f) = g\frac{\pi}{A} r^2 + O(r)$.)

2.1.6 The Ahlfors–Shimizu Formula

Let f be non-constant meromorphic with mutually distinct poles b_ν. We assume that f has no poles on $|z| = r$; the poles in $D = \{z : |z| < r\}$ are separated by mutually disjoint discs $\Delta_\nu = \{z : |z - b_\nu| \leq \rho\} \subset D$. On $D_\rho = \{z : |z| < r\} \setminus \bigcup_{|b_\nu|<r} \Delta_\nu$ with outer normal $\mathfrak{n}$, Green's formula

$$\int_{D_\rho} \Delta u \, d(x,y) = \int_{\partial D_\rho} \nabla u \cdot \mathfrak{n} \, |dz|$$

applies to $u(z) = \log(1 + |f(z)|^2)$ with Laplacian[4] $\Delta u(z) = 4 f^\sharp(z)^2$, where $f^\sharp$ denotes the spherical derivative of f. We thus obtain

$$\begin{aligned} 4 \int_{D_\rho} f^\sharp(z)^2 \, d(x,y) &= \int_0^{2\pi} r \frac{\partial}{\partial r} \log(1 + |f(re^{i\theta})|^2) \, d\theta \\ &\quad + \sum_{|b_\nu|<r} \int_0^{2\pi} -\rho \frac{\partial}{\partial \rho} \log(1 + |f(b_\nu + \rho e^{i\theta})|^2) \, d\theta \end{aligned}$$

(note $\partial/\partial\mathfrak{n} = \partial/\partial r$, $|dz| = r d\theta$ on $|z| = r$, and $\partial/\partial\mathfrak{n} = -\partial/\partial\rho$, $|dz| = \rho d\theta$ on $|z - b_\nu| = \rho$). The left-hand side tends to $4 \int_{|z|<r} f^\sharp(z)^2 \, d(x,y)$ as $\rho \to 0$, while the first integral on the right-hand side equals $r \frac{d}{dr} \int_0^{2\pi} \log(1 + |f(re^{i\theta})|^2) \, d\theta$. At every pole b_ν with multiplicity k_ν we have $\log(1 + |f(z)|^2) = -2k_\nu \log|z - b_\nu| + U_\nu(z)$,

[4] $U(w) = \log(1 + |w|^2)$ has Laplacian $\Delta U = \dfrac{4}{(1 + |w|^2)^2}$, hence $\Delta u = (\Delta U \circ f)|f'|^2 = 4 f^{\sharp 2}$ holds for $u = U \circ f$ by the chain-rule for the Laplacian.

where U_ν is real analytic on $|z - b_\nu| \le 2\rho$, say, hence it follows that

$$-\frac{\partial}{\partial\rho}\log(1+|f(b_\nu+\rho e^{i\theta})|^2) = \frac{2k_\nu}{\rho} + O(1) \quad (\rho\to 0),$$
$$\lim_{\rho\to 0}\int_0^{2\pi} -\frac{\partial}{\partial\rho}\log(1+|f(b_\nu+\rho e^{i\theta})|^2)\,\rho d\theta = 4k_\nu\pi, \quad \text{and}$$
$$4\int_{|z|<r} f^\sharp(z)^2\,d(x,y) = r\frac{d}{dr}\int_0^{2\pi}\log(1+|f(re^{i\theta})|^2)\,d\theta + 4\pi n(r,f).$$

Dividing by $4\pi r$, integrating over $\epsilon < r < R$, and passing to the limit $\epsilon \to 0$ gives

$$\frac{1}{\pi}\int_0^R\int_{|z|<r} f^\sharp(z)^2\,d(x,y)\,\frac{dr}{r} = \frac{1}{4\pi}\int_0^{2\pi}\log(1+|f(Re^{i\theta})|^2)\,d\theta + N(R,f) - c(f),$$

where $c(f) = \frac{1}{2}\log(1+|f(0)|^2)$ if f is regular, and $c(f) = \log|c|$ if $f(z) = cz^{-p}+\cdots$ has a pole of order p at $z=0$. We set

$$A(r,f) = \frac{1}{\pi}\int_{|z|<r} f^\sharp(z)^2\,d(x,y) \quad \text{and} \quad m_*(r,f) = \frac{1}{4\pi}\int_0^{2\pi}\log(1+|f(re^{i\theta})|^2)\,d\theta,$$

and define the *spherical characteristic* and *proximity function*

$$T_0(r,f) = \int_0^r \frac{A(t,f)}{t}\,dt \quad \text{and} \quad m_0(r,f) = m_*(r,f) - c(f),$$

respectively. Finally, since $\max\{1,x\} \le \sqrt{1+x^2} \le \sqrt{2}\max\{1,x\}$ $(x \ge 0)$ implies $m(r,f) \le m_*(r,f) \le m(r,f) + \frac{1}{2}\log 2$, we obtain the

Spherical First Main Theorem ([1, 150]) *For every non-constant meromorphic function the following is true:*

- $T_0(r,f) = m_0(r,f) + N(r,f) = T(r,f) + O(1)$;
- $T_0(r, S\circ f) = T_0(r,f)$ *for every rotation of the sphere* $S(w) = \frac{aw+b}{-\bar{b}w+\bar{a}}$.

2.1.7 The Order of Growth

The growth of an entire function f is measured by comparing its *maximum modulus* $M(r,f) = \max_{|z|=r}|f(z)|$ with e^{r^ϱ}. The number

$$\varrho(f) = \limsup_{r\to\infty}\frac{\log^+\log^+ M(r,f)}{\log r} \tag{2.7}$$

$(0 \le \varrho(f) \le \infty)$ is called the *order of growth* of f.

Exercise 2.10 For every non-constant entire function f and $s < r$ prove

$$m(s,f) \le \log^+ M(s,f) \le \frac{r+s}{r-s} m(s,f) \quad \text{and} \quad \varrho(f) = \limsup_{r\to\infty} \frac{\log^+ m(r,f)}{\log r}.$$

(**Hint.** From (2.1) deduce $\log|f(z)| \le \frac{1}{2\pi}\int_0^{2\pi} \log^+|f(re^{i\theta})|\frac{r+|z|}{r-|z|}\,d\theta$.)

It is quite natural to define the order of growth of any meromorphic function by

$$\varrho(f) = \limsup_{r\to\infty} \frac{\log^+ T(r,f)}{\log r}. \tag{2.8}$$

Exercise 2.11 Prove that the functions $f_1 \pm f_2$, $f_1 f_2$ and $f = f_1/f_2$ have order of growth at most $\max\{\varrho(f_1), \varrho(f_2)\}$, and equality certainly holds if $\varrho(f_1) \ne \varrho(f_2)$.

Remark For entire functions $f = f_1 f_2$ with $\varrho(f_1) > \varrho(f_2)$, say, it is quite laborious to prove $\varrho(f) = \varrho(f_1)$ within the theory of entire functions.

Exercise 2.12 Prove that the transcendental solutions to the Riccati differential equation $w' = a(z) + b(z)w + c(z)w^2$ with polynomial coefficients have finite order of growth. (**Hint.** Estimate $w^\sharp$ in terms of $|a|$, $|b|$, and $|c|$ as $z \to \infty$.)

2.1.8 Canonical Products

Let (a_μ), $a_\mu \ne 0$, be any sequence tending to infinity, and let $E(u,0) = 1-u$ and $E(u,q) = (1-u)\exp(u + \frac{1}{2}u^2 + \cdots + \frac{1}{q}u^q)$ denote the *Weierstraß prime factors*. Then the infinite product

$$\prod_{\mu=1}^{\infty} E\Big(\frac{z}{a_\mu}, q\Big) \tag{2.9}$$

converges absolutely and locally uniformly on the complex plane if and only if the series $\sum_{\mu=1}^{\infty} |a_\mu|^{-q-1}$ converges. If q is chosen as small as possible we speak of a *canonical product* of *genus* q. The infimum of those τ such that the series

$$\sum_{\mu=1}^{\infty} |a_\mu|^{-\tau}$$

converges is called the *exponent of convergence* of the zeros, and denoted by $\varrho_0(\{a_\mu\})$; $q \le \varrho_0 \le q+1$ holds by definition of the genus.

Exercise 2.13 Set $n(r) = \sum_{|a_\mu|\le r} 1, N(r) = \int_0^r n(t)t^{-1}\,dt$, and prove that

$$\sum_{\mu=1}^{\infty} |a_\mu|^{-\tau}, \quad \int_0^\infty n(t)t^{-\tau-1}\,dt, \quad \text{and} \quad \int_0^\infty N(t)t^{-\tau-1}\,dt$$

converge or diverge simultaneously. (**Hint.** Use Exercise 2.2 with $\phi(t) = t^{-\tau}$.)

Exercise 2.14 The inequality $\log|E(u,0)| \le \log(1+|u|)$ is obvious. For $q \ge 1$ prove $\log|E(u,q)| \le \frac{|u|^{q+1}}{q+1}(1+o(1))$ as $u \to 0$ and $\log|E(u,q)| \le \frac{|u|^q}{q}(1+o(1))$ as $u \to \infty$, and deduce that

$$\log|E(u,q)| \le C(q)\frac{|u|^{q+1}}{1+|u|} \quad (u \in \mathbb{C}).$$

Exercise 2.15 (Continued) Use Exercise 2.2 to prove that a canonical product $f(z) = \prod_{\mu=1}^\infty E(z/a_\mu, q)$ of genus $q \ge 0$ satisfies

$$\log M(r,f) \le C(q)r^{q+1}\int_0^\infty \frac{n(t)}{t^{q+1}(t+r)}\,dt.$$

Theorem 2.1 *Canonical products have order of growth $\varrho = \varrho_0(\{a_\mu\})$.*

Proof From Exercise 2.15 it follows that $\log M(r,f) = o(r^{q+1})$ holds as $r \to \infty$ (note that the integrand $\frac{n(t)}{t^{q+1}(t+r)}$ is dominated by $\frac{n(t)}{t^{q+1}(t+1)}$), hence $\varrho \le q+1$, and $\varrho = \varrho_0$ if $\varrho_0 = q+1$. We now assume $\varrho_0 < q+1$, hence $n(t) = O(t^k)$ for every $k \in (\varrho_0, q+1)$. This implies (note that $q+1-k < 1$)

$$\log M(r,f) = O(r^{q+1})\int_0^\infty \frac{t^k}{t^{q+1}(t+r)}\,dt = O(r^k),$$

from which the final assertion $\varrho \le \varrho_0$ follows since $k > \varrho_0$ was arbitrary.

2.2 The Second Main Theorem

The Second Main Theorem may be stated in several more or less equivalent ways. Its proof is based on the inequality (2.10) below and an appropriate estimate of the proximity function $m(r,f'/f)$ of the *logarithmic derivative*.

Exercise 2.16 Let f be non-constant meromorphic and a_ν $(1 \le \nu \le q)$ mutually distinct values. For $F = \displaystyle\sum_{\nu=1}^{q}\frac{1}{f-a_\nu}$ prove $\log^+|F(z)| = \displaystyle\sum_{\nu=1}^{q}\log^+\frac{1}{|f(z)-a_\nu|} +$

$O(1)$ as $z \to \infty$ to obtain

$$\sum_{\nu=1}^{q} m\Big(r, \frac{1}{f - a_\nu}\Big) = m(r, F) + O(1) \le m\Big(r, \frac{1}{f'}\Big) + \sum_{\nu=1}^{q} m\Big(r, \frac{f'}{f - a_\nu}\Big) + O(1). \tag{2.10}$$

(**Hint.** Consider $E_\nu = \{z : |f(z) - a_\nu| < \delta\}$ with $\delta = \frac{1}{2}\min_{\mu \ne \nu} |a_\mu - a_\nu|$.)

2.2.1 *The Lemma on the Logarithmic Derivative*

To estimate $m(r,f'/f)$ in terms of $\log T(r,f)$ we first prove a rather technical result due to Ngoan and Ostrovskii [129]:

Lemma 2.1 *Let f be non-constant meromorphic with $f(0) = 1$. Then*

$$\frac{1}{2\pi}\int_0^{2\pi} \left|\frac{f'(re^{i\theta})}{f(re^{i\theta})}\right|^\alpha d\theta \le 8RT(R,f)\Big\{\frac{2}{(R-r)^{2\alpha}} + \frac{1}{(R-r)r^\alpha(1-\alpha)}\Big\} \tag{2.11}$$

holds for $0 < \alpha < 1$ and $r < R$, provided $T(r,f) \ge 1$.

Proof For $r < s < R$, $|z| = r$, and $z \ne a_\mu, b_\nu$ the Poisson–Jensen formula yields

$$\begin{aligned}\log f(z) = {} & \frac{1}{2\pi}\int_0^{2\pi} \log|f(se^{i\theta})| \frac{se^{i\theta} + z}{se^{i\theta} - z}\, d\theta \\ & + \sum_{|a_\mu| < s} \log \frac{s(z - a_\mu)}{s^2 - \bar a_\mu z} - \sum_{|b_\nu| < s} \log \frac{s(z - b_\nu)}{s^2 - \bar b_\nu z} + 2k\pi i,\end{aligned}$$

where $k \in \mathbb{Z}$ is locally constant; this is true since both sides have the same real part. Differentiation yields

$$\begin{aligned}\frac{f'(z)}{f(z)} = {} & \frac{1}{2\pi}\int_0^{2\pi} \log|f(se^{i\theta})| \frac{2se^{i\theta}}{(se^{i\theta} - z)^2}\, d\theta \\ & + \sum_{|a_\mu| < s} \Big(\frac{1}{z - a_\mu} + \frac{\bar a_\mu}{s^2 - \bar a_\mu z}\Big) - \sum_{|b_\nu| < s} \Big(\frac{1}{z - b_\nu} + \frac{\bar b_\nu}{s^2 - \bar b_\nu z}\Big).\end{aligned}$$

We set $\{c_\kappa\} = \{a_\mu\} \cup \{b_\nu\}$ and note that

$$\left|\frac{1}{z - c} + \frac{\bar c}{s^2 - \bar c z}\right| = \frac{1}{|z - c|}\left|1 + \frac{\bar c}{s}\frac{s(z - c)}{s^2 - \bar c z}\right| \le \frac{2}{|z - c|}$$

holds for $|c| < s$, $|z| = r < s$, $z \neq c$. This yields

$$\left|\frac{f'(z)}{f(z)}\right| \leq \frac{1}{2\pi}\int_0^{2\pi} |\log|f(se^{i\theta})|| \frac{2s}{(s-r)^2}\,d\theta + \sum_{|c_\kappa|<s} \frac{2}{|z-c_\kappa|}.$$

The integral is at most $\frac{2s}{(s-r)^2}(m(s,f)+m(s,1/f) \leq \frac{4s}{(s-r)^2}T(s,f)$, and on combination with $|z-c|^2 = |re^{i\theta} - |c|e^{i\gamma}|^2 \geq r^2\sin^2(\theta-\gamma)$ we obtain

$$\left|\frac{f'(re^{i\theta})}{f(re^{i\theta})}\right| \leq \frac{4s}{(s-r)^2}T(s,f) + \sum_{|c_\kappa|<s} \frac{2}{r|\sin(\theta-\gamma_\kappa)|} \quad (\gamma_k = \arg c_k).$$

Since $\left|f'(re^{i\theta})/f(re^{i\theta})\right|$ is not integrable over $[0,2\pi]$ if f has zeros or poles on $|z| = r$, we have to pass to $\left|f'(re^{i\theta})/f(re^{i\theta})\right|^\alpha$. This is done by

Exercise 2.17 Prove the inequality $\left(\sum_{\nu=1}^p x_\nu\right)^\alpha \leq \sum_{\nu=1}^p x_\nu^\alpha$ $(x_\nu \geq 0, 0 < \alpha < 1)$, first assuming $\sum_{\nu=1}^p x_\nu = 1$.

Noting that $T(s,f) \geq T(r,f) \geq 1$ and $\int_0^{2\pi} \frac{d\phi}{|\sin\phi|^\alpha} < \frac{2\pi}{1-\alpha}$, this implies

$$\frac{1}{2\pi}\int_0^{2\pi} \left|\frac{f'(re^{i\theta})}{f(re^{i\theta})}\right|^\alpha d\theta \leq \frac{(4s)^\alpha}{(s-r)^{2\alpha}}T(s,f) + \frac{2n(s)}{r^\alpha(1-\alpha)}$$

with $n(s) = n(s,f) + n(s,1/f)$. To estimate $n(s)$, set $s = (R+r)/2$ to obtain

$$2T(R,f) \geq \int_s^R \frac{n(t)}{t}\,dt \geq \frac{R-s}{R}n(s) = \frac{R-r}{2R}n(s)$$

from the First Main Theorem, hence the assertion (2.11).

There is just one small step to the key result in Nevanlinna Theory, known as the *Lemma on the Logarithmic Derivative.* Its significance is far from being limited to the proof of the Second Main Theorem. It is rather indispensable in applications to algebraic differential equations and many other fields outside Nevanlinna Theory.

Lemma on the Logarithmic Derivative *For every meromorphic function f,*

$$m\left(r, \frac{f'}{f}\right) = O\big(\log(rT(r,f))\big) \quad (r \to \infty)$$

holds outside some set of finite measure $E \subset (0,\infty)$. If f has finite order ϱ, then

$$m\left(r, \frac{f'}{f}\right) \leq (\varrho - 1 + o(1))^+ \log r + O(1) \quad (r \to \infty)$$

is true without exception.

Proof We may assume $f(0) = 1$, otherwise consider $g(z) = cz^p f(z)$ with $g(0) = 1$ and note that $m(r, f'/f) = m(r, g'/g) + O(1)$ and $T(r,f) = T(r,g) + O(\log r)$; thus the result for g implies the same result for f. To estimate $m(r,f'/f)$ we have to pass from $|f'/f|^\alpha$ to $\log^+ |f'/f|$, and this may be done as follows.

Exercise 2.18 Prove *Jensen's inequality*

$$\frac{1}{b-a}\int_a^b \log\phi(\theta)\,d\theta \le \log\left(\frac{1}{b-a}\int_a^b \phi(\theta)\,d\theta\right),$$

where ϕ is positive and integrable, and deduce

$$\frac{1}{b-a}\int_a^b \log^+\phi(\theta)\,d\theta \le \log^+\left(\frac{1}{b-a}\int_a^b \phi(\theta)\,d\theta\right) + \log 2.$$

(**Hint.** Set $\mu = \frac{1}{b-a}\int_a^b \phi(\theta)\,d\theta$ and use $\log x \le \log\mu + (x-\mu)/\mu$.)

From Lemma 2.1 we thus obtain

$$\alpha\, m\left(r, \frac{f'}{f}\right) \le \log^+\left[8RT(R,f)\left\{\frac{2}{(R-r)^{2\alpha}} + \frac{1}{(R-r)r^\alpha(1-\alpha)}\right\}\right] + \log 16. \tag{2.12}$$

If f has finite order ϱ we set $R = 2r$ and $\alpha = 1-\epsilon$, where $0 < \epsilon < 1$ is arbitrary. Then $T(r,f) = O(r^{\varrho+\epsilon})$, and the right-hand side in (2.12) is $\log^+ r^{\varrho+3\epsilon-1} + O(\log\frac{1}{\epsilon})$ ($\epsilon \to 0$, $r \to \infty$), hence

$$m\left(r, \frac{f'}{f}\right) \le \frac{(\varrho + 3\epsilon - 1)^+}{1-\epsilon}\log r + O\left(\log\frac{1}{\epsilon}\right),$$

from which the assertion follows, since $0 < \epsilon < 1$ was arbitrary. In the general case we set $\alpha = \frac{1}{2}$, $R = r + 1/T(r,f)$, and apply Exercise 2.19 below with $h(r) = T(r,f)$ to obtain the assertion.

Exercise 2.19 Let $h : [r_0, \infty) \longrightarrow [1, \infty)$ be continuous and increasing. Prove that

$$h(r + 1/h(r)) \le 2h(r) \tag{2.13}$$

holds, possibly with the exception of some set E of measure at most two, as follows:

- If $E \ne \emptyset$ set $r_1 = \inf E$ and $r_1' = r_1 + 1/h(r_1)$, and note that $h(r_1') \ge 2h(r_1)$.
- Again if $E_1 = E \cap [r_1', \infty) \ne \emptyset$ set $r_2 = \inf E_1$ and $r_2' = r_2 + 1/h(r_2)$ etc.
- Prove that $r_n \to \infty$ if there are infinitely many r_n.
- Prove that $E \subset \bigcup_n [r_n, r_n']$ holds in any case, and $\operatorname{meas} E \le \sum_n (r_n' - r_n) \le 2$.

2.2.2 *The Second Main Theorem*

To some extent, $m(r,f'/f)$ is a *quantité négligeable*. We will write $\phi(r) = S(r,f)$ to indicate that $\phi(r) = O(\log(rT(r,f)))$ or sometimes just $\phi(r) = o(T(r,f))$ holds as $r \to \infty$, possibly outside a set of finite measure, not always the same at every occurrence; thus $m(r, f'/f) = S(r,f)$. We also denote by

$$\overline{N}(r,f) = \bar{n}(0,f)\log r + \int_0^r \frac{\bar{n}(t,f) - \bar{n}(0,f)}{t}\,dt$$

the counting function of poles, where this time $\bar{n}(t,f)$ counts each pole only once despite its multiplicity, and set

$$N_{\rm crit}(r) = N(r,f) - \overline{N}(r,f) + N(r, 1/f');$$

$N_{\rm crit}(r)$ is the counting function of the critical points of f. It 'counts' each point z_0 with multiplicity $p-1$ where f assumes the value $f(z_0)$ with multiplicity p. With this notation the Second Main Theorem may be stated as follows:

Second Main Theorem *Let f be non-constant meromorphic and a_ν $(1 \le \nu \le q)$ mutually distinct finite values. Then*

$$\sum_{\nu=1}^{q} m\Big(r, \frac{1}{f-a_\nu}\Big) \le T(r,f) + N(r,f) - N_{\rm crit}(r,f) + S(r,f) \tag{2.14}$$

holds.

Proof The inequality (2.14) follows from (2.10),

$$\begin{aligned} m(r,1/f') &= T(r,f') - N(r,1/f') + O(1),\\ m(r,f') &\le m(r,f) + m\Big(r,\frac{f'}{f}\Big) = m(r,f) + S(r,f), \end{aligned}$$

and $N(r,f') = N(r,f) + \overline{N}(r,f)$. ☕

Exercise 2.20 Prove the following versions of the Second Main Theorem.

$$\begin{aligned} m(r,f) + \sum_{\nu=1}^{q} m\Big(r,\frac{1}{f-a_\nu}\Big) &\le 2T(r,f) - N_{\rm crit}(r) + S(r,f),\\ (q-2)T(r,f) &\le \sum_{\nu=1}^{q} \overline{N}\Big(r,\frac{1}{f-a_\nu}\Big) + S(r,f),\\ (q-1)T(r,f) &\le \overline{N}(r,f) + \sum_{\nu=1}^{q} \overline{N}\Big(r,\frac{1}{f-a_\nu}\Big) + S(r,f). \end{aligned} \tag{2.15}$$

The right-hand side in the first inequality (2.15) has an interesting interpretation.

$$\begin{aligned}-\tfrac{1}{2\pi}\int_0^{2\pi}\log f^\sharp(re^{i\theta})\,d\theta &= m(r,1/f') - m(r,f') + \tfrac{1}{2\pi}\int_0^{2\pi}\log(1+|f(re^{i\theta}|^2)\,d\theta\\ &= N(r,f') - N(r,1/f') + 2m(r,f) + O(1)\\ &= N(r,f) + \overline{N}(r,f) - N(r,1/f') + 2m(r,f) + O(1)\\ &= 2T(r,f) - N_{\rm crit}(r) + O(1) (\text{Yosida [210]}).\end{aligned}$$

2.2.3 *The Deficiency Relation*

The number $\delta(\infty,f) = \liminf\limits_{r\to\infty}\dfrac{m(r,f)}{T(r,f)} = 1 - \limsup\limits_{r\to\infty}\dfrac{N(r,f)}{T(r,f)}$ is called the *deficiency* of ∞; $\delta(a,f)$ is defined similarly, and the value a is called *deficient* if $\delta(a,f) > 0$. From the Second Main Theorem it follows that

$$\delta(\infty,f) + \sum_{\nu=1}^{q}\delta(a_\nu,f) \le 2 - \liminf_{r\to\infty}\frac{N_{\rm crit}(r)}{T(r,f)} \le 2,$$

hence the *deficiency relation*

$$\sum_{a\in\hat{\mathbb{C}}}\delta(a,f) \le 2.$$

In particular, there are at most countably many deficient values. Since Picard exceptional values of f have deficiency $\delta(a,f) = 1$, we obtain Picard's Theorem.[5]

2.2.4 *Higher Derivatives*

The derivative $f^{(p)}$ has counting function of poles $N(r,f^{(p)}) = N(r,f) + p\overline{N}(r,f)$. Writing $\dfrac{f^{(p)}}{f} = \dfrac{f^{(p)}}{f^{(p-1)}}\dfrac{f^{(p-1)}}{f^{(p-2)}}\cdots\dfrac{f'}{f}$ and $f^{(p)} = \dfrac{f^{(p)}}{f}f$ it is easily proved by induction that $T(r,f^{(p)}) \le T(r,f) + p\overline{N}(r,f) + S(r,f) \le (p+1)T(r,f) + S(r,f)$ and also $m\big(r,f^{(p)}/f^{(q)}\big) = S(r,f)$ holds for $p > q \ge 0$.

[5]One of Nevanlinna's main goals was to find an 'elementary' proof of Picard's Theorem, that is, a proof that avoids the elliptic modular function. As a consequence he proved the Second Main Theorem for $q = 3$(!) only.

2.2.5 Generalisations of the Second Main Theorem

For $q = 3$ the second inequality in (2.15) remains true if the constants a_ν are replaced with meromorphic functions with characteristic $T(r, a_\nu) = S(r,f)$, so-called 'small functions' (with respect to f). One has just to replace f with the cross-ratio $M(f) = (f, a_1, a_2, a_3)$; the Möbius transformation M maps the functions a_1, a_2, a_3 onto the constant values $1, 0, \infty$. For every $q \geq 3$ we can prove the following generalisation with moderate effort:

Theorem 2.2 *Let f be meromorphic and let a_ν $(1 \leq \nu \leq q)$ be small meromorphic functions. Then*

$$m(r,f) + \sum_{\nu=1}^{q} m\Big(r, \frac{1}{f - a_\nu}\Big) \leq (2 + \epsilon)T(r,f) + S(r,f) \tag{2.16}$$

holds for every $\epsilon > 0$, possibly outside some set E_ϵ of finite measure.

Proof Let $\mathscr{L}(s)$ denote the linear space spanned by the functions

$$a_1^{n_1} a_2^{n_2} \cdots a_q^{n_q} \quad (n_\nu \geq 0,\ n_1 + n_2 + \cdots + n_q = s),$$

and let $\beta_1, \ldots, \beta_k$ and $b_1, \ldots, b_n$ be any basis of $\mathscr{L}(s)$ and $\mathscr{L}(s+1)$, respectively. Then

$$P[f] = W(b_1, \ldots, b_k, \beta_1 f, \ldots, \beta_n f),$$

where $W(g_1, \ldots, g_m)$ denotes the Wronskian determinant of the meromorphic functions $g_1, \ldots, g_m$, is a homogeneous polynomial in $f, f', \ldots, f^{(n+k-1)}$ of degree n over the field $\mathfrak{F}$ generated by the functions a_ν and their successive derivatives. It has the following properties.

Exercise 2.21 Use the rules in Exercise 1.13 to prove that

- $P[f]/f^n = Q[f'/f]$; $Q[h]$ is a polynomial in $h, h', h'', \ldots$[6] over the same field $\mathfrak{F}$;
- $P[f] \not\equiv 0$;
- $P[f - a_j] = P[f]$ (use $a_j\mathscr{L}(s) \subset \mathscr{L}(s+1)$);
- $m(r, P[f]) \leq n\, m(r,f) + S(r,f)$;
- $N(r, P[f]) \leq (n + k)N(r,f) + S(r,f)$.

Writing $h_\nu = f - a_\nu$ and $Q_\nu(z) = Q[h'_\nu/h_\nu](z)$ we obtain

$$(f(z) - a_\nu(z))^n = \frac{P[f - a_\nu](z)}{Q_\nu(z)} = \frac{P[f](z)}{Q_\nu(z)}.$$

[6] Like P and similar expressions (not necessarily homogeneous) Q is called a *differential polynomial*.

Since the sets $E_\nu = \{z : |f(z) - a_\nu(z)| \le \delta(z)\}$ with $\delta(z) = \frac{1}{2}\min_{\mu\neq\nu}|a_\mu(z) - a_\nu(z)|$ are mutually disjoint, this yields

$$\sum_{\nu=1}^{q}\log^+\frac{1}{|f(z)-a_\nu(z)|} \le \frac{1}{n}\log^+\frac{1}{|P[f](z)|} + n\log^+\frac{1}{\delta(z)} + \frac{1}{n}\sum_{\nu=1}^{q}\log^+|Q_\nu(z)| \tag{2.17}$$

everywhere, and from $1/\delta(z) \le \sum_{\mu<\nu} 2/|a_\mu(z) - a_\nu(z)|$ we finally obtain

$$\begin{aligned} m(r,f) + \sum_{\nu=1}^{q} m\Big(r, \frac{1}{f-a_\nu}\Big) &\le \frac{n+k}{n}T(r,f) + n\sum_{\mu<\nu}T(r, a_\mu - a_\nu) + O(1) \\ &= \frac{n+k}{n}T(r,f) + S(r,f). \end{aligned}$$

It remains to show that we can choose s in such a way that $k/n < 1+\epsilon$ holds. This, however, follows from $\dim\mathscr{L}(s) \le \binom{q+s-1}{s}$,[7] which implies $\liminf\limits_{s\to\infty}\frac{\dim\mathscr{L}(s+1)}{\dim\mathscr{L}(s)} = 1$. The exceptional set consists of the exceptional set for f and the interval $(0, r_\epsilon)$. ☕

2.2.6 *Zeros of Linear Differential Polynomials*

In a bid to provide a new proof of the generalised Second Main Theorem, Frank and Weissenborn [49], see also Langley [106], proved an equivalent form of

Theorem 2.3 *Suppose f is meromorphic and $a_1, \ldots, a_q$ are linearly independent small meromorphic functions. Then $L[f] = W(a_1, \ldots, a_n, f)$ satisfies*

$$m(r, 1/L[f]) \le m(r, L[f]) + (2+\epsilon)N(r,f) + S(r,f); \tag{2.18}$$

$S(r,f)$ depends on $\epsilon > 0$.

Exercise 2.22 Frank and Weissenborn considered $L[y] = y''' + 4y'$ and $f(z) = \tan z$ as a counterexample if the functions a_ν are not necessarily small; L is generated by $y_1 = 1, y_2 = e^{2iz}, y_3 = e^{-2iz}$. Prove that $L[f](z) = 6(\cos z)^{-4}$, hence $m(r, L[f]) = O(1)$ and $m(r, 1/L[f]) = 8r/\pi + O(1) = 4T(r,f) + O(1)$. Thus (2.18) is no longer true, and the functions y_ν are not small with respect to f. Note, however, that f is a rational function of y_1, y_2, y_3! ☕

[7] This is just the number of 'ordered words' $x_1^{n_1}x_2^{n_2}\cdots x_q^{n_q}$ of length s over the 'alphabet' $\{x_1, \ldots, x_q\}$.

The above example describes the typical situation which may occur if the hypothesis $T(r, a_\nu) = S(r,f)$ is dropped. In this direction we will prove Theorem 2.4 below. To avoid technical complications we will restrict ourselves to linear operators

$$L[y] = y^{(q)} + c_{q-1}y^{(q-1)} + \cdots + c_0 y \tag{2.19}$$

with constant rather than rational coefficients, which are admitted in [169]; $y_1, \ldots, y_q$ denotes the canonical fundamental set to $L[y] = 0$.

Theorem 2.4 *For L given by* (2.19) *and any transcendental meromorphic function f, either* (2.18) *holds or else f is a rational function of* $y_1, \ldots, y_q$.

Proof Again we denote by $\mathscr{L}(s)$ the linear space spanned by the functions $y_1^{n_1} \cdots y_q^{n_q}$ with $n_1 + \cdots + n_q = s$, choose any basis $u_1, \ldots, u_n$ and $U_1, \ldots, U_k$ of $\mathscr{L}(s)$ and $\mathscr{L}(s+1)$, respectively, and set

$$\Omega[w] = \frac{W(U_1, \ldots, U_k, u_1 w, \ldots, u_n w)}{W(U_1, \ldots, U_k) W(u_1, \ldots, u_n)},$$

$$M[w] = \frac{W(U_1, \ldots, U_k, w)}{W(U_1, \ldots, U_k)} \quad \text{and} \quad M_j[w] = M[u_j w]/u_j.$$

Then Ω and M are independent of the special choice of the bases, and we may assume that $u_1 \cdots u_n$ and $U_1 \cdots U_k$ have only finitely many zeros. In particular,

$$\phi = \frac{u_1 \cdots u_n}{W(u_1, \ldots, u_n)} \tag{2.20}$$

is a rational function since it has only finitely many zeros and $m(r, 1/\phi) = O(\log r)$ holds; also M_j has rational coefficients. From Exercise 1.13 it follows that

$$\Omega[w] = \frac{W(M[u_1 w], \ldots, M[u_n w])}{W(u_1, \ldots, u_n)} = W(M_1[w], \ldots, M_n[w]) \frac{u_1 \cdots u_n}{W(u_1, \ldots, u_n)}.$$

Since M_j annihilates the functions y_ν (note that $y_\nu u_j \in \mathscr{L}(s+1)$), there exist linear operators K_j such that $M_j = K_j \circ L$; this follows from an elimination process in [94], p. 126 (see also Sect. 1.3.4), which is in some sense related to the euclidean algorithm. We thus obtain

$$\Omega[w] = \phi(z) W(K_1[L[w]], \ldots, K_n[L[w]]) = Q[L[w]], \tag{2.21}$$

where Q is a homogeneous differential polynomial of degree n with rational coefficients. If $Q[L[f]](z) = \Omega[f](z)$ vanishes identically, then f is the ratio of linear combinations of $U_1, \ldots, U_k$ and $u_1, \ldots, u_n$, that is, f is a rational function of

$y_1, \ldots, y_q$. Otherwise we set $F = L[f]$ to obtain

$$\begin{aligned} nm(r, 1/F) &\leq m(r, 1/Q[F]) + m(Q[F]/F^n) \\ &\leq T(r, Q[F]) + S(r,f) \\ &\leq nm(r, F) + (k+n)N(r,f) + S(r,f), \end{aligned}$$

hence (2.18) if s is chosen such that $k/n < 1 + \epsilon$ holds.

Remark 2.1 It is not hard to prove Theorem 2.3 (with $a_\nu = y_\nu$ and $T(r, a_\nu) = S(r,f)$) by modifying the proof of Theorem 2.4 as follows.

- The operator L in Theorem 2.3 may be replaced with $\dfrac{W(a_1, \ldots, a_q, f)}{W(a_1, \ldots, a_q)}$ without changing the claim. The coefficients in L, Ω, Q, and various differential operators are no longer constants but meromorphic functions that are small even with respect to $\max_{1 \leq \nu \leq n} T(r, a_\nu)$; also ϕ is not a rational but a small function.
- $\Omega[f]$ does not vanish identically if the a_ν are small.

Also Theorem 2.4 remains true if the coefficients in L are rational functions. Then the coefficients in Ω are also rational (see Sect. 1.3.4 with $\mathfrak{F} = \mathbb{C}(z)$). The only obstacle is the function ϕ given in (2.20), which, however, is an artefact since Ω and Q are independent of the choice of the basis $u_1, \ldots, u_n$. Again Q has rational coefficients.

2.2.7 *Yamanoi's Work*

The first attempts to prove Theorem 2.2 were due to Chuang [24], who considered *entire* functions f, and Frank and Weissenborn [48], who settled the case of *rational* a_ν and transcendental meromorphic f, see also [49, 169]. They also introduced Wronskian determinants as a decisive tool. In the present form, the statement of Theorem 2.2 is due to Osgood [133] and the author [164]; the proof follows [164]. The true form of the Second Main Theorem for small functions, namely

$$(q - 2 - \epsilon)T(r,f) \leq \sum_{\nu=1}^{q} \overline{N}\Big(r, \frac{1}{f - a_\nu}\Big)$$

outside some exceptional set that depends on $\epsilon > 0$, is due to Yamanoi [203]. Yamanoi's paper [204] contains a less technical proof in the case of rational functions a_ν. Yamanoi's proof is rather elaborate and technical, so lack of space regrettably prevents a detailed presentation here. A further breakthrough in Nevanlinna Theory is also due to Yamanoi [205], who succeeded in proving

Gol'dberg's Conjecture $\quad \overline{N}(r,f) \le N\Big(r,\frac{1}{f''}\Big) + S(r,f)$

in the following more general and even stronger form.

Theorem 2.5 *For $k \ge 2$ and mutually distinct complex numbers a_ν $(1 \le \nu \le q)$,*

$$(k-1)\overline{N}(r,f) + \sum_{\nu=1}^{q} N_1\Big(r,\frac{1}{f-a_\nu}\Big) \le N\Big(r,\frac{1}{f^{(k)}}\Big) + \epsilon T(r,f) \tag{2.22}$$

holds outside some set E (depending on f, k, $\epsilon > 0$, and the a_ν) of logarithmic density zero; $N_1(r,h) = N(r,h) - \overline{N}(r,h)$ 'counts' the multiple poles of h.

We note that in Gol'dberg's Conjecture equality holds for $f(z) = \tan z$. The validity of the so-called

Mues Conjecture ([117]) $\quad \sum_{c\in\mathbb{C}} \delta(c,f') \le 1$

follows at once. Given n values $c_1,\ldots,c_n$ and $\epsilon > 0$, Gol'dberg's Conjecture and the Second Main Theorem for f' yield

$$\sum_{\nu=1}^{n} m\Big(r,\frac{1}{f'-c_\nu}\Big) \le m\Big(r,\frac{1}{f''}\Big) + S(r,f') \le T(r,f') + S(r,f') + \epsilon T(r,f).$$

Since $T(r,f)/T(r,f')$ is bounded (even by a universal constant) outside some set of upper logarithmic density less than 1 (a non-trivial fact), the assertion follows.

2.3 Applications of Nevanlinna Theory

2.3.1 Theorems of Hadamard and Borel

The zeros of any meromorphic function f of finite order are finite in number or form a sequence having exponent of convergence $\varrho_0(f) \le \varrho(f)$. By Π_0 we denote the corresponding (polynomial or) canonical product; in any case Π_0 has order of growth at most $\varrho(f)$, and the same is true for the (polynomial or) canonical product Π_∞ that corresponds to the (non-zero) poles of f. The Lemma on the Logarithmic Derivative yields

Hadamard's Factorisation Theorem *Any meromorphic function f of finite order may be written as*

$$f(z) = z^m e^{Q(z)} \frac{\Pi_0(z)}{\Pi_\infty(z)},$$

where m resp. $-m$ is the order of the zero resp. pole at $z = 0$, Π_0 and Π_∞ are canonical products or polynomials having their zeros at the (non-zero) zeros and poles of f, respectively, and Q is a polynomial with

$$\varrho(f) = \max\{\varrho(\Pi_0), \varrho(\Pi_\infty), \deg Q\} = \max\{\varrho_0(f), \varrho_\infty(f), \deg Q\}.$$

The integer $\max\{\deg Q, \mathrm{genus}(\Pi_0), \mathrm{genus}(\Pi_\infty)\}$ is called the *genus* of f.

Proof The inequality $\varrho(f) \le \max\{\deg Q, \varrho(\Pi_0), \varrho(\Pi_\infty)\}$ is known. On the other hand, $f(z)z^{-m}\Pi_\infty(z)/\Pi_0(z) = e^{Q(z)}$ has order of growth at most ϱ, and from

$$T(r, Q') = m(r, Q') \le (\varrho - 1 + o(1))^+ \log r$$

it follows that Q is a polynomial of degree at most ϱ.

Any value a with exponent of convergence $\varrho_a(f) < \varrho(f)$ is called *Borel exceptional* for f. By Hadamard's Theorem the order of growth of functions with two Borel exceptional values is infinite or an integer.

Borel's Theorem *A non-constant meromorphic function can have at most two Borel exceptional values, that is, $\varrho(f) = \max_{1\le\nu\le3} \varrho_{a_\nu}(f)$ holds for every triple a_1, a_2, a_3. Functions with two Borel exceptional values have the form $f = he^g$, where h has finite order and g is either entire transcendental ($\varrho(f) = \infty$) or else a polynomial with $\deg g = \varrho(f) > \varrho(h)$.*

Proof We assume that the values a_ν are finite and have finite exponent of convergence, and choose $\sigma > \max_{1\le\nu\le3} \varrho_{a_\nu}(f)$. The Second Main Theorem yields

$$T(r,f) \le \sum_{\nu=1}^{3} N\Big(r, \frac{1}{f - a_\nu}\Big) + S(r,f) = O(r^\sigma) + S(r,f),$$

hence $T(r,f) = O(r^\sigma)$ outside some exceptional set E of measure $\ell < \infty$. Given $r > 0$ there exists an $\tilde{r} \in (r, r + \ell) \setminus E$, and since T is increasing this implies

$$T(r,f) \le T(\tilde{r},f) = O(\tilde{r}^\sigma) = O(r^\sigma).$$

Remark 2.2 Even for entire functions both proofs are much more elaborate within the theory of entire functions. It is hardly an exaggeration to state that to some extent the Lemma on the Logarithmic Derivative exceeds the strength of the theory of entire functions.

2.3.2 The Tumura–Clunie Theorem

The inequality (2.16) in the Generalised Second Main Theorem may be written as

$$m(r,f) + m\Big(r, \frac{1}{Q(z,f(z))}\Big) \le (2+\epsilon)T(r,f) + S(r,f),$$

with $Q(z,w) = \prod_{\nu=1}^{q}(w - a_\nu(z))$. Without going into details we will indicate how the proof of Theorem 2.2 can be adapted to the case when $Q(z,w)$ is any monic polynomial in w over the field of small functions w.r.t. f without multiple roots, which now may be algebraic or 'small' algebroid. We refer to the notation in the proof of Theorem 2.2. First of all, local bases $\beta_1 \ldots, \beta_n$ and $b_1 \ldots, b_k$ remain bases under analytic continuation. Also analytic continuation along some closed curve changes $P[f]$ into $CP[f]$, and returns to $P[f]$ after a suitable number of continuations along the same curve. Thus we have $|C| = 1$, and $|P[f]|$ is well defined. The same is true for $\delta(z)$ and $|h'_\nu/h_\nu|$, hence (2.17) holds in the form

$$\log^+ \frac{1}{|Q(z,f(z))|} \le \frac{1}{n}\log^+ \frac{1}{|P[f](z)|} + n\log^+ \frac{1}{\delta(z)} + \frac{1}{n}\sum_{\nu=1}^{q} \log^+ |Q_\nu(z)|.$$

Integrating over $[0,2\pi]$ then yields the desired result. The estimate of the terms involving h'_ν/h_ν is now based on the Selberg–Valiron Theory of Algebroid Functions, which will be discussed in some detail in Sect. 2.5. This result is used to generalise the so-called Tumura–Clunie Theorem for meromorphic functions [122, 198] as follows.

Theorem 2.6 *Let f be meromorphic and $Q(z,w) = w^q + c_{q-1}(z)w^{q-1} + \cdots + c_0(z)$ any polynomial over the field of small functions such that $Q(z,w) = 0$ has $q \ge 3$ distinct (algebroid or meromorphic) solutions. Then for every $\epsilon > 0$,*

$$N\Big(r, \frac{1}{Q(z,f(z))}\Big) \ge (q-2-\epsilon)T(r,f)$$

holds outside some set E_ϵ of finite measure.

2.3.3 The Order of the Derivative

From $T(r,f') \le 2T(r,f) + S(r,f)$ it follows that the order of growth of f' satisfies $\varrho(f') \le \varrho(f)$. Whittaker [199] proved that the converse is also true (see also Tsuji [189] and Clunie [27]).

Theorem 2.7 *The order of growth of f and f' coincide, $\varrho(f) = \varrho(f')$.*

Proof To prove $\varrho(f) \le \varrho(f') = \varrho'$ we may assume that ϱ' is finite, so that the poles of f have exponent of convergence at most ϱ'. In particular,

$$N(r,f) \le N(r,f') = O(r^{\varrho'+\epsilon})$$

holds for every $\epsilon > 0$. To estimate the proximity function $m(r,f)$ we denote by (b_ν) the sequence of distinct poles of f and consider the union M of closed discs $\{z : |z - b_\nu| \le |b_\nu|^{-\kappa}\}$. We may assume that $|b_\nu| > 2$ holds for every ν (otherwise consider $f - R$, where R is an appropriate rational function). To proceed we need

Exercise 2.23 Prove that $\log |f'(z)| = O(|z|^{\varrho'+\epsilon})$ ($\epsilon > 0$ arbitrary) holds on $\mathbb{C} \setminus M$. (**Hint.** Employ the Poisson–Jensen formula for f' to obtain

$$\log |f'(z)| \le 3m(r,f') + O(n(r,f')\log r) \quad (|z| = r/2, z \notin M).)$$

For $\kappa > \varrho'$ the series $\sum_\nu |b_\nu|^{-\kappa}$ converges, hence the circular projection E of M onto the positive axis has measure at most $2\sum_\nu |b_\nu|^{-\kappa}$, and the radial projection onto the unit circle has angular measure at most $2\sum_\nu |b_\nu|^{-\kappa-1}$, which is less than 2π if κ is chosen sufficiently large. Thus there exists some ray $\arg z = \alpha$ not intersecting M, and given z with $|z| \notin E$ also a path Γ_z of length $O(|z|)$ joining the origin with z (for example, we can take the straight line segment from 0 to $|z|e^{i\alpha}$ followed by a circular arc from $|z|e^{i\alpha}$ to z). In combination with Exercise 2.23 this yields

$$|f(z)| \le (A + B|z|)e^{C|z|^{\varrho'+\epsilon}} \quad (A, B, C > 0),$$

hence $\log^+ |f(z)| = O(|z|^{\varrho'+\epsilon})$ for $|z| \notin E$, $m(r,f) = O(r^{\varrho'+\epsilon})$, and

$$T(r,f) = m(r,f) + O(r^{\varrho'+\epsilon}) = O(r^{\varrho'+\epsilon}) \quad (r \notin E).$$

Since E has finite measure and $T(r,f)$ is increasing, this holds as $r \to \infty$ without exception.

Exercise 2.24 Prove that Theorem 2.6 may also be stated as follows: Let g and h be entire functions such that h and $-W(g,h) = g'h - gh'$ have order of growth at most ϱ. Then g also has order of growth at most ϱ.

2.3.4 *Ramified Values*

The number

$$\vartheta(\infty,f) = \liminf_{r\to\infty} \frac{N(r,f) - \overline{N}(r,f)}{T(r,f)}$$

'measures' the ramification of the value ∞; it is positive if and only if f has 'many' multiple poles counting multiplicities. The *ramification index* $\vartheta(a,f)$ of $a \in \mathbb{C}$ is defined in the same way. We also define

$$\Theta(\infty,f) = 1 - \limsup_{r\to\infty} \frac{\overline{N}(r,f)}{T(r,f)},$$

and similarly $\Theta(a,f)$. Since $\vartheta(a,f) + \delta(a,f) \le \Theta(a,f)$ holds, the Second Main Theorem implies

$$\sum_{a\in\hat{\mathbb{C}}} \vartheta(a,f) + \sum_{a\in\hat{\mathbb{C}}} \delta(a,f) \le \sum_{a\in\hat{\mathbb{C}}} \Theta(a,f) \le 2.$$

The value a is called *completely ramified* if f assumes a always with multiplicity at least two. From $\Theta(a,f) \ge 1/2$ we obtain

Theorem 2.8 *Non-constant meromorphic functions can have at most four completely ramified values.*

The number four is attained by the Weierstraß P-function, which has four completely ramified values e_1, e_2, e_3, and $e_4 = \infty$. This can easily be read off the differential equation $\wp'^2 = 4(\wp - e_1)(\wp - e_2)(\wp - e_3)$.

Exercise 2.25 It is obvious that elliptic functions f have only finitely many critical (= ramified!) values. Prove $\sum_{a\in\hat{\mathbb{C}}} \vartheta(a,f) = \sum_{a\in\hat{\mathbb{C}}} \Theta(a,f) = 2$. (**Hint.** First assume that f has only simple poles and prove $N(r,1/f') = 2T(r,f) + O(r)$.)

2.3.5 Parametrisation of Simple Algebraic Curves

The algebraic curve

$$F : x^n + y^m = 1 \quad (n \ge m \ge 2) \tag{2.23}$$

has genus zero if $n = m = 2$, genus one if $(n,m) \in \{(4,2),(3,3),(3,2)\}$, and genus greater than one otherwise. Valiron [191] proved that F has no parametrisation by entire functions (f,g) if $\frac{1}{m} + \frac{1}{n} < 1$. This, of course, is a special case of the Uniformisation Theorem, since the basic parametrisation is given by elliptic functions in the second case, and by functions meromorphic in the unit disc in the third. Without reference to the Uniformisation Theorem we will prove

Theorem 2.9 *Suppose that non-constant meromorphic functions f and g parametrise the algebraic curve* (2.23) *with $\frac{1}{m} + \frac{1}{n} < 1$. Then (m,n) equals* $(4,2)$ *or* $(3,3)$ *or* $(3,2)$. *In any case f and g are given by*

$$f = E \circ \psi \quad \text{and} \quad g = \sqrt[m-1]{E'} \circ \psi,$$

where E is an elliptic function satisfying

$$E'^2 = 1 - E^4, \quad E'^3 = (1 - E^3)^2 \quad \text{and} \quad E'^2 = 1 - E^3,$$

respectively, and ψ is any non-constant entire function.

Proof At every zero of g, f assumes one of the values $a_\nu = e^{2\nu\pi i/n}$, and *vice versa*. The multiplicity is always $\geq m \geq 2$, this implying $\Theta(a_\nu, f) \geq 1 - 1/m \geq 1/2$, hence $n \leq 4$, and $m = 2$ if $n = 4$. In this case, the meromorphic function

$$\phi = \frac{f'^2}{1 - f^4} = \frac{f'}{(1-f)(1+f)} \frac{f'}{(1-if)(1+if)}$$

satisfies $m(r, \phi) = S(r, f)$ by the Lemma on the Logarithmic Derivative. Since, however, the zeros of $1 - f^4$ and the poles of f'^2 are cancelled by the zeros of f'^2 and the poles of $1 - f^4$, respectively, ϕ is entire, and f is a solution to

$$f'^2 = \phi(z)(1 - f^4) = \psi'^2(z)(1 - f^4),$$

this following from $\phi = \left(f'/g\right)^2$. Thus f may be written as $f = E \circ \psi$, where E satisfies $E'^2 = 1 - E^4$, hence is closely related to (and may be expressed by) some Jacobi elliptic function of elliptic order two. In the second and third case the proof runs along the same lines with

$$\phi = \frac{f'^3}{(1-f^3)^2} = \left(\frac{f'}{g^2}\right)^3 = \psi'^3 \quad \text{and} \quad \phi = \frac{f'^2}{1-f^3} = \left(\frac{f'}{g}\right)^2 = \psi'^2,$$

respectively. The third case, $(n, m) = (3, 2)$, differs from the others insofar as the poles of f and g enter the stage: a pole of f of order p requires a pole of g of order q such that $3p = 2q$, and *vice versa*.

Exercise 2.26 The equation $f^2 + g^2 = 1$ is solved by $f(z) = \cos z$ and $g(z) = \sin z$. Prove that the general *entire* solution is given by $f(z) = \cos \psi(z)$ and $g(z) = \sin \psi(z)$, where ψ is any non-constant entire function. There are, however, many other *meromorphic* solutions ($f(z) = i \tan z$ and $g(z) = 1/\cos z$, for example). Find some more among the Jacobi elliptic functions (see Exercise 1.18). The general solution has the form $f(z) = \frac{1-h(z)^2}{1+h(z)^2}$, $g(z) = \frac{2h(z)}{1+h(z)^2}$, h any non-constant meromorphic function. To determine h if $f(z) = \text{cn}\, z$ and $g(z) = \text{sn}\, z$ is non-trivial. Prove that h satisfies $h'^2 = \frac{1}{4}(h^2 + 2\kappa h + 1)(h^2 - 2\kappa h + 1)$, $h(0) = 0$, $h'(0) = \frac{1}{2}$, $\kappa^2 \neq 0, 1$.

Remark 2.3 Gross [61, 62] and Gross and Osgood [63] considered $f^n + g^n = 1$. Gundersen [69, 71] constructed several interesting solutions to $f^n + g^n + h^n = 1$. For general results on Waring's problem for analytic functions, see Hayman [80].

2.4 Cartan's Theory of Entire Curves

Let $g_1, \ldots, g_p$ be linearly independent entire functions without common zeros. Then

$$\mathfrak{g} = (g_1, \ldots, g_p) : \mathbb{C} \longrightarrow \mathbb{C}^p$$

is called an *entire curve*. It is just a matter of taste to look upon $\mathfrak{g}$ as a map from $\mathbb{C}$ into $\mathbb{C}^p$ or into projective space. In this section we will discuss Cartan's generalisation of Nevanlinna Theory to entire curves.

2.4.1 *The Characteristic of an Entire Curve*

To make a definite choice we assume that $\mathbb{C}^p$ is endowed with the euclidean norm $\|\mathfrak{a}\| = \sqrt{|a_1|^2 + \cdots + |a_p|^2}$. Then

$$T_C(r, \mathfrak{g}) = \frac{1}{2\pi} \int_0^{2\pi} \log \|\mathfrak{g}(re^{i\theta})\| \, d\theta$$

is called the *Cartan characteristic* of $\mathfrak{g}$. A change of the norm or a change of the basis of the space spanned by $g_1, \ldots, g_p$ results in an additional bounded term; actually T_C may be viewed as the characteristic of the linear space spanned by $g_1, \ldots, g_p$.

Example 2.1 The quotient $f = g_\mu/g_\nu$ is a meromorphic function, and from $\log \|\mathfrak{g}\| \ge \frac{1}{2} \log(1 + |f|^2) + \log |g_\nu| = \log^+ |f| + \log |g_\nu| + O(1)$ it follows that

$$\begin{aligned} T_C(r, \mathfrak{g}) &\ge m(r,f) + m(r, g_\nu) - m(r, 1/g_\nu) + O(1) \\ &= m(r,f) + N(r, 1/g_\nu) + O(1) \\ &\ge T(r,f) + O(1) \end{aligned}$$

holds. Equality, $T_C(r, \mathfrak{g}) = T(r,f) + O(1)$, holds for $p = 2$. ☕

2.4.2 *The Ahlfors–Shimizu Formula*

The function $u(z) = \log \|\mathfrak{g}(z)\|$ is regular subharmonic with Laplacian

$$\Delta u = \frac{2}{\|\mathfrak{g}\|^4} \sum_{1 \le \mu < \nu \le p} |g'_\mu g_\nu - g_\mu g'_\nu|^2 = 2\|\mathfrak{g}'\|^2,$$

this defining $\|\mathfrak{g}'\|$. Green's formula

$$\int_{|z|<r} \Delta u(z)\, d(x,y) = \int_{|z|=r} \frac{\partial}{\partial \mathfrak{n}} u(z)\, |dz| = r\frac{d}{dr}\int_0^{2\pi} u(re^{i\theta})\, d\theta$$

yields the analogue to the *Ahlfors–Shimizu formula* in Nevanlinna Theory, namely

$$T_C(r,\mathfrak{g}) = \log\|\mathfrak{g}(0)\| + \int_0^r A(t,\mathfrak{g})\,\frac{dt}{t} \quad \text{with} \quad A(t,\mathfrak{g}) = \frac{1}{\pi}\int_{|z|<t} \|\mathfrak{g}'(z)\|^2\, d(x,y).$$

We note that $\|\mathfrak{g}'\| = f^\sharp$ holds for $\mathfrak{g} = (g_1, g_2)$ and $f = g_1/g_2$.

2.4.3 Cartan's First and Second Main Theorem

From Jensen's formula applied to the entire function $\mathfrak{a}{\cdot}\mathfrak{g} = a_1g_1 + \cdots + a_pg_p$ ($\mathfrak{a} \neq \mathfrak{o}$) it follows that

$$\frac{1}{2\pi}\int_0^{2\pi} \log|\mathfrak{a}{\cdot}\mathfrak{g}(re^{i\theta})|\, d\theta = N\Big(r, \frac{1}{\mathfrak{a}{\cdot}\mathfrak{g}}\Big) + \log^+|\mathfrak{a}{\cdot}\mathfrak{g}(0)| = N_C(r,\mathfrak{a}),$$

this defining the counting function $N_C(r,\mathfrak{a})$; it 'counts' how often $\mathfrak{g}(z)$ intersects the hyperplane $\mathfrak{a}\cdot\mathfrak{w} = 0$ on $|z| < r$. With

$$m_C(r,\mathfrak{a}) = \frac{1}{2\pi}\int_0^{2\pi} \log\frac{\|\mathfrak{a}\|\,\|\mathfrak{g}(re^{i\theta})\|}{|\mathfrak{a}{\cdot}\mathfrak{g}(re^{i\theta})|}\, d\theta$$

we obtain Cartan's

First Main Theorem *For every $\mathfrak{a} \neq \mathfrak{o}$ we have*

$$m_C(r,\mathfrak{a}) + N_C(r,\mathfrak{a}) = T_C(r,\mathfrak{g}) + \log\|\mathfrak{a}\|.$$

To state Cartan's Second Main Theorem we introduce the Wronskian determinant $W(\mathfrak{g}) = W(g_1,\ldots,g_p)$ and set $N_{\rm crit}(r,\mathfrak{g}) = N(r, 1/W(\mathfrak{g}))$, noting that $N_{\rm crit}(r,\mathfrak{g}) = N_{\rm crit}(r,f)$ if $p = 2$ and $f = g_1/g_2$. The vectors $\mathfrak{a}_1,\ldots,\mathfrak{a}_q \in \mathbb{C}^p$ ($q > p$) are said to be in *general position* if any p mutually distinct vectors $\mathfrak{a}_{j_1},\ldots,\mathfrak{a}_{j_p}$ are linearly independent. Then the role of $\mathfrak{g}$ may be taken by $\tilde{\mathfrak{g}} = (\mathfrak{a}_{j_1}\cdot\mathfrak{g},\ldots,\mathfrak{a}_{j_p}\cdot\mathfrak{g})$, since $T_C(r,\mathfrak{g}) = T_C(r,\tilde{\mathfrak{g}}) + O(1)$ and $N(r,1/W(\mathfrak{g})) = N(r,1/W(\tilde{\mathfrak{g}}))$ holds.

Second Main Theorem *Given $q > p$ vectors $\mathfrak{a}_\nu$ in general position,*

$$\sum_{\nu=1}^q m_C(r,\mathfrak{a}_\nu) \le pT_C(r,\mathfrak{g}) - N_{\rm crit}(r,\mathfrak{g}) + S(r,\mathfrak{g})$$

holds, with $S(r,\mathfrak{g}) = O(\log(rT_C(r,\mathfrak{g})))$ *outside some set* $E \subset (0,\infty)$ *of finite measure. For entire curves* $\mathfrak{g}$ *of finite order* $\varrho(\mathfrak{g}) = \limsup_{r\to\infty} \frac{\log^+ T_C(r,\mathfrak{g})}{\log r}$, $S(r,\mathfrak{g}) = O(\log r)$ *holds without exception.*

Proof There is no loss of generality to assume $\mathfrak{a}_\nu = \mathfrak{e}_\nu$ (the ν-th unit vector) for $1 \le \nu \le p$. We set $g_\nu = \mathfrak{a}_\nu \cdot \mathfrak{g}$ for $p+1 \le \nu \le q$ and, following Eremenko [37],

$$\mathfrak{g}_J = (g_{j_1},\ldots,g_{j_p}) \quad \text{and} \quad W_J = W(g_{j_1},\ldots,g_{j_p})$$

for any set $J = \{j_1,\ldots,j_p\} \subset \{1,\ldots,q\}$ of cardinality p, and $W = W(\mathfrak{g})$. Then

$$\log|W_J| = \log|W| + O(1) \quad \text{and} \quad \max_{j\in J} \log|g_j| = \log\|\mathfrak{g}\| + O(1)$$

hold, and at least $q-p$ of the functions g_ν satisfy $\log|g_\nu| = \log\|\mathfrak{g}\| + O(1)$. This has to be understood point-wise as $z \to \infty$; note that there are only finitely many $O(1)$-terms. Then

$$\begin{aligned}
&\textstyle\sum_{\nu=1}^q \log|g_\nu| \\
&\ge (q-p)\log\|\mathfrak{g}\| + \min_J \Big(\textstyle\sum_{j\in J}\log|g_j| - \log|W_J|\Big) + \log|W| + O(1)\\
&\ge (q-p)\log\|\mathfrak{g}\| + \log|W| - \textstyle\sum_J \Big(\log|W_J| - \sum_{j\in J}\log|g_j|\Big)^+ + O(1)\\
&= (q-p)\log\|\mathfrak{g}\| + \log|W| - \textstyle\sum_J \log^+ \frac{|W_J|}{\prod_{j\in J}|g_j|} + O(1)
\end{aligned}$$

also holds. Jensen's formula for g_ν yields

$$\sum_{\nu=1}^q N\Big(r,\frac{1}{g_\nu}\Big) \ge (q-p)T_C(r,\mathfrak{g}) + N_{\text{crit}}(r,\mathfrak{g}) - \sum_J m\Big(r,\frac{W_J}{\prod_{j\in J} g_j}\Big) + O(1).$$

Since $N(r,1/g_\nu) = N_C(r,\mathfrak{a}_\nu) + O(1)$ we have to estimate $m\Big(r,\frac{W_J}{\prod_{j\in J} g_j}\Big)$. The standard estimate $O(\log \max_{1\le\nu\le p} rT(r,g_\nu))$, however, is too crude. For $J = \{1,\ldots,p\}$, say,

$$\frac{W(g_1,\ldots,g_p)}{g_1\cdots g_p} = (-1)^p \frac{W(f_1',\ldots,f_{p-1}')}{f_1\cdots f_{p-1}} \quad (f_\nu = g_\nu/g_p)$$

holds, and $T(r,f_\nu) = T(r,g_\nu/g_p) \le T_C(r,\mathfrak{g}) + O(1)$ implies

$$m\Big(r,\frac{W(g_1,\ldots,g_p)}{g_1\cdots g_p}\Big) = m\Big(r,\frac{W(f_1',\ldots,f_{p-1}')}{f_1\cdots f_{p-1}}\Big) = \sum_{\nu=1}^{p-1} S(r,f_\nu) = S(r,\mathfrak{g}).$$

Example 2.2 The solutions to linear differential equations

$$w^{(n)} + c_{n-1}(z)w^{(n-1)} + \cdots + c_0(z)w = 0$$

with polynomial coefficients are entire functions of finite order, and every fundamental system defines an entire curve $\mathfrak{w} = (w_1, \ldots, w_n)$. If

$$N(r, 1/w_\nu) = o(T(r)) = o(\max_{1\le\mu\le n} T(r, w_\mu))$$

holds, then every linear combination $w = \sum_{\nu=1}^n b_\nu w_\nu$ with $b_1 \cdots b_n \neq 0$ satisfies

$$N(r, 1/w) = T_C(r, \mathfrak{w}) + o(T(r)).$$

Remark 2.4 Cartan developed his theory in [22]. His Main Theorems are obvious generalisations of Nevanlinna's Main Theorems. We note, however, that the Lemma on the Logarithmic Derivative and Jensen's formula are indispensable. For further generalisations, see Eremenko [38]. There are good reasons for the notation $N_{\rm crit}$: for $p = 2$, $f = g_1/g_2$ has the derivative $f' = -W(g_1, g_2)/g_2^2$, and the zeros of $W(g_1, g_2)$ are exactly the critical points of f including the multiple poles (with the same multiplicities). In the general case, the coefficients of the linear differential operator $L[w] = \frac{W(g_1,\ldots,g_p,w)}{W(g_1,\ldots,g_p)}$ have poles exactly where $W(g_1, \ldots, g_p)$ vanishes; these points are critical for the linear equation $L[w] = 0$. For applications of Cartan's theory, see the paper [73] by Gundersen and Hayman.

2.4.4 Borel Identities

Picard's Theorem can be deduced from the fact that zero-free entire functions $e^{g_1}, e^{g_2}, e^{g_3}$ are linearly independent unless all differences $g_\mu - g_\nu$ are constants. For if $f = e^g$ never assumes the value 1, then $f = e^h + 1$ and $e^g - e^h - e^0 = 0$ holds, hence e^g, e^h, and e^0 are linearly dependent. This was generalised by Borel, who proved the following.

Theorem 2.10 *Suppose that the entire functions $f_0, f_1, \ldots, f_n$ are linearly dependent and zero-free. Then there exist fewer than $n+1$, and thus two of these functions that are linearly dependent.*

Proof We shall deduce the proof from Theorem 2.11 below, which is sometimes called Nevanlinna's Third Main Theorem. We start with $\sum_{\nu=0}^n c_\nu f_\nu \equiv 0$ (non-trivial), and are done if some c_ν vanishes. So we may assume $c_\nu \neq 0$ $(0 \le \nu \le n)$, and will derive a contradiction to the assumption that the functions $f_1, \ldots, f_n$ are linearly independent. Assuming this, the non-constant *entire* functions $g_\nu = -\dfrac{c_\nu f_\nu}{c_0 f_0}$

are linearly independent and satisfy

$$g_1 + g_2 + \cdots + g_n \equiv 1. \tag{2.24}$$

Since g_ν has no zeros and $W = W(g_1, \ldots, g_n)$ is entire, Theorem 2.11 gives

$$T(r, g_\mu) = m(r, g_\mu) \le -N\big(r, 1/W\big) + S(r) \le S(r) \quad (1 \le \mu \le n),$$

where $S(r)$ is a remainder term with respect to $T(r) = \max_{1\le\nu\le n} T(r, g_\nu)$. This, however, is impossible for non-constant entire functions. ☕

Theorem 2.11 ([126]) *Suppose that linearly independent meromorphic functions $g_1, \ldots, g_n$ satisfy* (2.24). *Then*

$$m(r, g_\mu) \le \sum_{\nu=1}^{n} N\Big(r, \frac{1}{g_\nu}\Big) + N(r, W) - \sum_{\nu=1}^{n} N(r, g_\nu) - N\Big(r, \frac{1}{W}\Big) + S(r)$$

($1 \le \mu \le n$) *holds, where again W denotes the Wronskian determinant of $g_1, \ldots, g_n$.*

Proof Let W_1 denote the Wronskian determinant of the functions $g_2', \ldots, g_n'$, and set $G = g_1 \cdots g_n$ and $G_1 = g_2 \cdots g_n$. Then $W = W_1$ holds by Cramer's rule applied to the linear system that consists of (2.24) and $g_1^{(\nu)} + \cdots + g_n^{(\nu)} \equiv 0$ $(1 \le \nu < n)$. From $g_1 = \Delta_1/\Delta$ with $\Delta_1 = W_1/G_1$ and $\Delta = W/G$, Jensen's formula, and $m(r, \Delta) + m(r, \Delta_1) = S(r)$ it follows that

$$\begin{aligned}
m(r, g_1) &\le m(r, 1/\Delta) + S(r)\\
&= m(r, 1/\Delta) - m(r, \Delta) + S(r)\\
&= -\frac{1}{2\pi}\int_0^{2\pi} \log|\Delta(re^{i\theta})|\, d\theta + S(r)\\
&= \frac{1}{2\pi}\int_0^{2\pi} \big(\log|G(re^{i\theta})| - \log|W(re^{i\theta})|\big)\, d\theta + S(r)\\
&= N(r, 1/G) - N(r, G) + N(r, W) - N(r, 1/W) + S(r).
\end{aligned}$$

This proves Theorem 2.11, since each g_μ can take the role of g_1. ☕

2.5 The Selberg–Valiron Theory of Algebroid Functions

2.5.1 *Algebroid Functions*

We will consider irreducible polynomials[8]

$$P(z, w) = \sum_{\kappa=0}^{k} A_\kappa(z) w^\kappa$$

of degree $k > 1$ in w over the ring of entire functions. It is assumed that the coefficients A_κ have no common zeros and that at least one ratio A_κ/A_k is transcendental. The local solutions to the algebraic equation $P(z, w) = 0$ are denoted by $f_1, \ldots, f_k$. They admit analytic continuation along any curve that avoids the singularities. The singularities are algebraic and are zeros of the entire function $A_k \Delta_P$, where

$$\Delta_P = A_k^{2k-2} \prod_{1 \le \kappa < \nu \le k} (f_\kappa - f_\nu)^2 \tag{2.25}$$

denotes the discriminant of P with respect to w. We note *Vieta's rules*

$$\frac{A_{k-\kappa}}{A_k} = (-1)^\kappa \sum_{\nu_1 < \nu_2 < \cdots < \nu_\kappa} f_{\nu_1} f_{\nu_2} \cdots f_{\nu_\kappa} \quad (0 < \kappa \le k). \tag{2.26}$$

The union of branches $\mathfrak{f} = \{f_1, \ldots, f_k\}$ is called the *Algebroid Function* of degree k with *minimal polynomial* P; P is uniquely determined up to a factor e^ϕ, where ϕ may be any entire function. We note that analytic continuation along any closed curve γ effects a permutation of the branches f_κ, and every permutation may be achieved by choosing γ appropriately (if $P = P_1 \cdots P_n$ is the product of irreducible polynomials, the latter holds for the branches of $P_\nu = 0$ separately). Non-constant meromorphic functions are algebroid functions of degree 1, and *vice versa.*

Exercise 2.27 Let $f_1, \ldots, f_k$ be analytic on some domain D and suppose that

- each f_κ admits unrestricted analytic continuation to $\mathbb{C} \setminus S$, where S is a discrete subset of $\mathbb{C}$;
- for each f_κ every point in S is either regular or else an algebraic singularity or an algebraic pole;
- analytic continuation along any closed curve in $\mathbb{C} \setminus S$ effects a permutation of the branches f_κ.

[8]The reader will notice that irreducibility is not essential for the theory. In many places it suffices to assume that P is a product of *different* irreducible polynomials (P contains no repeated factor).

Prove that $f_1, \ldots, f_k$ form the branches of an (algebraic or) algebroid function. (**Hint.** Consider locally $\prod_{\kappa=1}^{k}(w - f_\kappa(z)) = w^k + \sum_{\kappa=0}^{k-1} a_\kappa(z) w^\kappa$.)

Exercise 2.28 Prove that if $\mathfrak{g}$ is any other algebroid function, then so are $\mathfrak{f} \pm \mathfrak{g}$ with branches $f_\kappa \pm g_\lambda$, $\mathfrak{f}\mathfrak{g}$, $\mathfrak{f}/\mathfrak{g}$ (defined in a similar way), and $\mathfrak{f}' = \{f_1', \ldots, f_k'\}$.
(**Hint.** The branches f_κ' have only algebraic singularities and poles, and f_κ' and f_κ are likewise permuted under analytic continuation.)

The algebroid functions form a field that is closed under differentiation. Suitable subfields may be defined by imposing certain growth conditions on the coefficients like $T(r, A_\kappa) = O(\Lambda(r))$ outside some exceptional set of finite measure, where Λ is fixed and increases to infinity faster than $\log r$.

Remark 2.5 The degree of $\mathfrak{f} + \mathfrak{g}$, $\mathfrak{f}\mathfrak{g}$, etc. is at most $k\ell$, but may be smaller. For example, $\mathfrak{f}^2$, $\mathfrak{f}'$, and $\mathfrak{f}'/\mathfrak{f}$ also have degree k (or less).

Exercise 2.29 Prove that if P is *irreducible* and some branch f_κ also satisfies an *irreducible* equation $Q(z, w) = \sum_{\lambda=0}^{\ell} B_\lambda(z) w^\lambda = 0$, then $k = \ell$ and $A_\kappa = B_\kappa e^\phi$ holds for $0 \le \kappa \le k$, with ϕ some entire function.

2.5.2 Two Equivalent Approaches

The analogue to Nevanlinna Theory for algebroid functions was developed by Selberg [149] and simplified by Valiron [192] using a different approach. Following Selberg, it is quite natural to define the *proximity function*, the *counting function* of poles, and the *(Selberg) characteristic* by

$$m(r, \mathfrak{f}) = \frac{1}{2k\pi} \int_0^{2\pi} \sum_{\kappa=1}^{k} \log^+ |f_\kappa(re^{i\theta})| \, d\theta,$$

$$N(r, \mathfrak{f}) = \frac{1}{k} N(r, 1/A_k), \quad \text{and} \quad T_S(r, \mathfrak{f}) = m(r, \mathfrak{f}) + N(r, \mathfrak{f}).$$

Exercise 2.30 Set $P(z, w) = A_k(z) \prod_{\kappa=1}^{k}(w - f_\kappa(z))$. Prove that for z fixed, Jensen's formula with $r = 1$ yields

$$U(z) = \frac{1}{2\pi} \int_0^{2\pi} \log |P(z, e^{i\phi})| \, d\phi = \log |A_k(z)| + \sum_{\kappa=1}^{k} \log^+ |f_\kappa(z)|$$

(compare the proof of Cartan's Identity), hence

$$T_S(r, \mathfrak{f}) = \frac{1}{2k\pi} \int_0^{2\pi} U(re^{i\theta}) \, d\theta + O(1) \tag{2.27}$$

holds by Jensen's formula for A_k.

The *Valiron characteristic* of $\mathfrak{f}$ is defined by

$$T_V(r,\mathfrak{f}) = \frac{1}{2k\pi}\int_0^{2\pi} \log A(re^{i\theta})\,d\theta \quad \text{with} \quad A(z) = \max_{0\le\kappa\le k} |A_\kappa(z)|. \tag{2.28}$$

The advantage of Valiron's approach is that it works with the known coefficients A_κ rather than the unknown branches f_κ; considering $e^\phi P$ in place of P results in an additive term $\operatorname{Re}\phi(0)/k$. On the other hand, Selberg's approach allows us to define expressions like $R\circ\mathfrak{f}$ simply by its branches $R\circ f_\kappa$.

Exercise 2.31 Compute $T_V(r,\mathfrak{f})$ for $\mathfrak{f}$ with minimal polynomial

$$P(z,w) = w^3\cos z + w\sinh z - e^{(1+i)z},$$

and compare $T_V(r,\mathfrak{f})$ with the Nevanlinna characteristic of the entire function

$$f(z) = c_3\cos z + c_1\sinh z + c_0e^{(1+i)z} \quad (c_0c_1c_3 \neq 0)$$

and the Cartan characteristic of the entire curve

$$\mathfrak{g}(z) = (\cos z, \sinh z, e^{(1+i)z}).$$

The equivalence of both approaches, namely

$$T_V(r,\mathfrak{f}) = T_S(r,\mathfrak{f}) + O(1),$$

follows from $U(z) = \log A(z) + O(1)$. To prove $\log A(z) \le U(z) + O(1)$ we employ Vieta's formula (2.26) to obtain

$$\begin{aligned}\log|A_{k-\kappa}(z)| &\le \log|A_k(z)| + \log^+ \sum_{\nu_1<\nu_2<\cdots<\nu_\kappa} |f_{\nu_1}(z)f_{\nu_2}(z)\cdots f_{\nu_\kappa}(z)|\\ &\le \log|A_k(z)| + \sum_{\nu=1}^{k}\log^+|f_\nu(z)| + O(1) = U(z) + O(1),\end{aligned}$$

while the reversed inequality $U(z) \le \log A(z) + O(1)$ follows from

$$|P(z,e^{i\phi})| \le \sum_{\kappa=0}^{k} |A_\kappa(z)| \le (k+1)A(z).$$

Henceforth we will not distinguish between both forms and just write $T(r,\mathfrak{f})$. The definition of the Valiron characteristic resembles the definition of the Cartan characteristic $T_C(r,\mathfrak{g})$ of $\mathfrak{g} = (A_0,\ldots,A_k)$. One cannot, however, expect that the coefficients A_κ are linearly independent. The sum $f+\mathfrak{g}$, the product $\mathfrak{f}\mathfrak{g}$, and the ratio

$\mathfrak{f}/\mathfrak{g}$ of $\mathfrak{f}$ and $\mathfrak{g}$ with minimal polynomials $P(z,w) = \sum_{\kappa=0}^{k} A_\kappa(z)w^\kappa$ and $Q(z,w) = \sum_{\lambda=0}^{\ell} B_\lambda(z)w^\lambda$, say, satisfy some algebraic equation

$$R(z,w) = \sum_{\nu=0}^{k\ell} C_\nu(z)w^\nu = 0 \quad \text{with } C_{k\ell} = A_k^\ell B_\ell^k.$$

For $\mathfrak{f}\mathfrak{g}$, say, $R(z,w)$ is the resultant of the polynomials $P(z,w/y)y^k$ and $Q(z,y)$ with respect to y. We note, however, that R need not be irreducible and its coefficients may have common zeros.

Exercise 2.32 In any case ($\mathfrak{h} = f + \mathfrak{g}, \mathfrak{f}\mathfrak{g}, \mathfrak{f}/\mathfrak{g}$) prove that

$$m(r,\mathfrak{h}) \le m(r,\mathfrak{f}) + m(r,\mathfrak{g}) + O(1) \quad \text{and} \quad N(r,\mathfrak{h}) \le N(r,\mathfrak{f}) + N(r,\mathfrak{g}).$$

Compute the corresponding functions for $\mathfrak{f}$, $\mathfrak{g}$ (with minimal polynomials $P(z,w) = w^3\cos z - \sin z$, $Q(z,w) = w^2\sin^2 z - \cos z$), and any $\mathfrak{h} = f + \mathfrak{g}, \mathfrak{f}\mathfrak{g}, \mathfrak{f}/\mathfrak{g}$.

2.5.3 The First Main Theorem for Algebroid Functions

Combining both definitions of the characteristic it becomes obvious that

$$m(r,\mathfrak{f}) = \frac{1}{2k\pi}\int_0^{2\pi} \log\frac{A(re^{i\theta})}{|A_k(re^{i\theta})|}\,d\theta + O(1),$$
$$m\Big(r,\frac{1}{\mathfrak{f}-c}\Big) = \frac{1}{2k\pi}\int_0^{2\pi} \log\frac{A(re^{i\theta})}{|P(re^{i\theta},c)|}\,d\theta + O(1)$$

hold. The First Main Theorem has the expected form.

First Main Theorem $T\Big(r,\dfrac{1}{\mathfrak{f}-a}\Big) = T(r,\mathfrak{f}) + O(1).$

Proof For $a = 0$, the assertion $T_V(r,\mathfrak{f}) = T_V(r,1/\mathfrak{f})$ follows from Valiron's definition and the fact that $1/\mathfrak{f}$ has minimal polynomial $w^kP(z,1/w) = \sum_{\kappa=0}^{k} A_{k-\kappa}(z)w^\kappa$, while $T_S(r,\mathfrak{f}-a) = T_S(r,\mathfrak{f}) + O(1)$ obviously follows from Selberg's definition and the elementary inequality $|\log^+|f_\kappa - a| - \log^+|f_\kappa|| \le \log^+|a| + \log 2$.

Exercise 2.33 From $|A_\kappa| \le A$ deduce that $N(r,1/A_\kappa) \le kT(r,\mathfrak{f}) + O(1)$. Prove also that $N(r,1/P(z,c)) \le kT(r,\mathfrak{f}) + O(1)$ ($c \in \mathbb{C}$) and the analogue to Cartan's Identity,

$$T_S(r,\mathfrak{f}) = \frac{1}{2k\pi}\int_0^{2\pi} N\Big(r,\frac{1}{P(z,e^{i\phi})}\Big)\,d\phi + C, \tag{2.29}$$

holds for some constant C. (**Hint.** Integrate $U(re^{i\theta}) = \frac{1}{2\pi}\int_0^{2\pi} \log|P(re^{i\theta}, e^{i\phi})|\,d\phi$ with respect to θ and reverse the order of integration, see Cartan's Identity.)

By virtue of (2.29), $T_S(r, \mathfrak{f})$ is continuous and increasing with respect to r, and a convex function of $\log r$. This is also true for $T_V(r, \mathfrak{f})$ and the Cartan characteristic $T_C(r, \mathfrak{g})$, since $\log A(z)$ and $\log \|\mathfrak{g}(z)\|$ are subharmonic.

2.5.4 The Characteristics of $\mathfrak{f}$ and A_ν/A_k

The Selberg–Valiron characteristic of $\mathfrak{f}$ and the Nevanlinna characteristics of $B_\nu = A_\nu/A_k$ are related as follows.

Lemma 2.2 *For $B_\nu = A_\nu/A_k$ the following is true:*

$$m(r, B_\nu) + O(1) \le km(r, \mathfrak{f}) \le \sum_{\kappa=0}^{k-1} m(r, B_\kappa) + O(1),$$
$$N(r, B_\nu) \le kN(r, \mathfrak{f}) \le \sum_{\kappa=0}^{k-1} N(r, B_\kappa).$$

Proof The very first inequality follows from $\log^+|B_\nu| \le \sum_{\kappa=1}^{k} \log^+|f_\kappa| + O(1)$. To prove the second we fix some (large) number $H > 0$. For z fixed, at least one of the intervals $I_0 = [0, H), \ldots, I_k = [kH, (k+1)H)$, I_j, say, contains none of the numbers $\log|f_\kappa(z)|$, while there might be indices such that $\log|f_\nu(z)| \ge (j+1)H$. If there is no such ν we get $\sum_{\nu=1}^{k} \log^+|f_\nu(z)| \le k(j+1)H \le k(k+1)H$. If, however,

$$\log|f_\nu(z)| \ge (j+1)H \ (1 \le \nu \le \ell) \quad \text{and} \quad \log|f_\nu(z)| < (j+1)H \ (\ell < \nu \le k),$$

say, holds, we obtain $|B_\ell| \ge \prod_{\nu=1}^{\ell}|f_\nu| - \binom{k}{\ell}e^{-H}\prod_{\nu=1}^{\ell}|f_\nu| \ge \frac{1}{2}\prod_{\nu=1}^{\ell}|f_\nu|$ if H is chosen sufficiently large. This implies

$$\sum_{\nu=1}^{k} \log^+|f_\nu| \le \sum_{\nu=1}^{\ell} \log^+|f_\nu| + O(1) \le \log^+|B_\ell| + O(1),$$

hence

$$\sum_{\nu=1}^{k} \log^+|f_\nu| \le \sum_{\kappa=0}^{k-1} \log^+|B_\kappa| + O(1)$$

is true in any case, and the second inequality holds. The second assertion is obvious since $N(r, B_\nu) \le N(r, 1/A_k)$ is trivially true, and to every zero of A_k there exists some function A_κ that does not vanish there.

Exercise 2.34 Prove that $T(r,\mathfrak{f}) = O(\log r)$ holds if and only if $\mathfrak{f}$ is algebraic.

Example 2.3 Choose mutually distinct complex numbers c_κ $(0 \le \kappa \le k)$ and set $B_\kappa(z) = P(z, c_\kappa)$. Assuming $Q(z,w) = \sum_{\kappa=0}^k B_\kappa(z)w^\kappa$ to be irreducible (which, however, is not necessary for our considerations), a new algebroid function $\mathfrak{g}$ is defined by $Q(z,w) = 0$. Obviously, $|B_\kappa(z)| = O(A(z))$ holds, and since the Vandermonde matrix $\left(c_\kappa^j\right)_{j,\kappa=0}^k$ is invertible, $|A_\kappa(z)| = O(B(z))$ with $B(z) = \max_{0\le\kappa\le k}|B_\kappa(z)|$ also holds. This implies $A(z) \asymp B(z)$, $T(r,\mathfrak{g}) = T(r,\mathfrak{f}) + O(1)$, and

$$T\Big(r, \frac{P(z,c_j)}{P(z,c_k)}\Big) \le \frac{1}{k}T(r,\mathfrak{f}) + O(1).$$

2.5.5 *The Logarithmic Derivative*

As in Nevanlinna Theory, the proximity function of the logarithmic derivative $m(r,\mathfrak{f}'/\mathfrak{f})$ plays a crucial role. The proof of the key inequality

$$m(r,\mathfrak{f}'/\mathfrak{f}) = O(\log T(r,\mathfrak{f}) + \log r) \tag{2.30}$$

possibly outside some set E of finite measure, which is empty if $\mathfrak{f}$ has finite order

$$\varrho(\mathfrak{f}) = \limsup_{r\to\infty} \frac{\log T(r,\mathfrak{f})}{\log r},$$

is quite technical and will be subdivided into several lemmas. As usual, (2.30) will be written as $m(r,\mathfrak{f}'/\mathfrak{f}) = S(r,\mathfrak{f})$. As a first corollary we note that

$$m(r,\mathfrak{f}') \le m(r,\mathfrak{f}) + S(r,\mathfrak{f}). \tag{2.31}$$

Lemma 2.3 *The entire function $\Phi = \Delta_P \prod_{A_\nu\not\equiv 0} A_\nu$ has counting function of zeros $N(r,1/\Phi) = O(T(r,\mathfrak{f}))$, hence has at most $O\Big(\dfrac{T(R,\mathfrak{f})}{R-r}\Big)$ zeros on $|z| \le r < R$.*

Proof The inequality $N(r,1/\Delta_P) = O(T(r,\mathfrak{f}))$ follows from (2.25), and Lemma 2.2 implies $N(r,1/A_\nu) \le N(r,1/B_\nu) + N(r,1/A_k) = O(T(r,\mathfrak{f}))$. The second assertion is obvious.

Lemma 2.4 *$\Phi(z) \ne 0$ on the annulus $r - 2h < |z| < r + 2h$ implies*

$$m(r,\mathfrak{f}'/\mathfrak{f}) = O\big(\log^+ T(r+2h,\mathfrak{f})\big) + O\Big(\log^+ \frac{r+2h}{h}\Big) + O\Big(\log^+\log^+ \frac{r}{h}\Big).$$

Proof The zeros of any polynomial $x^k + \sum_{\kappa=0}^{k-1} b_\kappa x^\kappa$ of degree k are contained in the disc $|x| \le k \max_{0\le\nu<k} |b_\nu| + 1$, this implying

$$|f_\kappa(z)| \le k \max_{0\le\nu<k} |B_\nu(z)| + 1 \quad (B_\nu = A_\nu/A_k).$$

To estimate the right-hand side we apply the Poisson–Jensen formula to B_ν on the disc of radius $s = r + 2h$ to obtain

$$|\log|B_\nu(z)|| \le \frac{s+|z|}{s-|z|}(m(s,B_\nu) + m(s,1/B_\nu)) + \sum_{|z_j|<s} \log\frac{s+|z|}{|z-z_j|} \quad (|z|<s);$$

the sum runs over all zeros z_j of Φ on $|\zeta| < s$, hence on $|\zeta| \le r - 2h$ (all zeros and poles of B_ν are included), and is independent of ν. This yields

$$\begin{aligned} |\log|B_\nu(z)|| &\le \frac{s+|z|}{s-|z|}(2T(s,B_\nu) + O(1)) + n(r-2h, 1/\Phi)\log\frac{2(s+|z|)}{h} \\ &= O\Big(\frac{T(s,\mathfrak{f})}{h}\Big(s + \log^+\frac{s}{h}\Big)\Big) = \mu(s,h) \end{aligned}$$

on $r - \frac{3}{2}h \le |z| \le r + \frac{3}{2}h$, hence also

$$|\log|f_\kappa(z)|| \le \mu(s,h) + \log 2k = \nu(s,h).$$

Given $z_0 = re^{i\theta}$ we consider $\psi(\zeta) = \log f_\kappa(\zeta + z_0)$ with $|\mathrm{Im}\,\psi(0)| \le \pi$ on the disc $|\zeta| < \frac{3}{2}h$, noting that f_κ has no zeros and poles nor algebraic singularities on the disc $|z - z_0| < \frac{3}{2}h$ (this was the proper reason for incorporating the zeros of Δ_P). To proceed we apply the Borel–Carathéodory inequality (see Exercise 2.35 below) to ψ with $R = \frac{3}{2}h$, $A = \nu(s,h)$ and $|\psi(0)| \le \nu(s,h) + \pi$ to obtain $|\psi(\zeta)| = O\big(\nu(s,h) + 1\big)$ on $|\zeta| < h$, hence $|\psi'(0)| = O\big((\nu(s,h)+1)/h\big)$ by the Schwarz Lemma and

$$\log^+\left|\frac{f_\kappa'(re^{i\theta})}{f_\kappa(re^{i\theta})}\right| = \log^+|\psi'(0)| \le \log^+\frac{\mu(s,h)}{h} + O(1) \quad (s = r + 2h \to \infty).$$

Integrating with respect to $0 \le \theta \le 2\pi$ then gives the desired result. ☕

Exercise 2.35 Let ψ be holomorphic, satisfying $\mathrm{Re}\,\psi(\zeta) < A$ on $|\zeta| < R$. Prove the *Borel–Carathéodory inequality* $|\psi(\zeta)| \le \frac{2A|\zeta|}{R-|\zeta|} + |\psi(0)|\frac{R+|\zeta|}{R-|\zeta|}$.
(**Hint.** First assume $\psi(0) = 0$ and consider $\phi(z) = \frac{\psi(Rz)}{2A-\psi(Rz)}$ on $|z| < 1$.) ☕

Lemma 2.5 *For* $0 < r < R \le 2r$ *the following is true:*

$$m(r, \mathfrak{f}'/\mathfrak{f}) = O(\log^+ T(R,\mathfrak{f})) + \log\frac{R}{R-r} + O(1). \tag{2.32}$$

Proof On $|z| \le \frac{1}{2}(R+r)$, the function Φ in Lemma 2.3 has $n \le O\Big(\dfrac{T(R,\mathfrak{f})}{R-r}\Big)$ zeros, and at least one of the annuli

$$r + 4jh < |z| < r + 4(j+1)h \quad \Big(h = \frac{R-r}{4(n+2)^2},\ 0 \le j \le n\Big)$$

contains no zeros of Φ. Lemma 2.4 applies with $s = r + 4(j+\frac{1}{2})h$ for some j in place of r. Since $s \le r + (4n+2)h < r + \frac{1}{2}(R-r) = \frac{1}{2}(R+r)$, $h \le \frac{1}{2}r$, $s + 2h < R$, and $R - r = O\big(h\,T(R,\mathfrak{f})^2\big)$, this yields

$$\begin{aligned} m(s,\mathfrak{f}'/\mathfrak{f}) &= O\big(\log T(R,\mathfrak{f})\big) + O\Big(\log\frac{(s+r)T(R,\mathfrak{f})^4}{(R-r)^2}\Big)\\ &\quad + O\Big(\log^+\log^+\frac{RT(R,\mathfrak{f})^2}{R-r}\Big)\\ &= O\big(\log^+ T(R,\mathfrak{f})\big) + O\Big(\log^+\frac{R}{R-r}\Big). \end{aligned} \tag{2.33}$$

From the First Main Theorem for $\mathfrak{f}'/\mathfrak{f}$ and monotonicity of the characteristic we obtain

$$\begin{aligned} m(r,\mathfrak{f}'/\mathfrak{f}) &\le m(s,\mathfrak{f}'/\mathfrak{f}) + N(s,\mathfrak{f}'/\mathfrak{f}) - N(r,\mathfrak{f}'/\mathfrak{f}) + O(1)\\ &\le m(s,\mathfrak{f}'/\mathfrak{f}) + \frac{n}{r}(s-r) + O(1). \end{aligned}$$

The final estimate of $m(r,\mathfrak{f}'/\mathfrak{f})$ then follows from (2.33) and

$$\frac{n}{r}(s-r) \le \frac{n}{r}(4n+2)\frac{R-r}{4(n+2)^2} < \frac{R-r}{r}.$$

The final step from (2.32) to (2.30) is the same as in Nevanlinna Theory: set $R = 2r$ if $\mathfrak{f}$ has finite order, and $R = r + 1/T(r,\mathfrak{f})$ otherwise.

2.5.6 The Second Main Theorem for Algebroid Functions

We will start with two examples to see what can be expected.

Example 2.4 Suppose that the coefficients $A_0,\ldots,A_k$ of the minimal polynomial $P(z,w)$ are linearly independent. Then $\mathfrak{g} = (A_0,\ldots,A_k)$ is an entire curve $\mathbb{C} \longrightarrow \mathbb{C}^{k+1}$, and $\mathfrak{a}\cdot\mathfrak{g} = P(z,a)$ holds for $\mathfrak{a} = (1,a,\ldots,a^k)$. Also the vectors $\mathfrak{a}_\nu =$

$(1, a_\nu, \ldots, a_\nu^k)$ $(1 \le \nu \le q, q > k+1)$ are in general position provided $a_\mu \neq a_\nu$ for $\mu \neq \nu$. From

$$\begin{aligned} N_C(r; \mathfrak{a}_\nu) &= N(r, 1/P(z, a_\nu)) = kN\big(r, 1/(\mathfrak{f} - a_\nu)\big),\\ T_C(r, \mathfrak{g}) &= kT(r, \mathfrak{f}) + O(1),\\ S(r, \mathfrak{g}) &= S(r, \mathfrak{f}), \end{aligned}$$

and Cartan's Second Main Theorem it follows that

$$(q - (k+1))T(r, \mathfrak{f}) \le \sum_{\nu=1}^{q} N\Big(r, \frac{1}{\mathfrak{f} - a_\nu}\Big) + S(r, \mathfrak{f}) \quad (q > k+1).$$

Example 2.5 The algebroid function with minimal polynomial $w^k \cos z + \sin z$ has $2k$ Picard values, namely the k-th roots of i and $-i$.

Thus the best one may hope to obtain is

$$(q - 2k)T(r, \mathfrak{f}) \le \sum_{\nu=1}^{q} N\Big(\frac{1}{\mathfrak{f} - a_\nu}\Big) + S(r, \mathfrak{f}) \quad (q > 2k). \tag{2.34}$$

Theorem 2.12 *For every algebroid function $\mathfrak{f}$ of degree k and $q > 2k$ mutually distinct complex numbers a_ν, the inequalities*

$$\sum_{\kappa=1}^{k} m\Big(r, \frac{1}{f - a_\nu}\Big) \le m\Big(r, \frac{1}{\mathfrak{f}'}\Big) + S(r, \mathfrak{f})$$

and (2.34) *hold.*

Proof Just as in the proof of Nevanlinna's Second Main Theorem one can prove

$$\sum_{\nu=1}^{q} \log^+ \frac{1}{|f_\kappa(z) - a_\nu|} = \log^+ \Big| \sum_{\nu=1}^{q} \frac{1}{f_\kappa(z) - a_\nu} \Big| + O(1) \quad (z = re^{i\theta},\ 1 \le \kappa \le k).$$

Integrating w.r.t. θ and summing w.r.t. κ yields

$$\begin{aligned} \sum_{\nu=1}^{q} m\Big(r, \frac{1}{\mathfrak{f} - a_\nu}\Big) &\le \sum_{\kappa=1}^{k} \frac{1}{2k\pi} \int_0^{2\pi} \log^+ \Big| \sum_{\nu=1}^{q} \frac{f_\kappa'(re^{i\theta})}{f_\kappa(re^{i\theta}) - a_\nu} \Big| \, d\theta \\ &\quad + \sum_{\kappa=1}^{k} \frac{1}{2k\pi} \int_0^{2\pi} \log^+ \frac{1}{|f_\kappa'(re^{i\theta})|} \, d\theta + O(1) \\ &= m(r, 1/\mathfrak{f}') + S(r, \mathfrak{f}). \end{aligned}$$

The estimate of $m(r, 1/\mathfrak{f}')$ is part of the subsequent exercises.

Exercise 2.36 Let $P(z, w)$ be the minimal polynomial of $\mathfrak{f}$. The branches $w_1 = f_\kappa'(z)$ of $\mathfrak{f}'$ solve $P_z(z, w) + P_w(z, w)w_1\big|_{w=f_\kappa(z)} = 0$, hence the minimal polynomial for $\mathfrak{f}'$ divides the resultant $P_1(z, w_1)$ of the polynomials $P(z, w)$ and $P_z(z, w) + P_w(z, w)w_1$ with respect to w. We note without proof that $P_1(z, w_1)$ has leading coefficient $A_k(z)^2 \triangle_P(z)$, but need not be irreducible, and its coefficients may have common zeros. Taking this into account, prove

$$kN(r, \mathfrak{f}') \le 2N(r, 1/A_k) + N(r, 1/\triangle_P) = 2kN(r, \mathfrak{f}) + N(r, 1/\triangle_P).$$

Example 2.6 For $P(z, w) = w^2 \cos^3 z - \sin z$ with $\triangle_P(z) = 4 \sin z \cos^3 z$ we obtain

$$P_1(z, w_1) = -\cos^4 z\big[4w_1^2 \sin z \cos^5 z - (1 + 2\sin^2 z)^2\big].$$

Exercise 2.37 The discriminant of an ordinary polynomial $p(w) = \sum_{\kappa=0}^{k} a_\kappa w^\kappa$ of degree k is a universal homogeneous polynomial in the coefficients $a_0, \ldots, a_k$ of degree $2k-2$. Prove that the discriminant w.r.t. w of the minimal polynomial $P(z, w)$ is an entire function with Nevanlinna characteristic

$$T(r, \triangle_P) \le k(2k-2)T(r, \mathfrak{f}) + O(1).$$

Deduce that $T(r, \mathfrak{f}') + m(r, \mathfrak{f}) \le 2kT(r, \mathfrak{f}) + S(r, \mathfrak{f})$.

Remark 2.6 The attentive reader will miss some term $-N_{\mathrm{crit}}(r, \mathfrak{f})$ on the right-hand side of (2.34). It is possible to define this term by

$$N_{\mathrm{crit}}(r, \mathfrak{f}) = 2N(r, \mathfrak{f}) + N(r, 1/\mathfrak{f}') - N(r, \mathfrak{f}'),$$

and the Second Main Theorem takes the more customary form

$$(q-2)T(r, \mathfrak{f}) \le \sum_{\nu=1}^{q} N\Big(r, \frac{1}{\mathfrak{f} - a_\nu}\Big) - N_{\mathrm{crit}}(r, \mathfrak{f}) + S(r, \mathfrak{f})$$

(see [149], Satz II). Occasionally $N_{\mathrm{crit}}(r, \mathfrak{f})$ will be negative, in any case with lower bound $-\frac{1}{k}N(r, 1/\triangle_P) \ge -(2k-2)T(r, \mathfrak{f}) + O(1)$. This again leads to (2.34).

Exercise 2.38 For $\mathfrak{f}$ with minimal polynomial $w^k \cos^2 z - \sin z$ prove that

- $\log A(re^{i\theta}) = 2r|\cos\theta| + O(1)$ and
- $T(r, \mathfrak{f}) \sim \dfrac{4r}{k\pi}$ and $N\Big(r, \dfrac{1}{\mathfrak{f} - c}\Big) \sim T(r, \mathfrak{f})$ $(c \ne 0)$ holds;
- $\mathfrak{f}$ has zeros at $z = \nu\pi$ of order $1/k$ with $N(r, 1/\mathfrak{f}) \sim m(r, 1/\mathfrak{f}) \sim \frac{1}{2}T(r, \mathfrak{f})$, and
- poles at $z = (\frac{1}{2} + \nu)\pi$ of order $2/k$ with $N(r, \mathfrak{f}) \sim T(r, \mathfrak{f})$.

Exercise 2.39 (Continued) Prove that the derivative $\mathfrak{f}' = \dfrac{1+\sin^2 z}{k\sin z\cos z}\mathfrak{f}$ has the minimal polynomial $w^k k^k \sin^{k-1} z\cos^{k+2} z - (1+\sin^2 z)^k = \tilde{A}_k(z)w^k + \tilde{A}_0(z)$, and the following holds:

- $\log\tilde{A}(re^{i\theta}) = \log\max_{\kappa=0,k}|\tilde{A}_\kappa(z)| = (2k+1)r|\cos\theta| + O(1)$;
- $T(r,\mathfrak{f}') \sim (k+\frac{1}{2})T(r,\mathfrak{f})$;
- $N(r,1/\mathfrak{f}') \sim kT(r,\mathfrak{f})$ and $m(r,1/\mathfrak{f}') \sim \frac{1}{2}T(r,\mathfrak{f}) \sim m(r,1/\mathfrak{f})$;
- $N(r,\mathfrak{f}') \sim T(r,\mathfrak{f}') \sim (k+\frac{1}{2})T(r,\mathfrak{f})$, and $N_{\rm crit}(r,\mathfrak{f}) \sim (2-k)T(r,\mathfrak{f})$.

Exercise 2.40 Prove Valiron's Lemma $T(R\circ\mathfrak{f}) = dT(r,\mathfrak{f}) + O(1)$, R rational of degree $d > 1$.

Remark 2.7 It is not hard to adapt the proof of Theorem 2.2 (generalised Second Main Theorem) to prove (2.34) for small meromorphic and even small algebroid functions $\mathfrak{a}_\nu$ in place of the constants a_ν.

Chapter 3
Selected Applications of Nevanlinna Theory

The present chapter is devoted to applications of Nevanlinna Theory to general questions in the theory of entire and meromorphic functions. This concerns algebraic differential and functional equations, uniqueness of meromorphic functions, and the value distribution of differential polynomials. We will always consider meromorphic functions in the plane, though more or less all results remain true if meromorphic functions on some punctured neighbourhood of infinity are considered. We have tried to avoid too much overlap with the books of Hayman [78], Laine [102], and Wittich [202], nevertheless we will present the basic results in the first section.

3.1 Algebraic Differential Equations

3.1.1 Valiron's Lemma Again

The composite function $F = R \circ f$, where R is nonlinear rational of degree d, has Nevanlinna characteristic $dT(r,f) + O(1)$. The generalisation to the case when R also depends on the independent variable, hence $F(z) = R(z,f(z))$, is due to Valiron [192] if the coefficients of R are rational, and due to Mokhon'ko [115] in the more general setting of small coefficients. Then f is also called admissible. The terms 'small' (coefficients) and 'admissible' (functions, solutions) are reciprocal to each other.

Theorem 3.1 *Let f be non-constant meromorphic and $R(z, w)$ any rational function in w with small coefficients w.r.t. f. Then*

$$T(r, R(z,f(z))) = dT(r,f) + S(r) \quad (d = \deg_w R)$$

holds; $S(r)$ is a remainder term with $S(r) = O(\log r)$ if R has rational coefficients.

N. Steinmetz, *Nevanlinna Theory, Normal Families, and Algebraic Differential Equations*, Universitext, DOI 10.1007/978-3-319-59800-0_3

Proof For polynomials $R(z,w)$ with respect to w the proof is almost the same as in the autonomous case (Exercise 2.4), and will be left to the reader. In the general case we first assume that R is rational over $\mathbb{C}(z)$. As in the proof of Valiron's Lemma in Chap. 2 we set $R = P/Q$, $F(z) = R(z,f(z))$, $p(z) = P(z,f(z))$, and $q(z) = Q(z,f(z))$, and assume $\deg_w P \le \deg_w Q$. In place of $\{z : |q(z)| < \delta\}$ we consider the set $E_0 = \{z : |q(z)| < |z|^{-L}\}$ ($L > 0$ some large number) and obtain

$$\begin{aligned} |z|^{-K} < |F(z)q(z)| = |p(z)| < |z|^K \quad & (z \in E_0, |z| > r_0), \\ |F(z)| + |1/q(z)| < |z|^K \quad & (z \notin E_0, |z| > r_0), \end{aligned}$$

where $K > 0$ is also some large real number. This leads to

$$\log^+ |F(z)| = \log^+ |1/q(z)| + O(\log |z|) \quad (|z| > r_0)$$

without restriction, hence $m(r,F) = m(r,1/q) + O(\log r)$, while

$$N(r,F) = N(r,1/q) + O(\log r)$$

is obvious. In the general case of small coefficients we will use functions

$$\sigma : \mathbb{C} \longrightarrow [0,\infty] \quad \text{with} \quad \int_0^{2\pi} |\log \sigma(re^{i\theta})|\, d\theta = S(r,f)$$

in place of $|z|^L$ and $|z|^K$ to define the set $E_0 = \{z : |q(z)| < 1/\sigma(z)\}$ and to control the growth of $F(z)$, $p(z)$, and $1/q(z)$ inside and outside E_0. The interested reader is encouraged to fill in the details.

3.1.2 *Valiron's Lemma and Malmquist's First Theorem*

Malmquist's theorem characterises the Riccati equations

$$w' = a(z) + b(z)w + c(z)w^2 \tag{3.1}$$

among the differential equations

$$w' = R(z,w) \quad (R \text{ rational in } w). \tag{3.2}$$

In the classical case (Malmquist [113]) it is assumed that R has rational coefficients and (3.2) has a transcendental meromorphic solution. The more general case ('small' coefficients and 'admissible' solution) is due to Gackstatter and Laine [52].

Theorem 3.2 *Suppose that the differential equation* (3.2) *has an admissible solution. Then* (3.2) *reduces to some Riccati equation* (3.1).

Proof From $T(r,w') \le 2T(r,w) + S(r,w)$ and Valiron's Lemma it follows that $d = \deg_w R \le 2$. Replacing w with $y = c + 1/w$ then yields

$$y' = -R\Big(z, \frac{1}{y-c}\Big)(y-c)^2 = R_1(z,y).$$

Obviously, $\deg_y R_1 \le 2$ is only possible if R is a polynomial of degree at most two over the field of small resp. rational functions.

Exercise 3.1 (Yosida [208, 211]) Suppose $w'^n = R(z,w)$ (R rational in w) has an admissible solution. Prove that R is a polynomial in w of degree at most $2n$.

3.1.3 Eremenko's Lemma and Malmquist's Second Theorem

Eremenko ([34], Theorem 5) obtained a far reaching generalisation of Valiron's Lemma, which among other things provides the main step in proving Malmquist's so-called Second Theorem. He considered meromorphic functions f and g that are algebraically dependent over the field of rational functions.

Eremenko's Lemma *Suppose that the transcendental meromorphic functions f and g are algebraically dependent over $\mathbb{C}(z)$, that is,*

$$P(z, f(z), g(z)) = 0 \tag{3.3}$$

holds, where $P(z,x,y)$ is an irreducible polynomial over the field $\mathbb{C}(z)$. Then

$$d_xT(r,f) = d_yT(r,g) + S(r) \quad (d_x = \deg_x P,\ d_y = \deg_y P) \tag{3.4}$$

holds, where $S(r)$ denotes the common remainder term.

Remark 3.1 In the original version of Eremenko's Lemma it is assumed that P is irreducible over the algebraically closed field $\mathfrak{A}$ of algebraic functions in place of $\mathbb{C}(z)$. Even under this hypothesis the proof is far beyond the methods, scope and even terminology of this text and will be omitted.[1] It is, however, not hard to prove that irreducibility over $\mathbb{C}(z)$ and $\mathfrak{A}$ is the same if (3.3) is taken into account.

Besides equations of type (3.2), Malmquist [114] also considered implicit first-order differential equations

$$P(z,w,w') = \sum_{\nu=0}^{n} P_\nu(z,w)w'^\nu = 0; \tag{3.5}$$

[1] It would be desirable to overcome these difficulties and provide a purely complex analytic proof.

$P(z, x, y)$ is assumed to be an irreducible polynomial over $\mathbb{C}(z)$. If (3.5) has some transcendental meromorphic solution, then Eremenko's Lemma applies with $f = w$ and $g = w'$. This yields the following necessary conditions for the existence of at least one transcendental meromorphic solution.

Theorem 3.3 ([114]) *Suppose that the irreducible differential equation* (3.5) *has a transcendental meromorphic solution. Then*

$$\deg_w P_\nu \leq 2n - 2\nu \tag{3.6}$$

holds. In particular, P_n is independent of w.

Proof From $\deg_{w'} P = n$, $T(r, w') \leq 2T(r, w) + S(r, w)$ and Eremenko's Lemma it follows that $\deg_w P_0 \leq \deg_w P \leq 2n$. Substituting $w = c + 1/y$ with $P_0(z, c) \not\equiv 0$ yields

$$Q(z, y, y') = \sum_{\nu=0}^{n} Q_\nu(z, y) y'^\nu = 0,$$

where $Q_\nu(z, y) = (-1)^\nu P_\nu(z, c + 1/y) y^{d-2\nu}$, $d = \max\{2\nu + \deg_w P_\nu\}$, and $Q_0(z, y) = P_0(z, c) y^d + \cdots$. Then $d \leq 2n$ holds by our first argument, this proving (3.6).

Exercise 3.2 Suppose that the coefficients in Eq. (3.5) are rational. Prove that the meromorphic solutions have finite order of growth by estimating the spherical derivative (Gol'dberg [56] in the general case, Yosida [209] for equations $w'^n = P(z, w)$). (**Hint.** The zeros of any polynomial $p(x) = x^n + p_{n-1}x^{n-1} + \cdots + p_0$ are contained in $|x| \leq \max_{0 \leq \nu < n} \sqrt[n-\nu]{n|p_\nu|}$. Set $x = w'$ and $p_\nu = P_\nu(z, w)$.)

The first-order equations (3.1) and (3.5) will be discussed in detail in Chap. 5.

3.1.4 Elementary Techniques

We will now prove some general results on the solutions to algebraic differential equations. Algebraic differential equations will be written in the form

$$\Omega(z, w, w', \ldots, w^{(n)}) = 0,$$

where Ω is a polynomial in the dependent variables, while the coefficients are assumed to be small with respect to the solutions in question. Sometimes we will also write

$$\Omega[w] = 0, \tag{3.7}$$

where now (the same) Ω will be regarded as a *differential polynomial*, that is, a finite sum of terms $a_M(z)M[w]$, where $a_M \not\equiv 0$ may be any small function and

$$M[w] = w^{\ell_0}(w')^{\ell_1}\cdots(w^{(n)})^{\ell_n}$$

is any *differential monomial*. The *degree* and *weight* of M and Ω are defined by

$$\begin{aligned} d_M &= \ell_0 + \ell_1 + \cdots + \ell_n, \; d_\Omega = \max_M d_M, \\ \mathsf{d}_M &= \ell_0 + 2\ell_1 + \cdots + (n+1)\ell_n, \; \mathsf{d}_\Omega = \max_M \mathsf{d}_M. \end{aligned} \tag{3.8}$$

At poles of w of order p, $M[w]$ has poles of order at most $pd_M + (\mathsf{d}_M - d_M) \le p\mathsf{d}_M$. This and $M[w] = w^{d_M}(w'/w)^{\ell_1}\cdots(w^{(n)}/w)^{\ell_n}$ yield

$$N(r, \Omega[w]) \le \mathsf{d}_\Omega N(r, w) + S(r, w) \quad \text{and} \quad m(r, \Omega[w]) \le d_\Omega m(r, w) + S(r, w) \tag{3.9}$$

for any admissible meromorphic function w.

Although only few algebraic differential equations are known to have admissible solutions, it is legitimate to introduce some basic techniques and elementary results, ignoring the question of global existence for a while. In other words, we just *assume* that the differential equation (3.7) in question has an admissible solution which has to be examined. One of the first papers to adopt this point of view was [52].

Exercise 3.3 Suppose that Ω in (3.7) contains some term $a(z)w^d$, while all other terms have degree less than d. Prove that every admissible solution satisfies

$$m(r, w) = S(r, w).$$

By a slight modification of Exercise 3.3 we obtain the following result, known as Clunie's Lemma. Wittich, who never systematised his observations, proved results like those in Exercises 3.3–3.5 only in the context of particular differential equations; Clunie [28] did not refer to differential equations.

Exercise 3.4 (Clunie's Lemma) Suppose w is an admissible solution to

$$w^q P[w] = Q[w],$$

where P and Q are differential polynomials with $d_Q \le q$. Prove $m(r, P[w]) = S(r, w)$. (**Hint.** Let M and N denote typical monomials occurring in P and Q, respectively. Then $|M[w]| \le |M[w]|/|w|^{d_M}$ resp. $|N[w]/w^q| \le |N[w]|/|w|^{d_N}$ holds for $|w| < 1$ resp. $|w| \ge 1$.) Exercise 3.3 is a particular case with $P[w] = a(z)w^{d-q}$.

In a similar way, we obtain the solution to the next problem due to Mokhon'ko.

Exercise 3.5 ([116]) For any admissible solution to (3.7) with $\Omega[0](z) \not\equiv 0$ prove

$$m(r, 1/w) = S(r, w).$$

(**Hint.** Set $P[w] = \Omega[w] - \Omega[0]$ and consider $\Omega[0]/w = -P[w]/w$.)

The hypothesis says that the constant 0 is not a solution. Obviously, $\Omega[\psi](z) \not\equiv 0$ for some small function ψ implies $m(r, 1/(w-\psi)) = S(r, w)$.

We will finally consider several particular differential equations.

Example 3.1 Assume that $w'' = a(z) + b(z)w + 2w^3$ with polynomial coefficients a and b has a transcendental meromorphic solution. By Exercise 3.3, w has infinitely many poles, each simple with residue 1 or -1. Trying to determine the coefficients in $w = \pm(z-p)^{-1} + \sum_{k=0}^{\infty} c_k(z-p)^k$ leads to the *resonance conditions* $a'(p) = b''(p) = 0$ (and c_4 remains undetermined). It follows that $a(z) = a_0$ and $b(z) = b_0 + b_1 z$ is necessary for the existence of a transcendental meromorphic solution. In other words, only $w'' = a_0 + (b_0 + b_1 z)w + 2w^3$ passes the so-called *Painlevé test* for poles. One may also speak of a *Theorem of Malmquist-type*.

Exercise 3.6 State and prove an analogous result for $w'' = a(z) + 6w^2$. (**Hint.** First show that the poles have multiplicity two.) **Solution.** $a'' = 0$.

Exercise 3.7 Prove that the solutions (assumed to be meromorphic on the plane) to Painlevé's first equation $w'' = z + 6w^2$ are transcendental and assume every value infinitely often.

Exercise 3.8 The solutions to the Riccati equation (3.1) with polynomial coefficients are known to be meromorphic in the plane. Prove that any transcendental solution has infinitely many poles if $c(z) \not\equiv 0$, and assumes each value κ with $a(z) + b(z)\kappa + c(z)\kappa^2 \not\equiv 0$ infinitely often.
What happens if $a(z) + b(z)\kappa + c(z)\kappa^2$ vanishes identically?

Meromorphic solutions to Riccati equations with rational coefficients have finite order of growth. This may be generalised as follows.

Theorem 3.4 *Let w be any meromorphic solution to the algebraic differential equation $w^k = \Omega[w]$, where $\Omega[w] = \sum a(z)w^{q_0}(w')^{q_1}\cdots(w^{(n)})^{q_n}$ is a differential polynomial with rational coefficients, degree $d_\Omega < k$ and weight $\mathsf{d}_\Omega \le k$. Assume also that up to finitely many the poles z_0 of w are simple with residues $\mu\rho(z_0)$, where $\mu = \mu(z_0)$ is an integer and ρ is some rational function. Then w has finite order.*

Proof We may assume that the residue condition holds with $\rho \equiv 1$ for every pole, otherwise introduce a new dependent variable $\tilde{w} = w/\rho + \sigma$ with suitably chosen rational function σ. Then $w = g'/g$ holds for some entire function g. We will show that the central index of g satisfies $\nu(r) = O(r^\lambda)$ for some $\lambda > 0$, possibly outside some exceptional set E of finite logarithmic measure. Since $\nu(r)$ is increasing we can disregard E, hence g and *a fortiori* w have finite order of growth.

The derivatives of w have the form

$$w^{(j)} = \frac{P_j(g, g', \ldots, g^{(j+1)})}{g^{j+1}} = P_j(1, g'/g, \ldots, g^{(j+1)}/g) \quad (j \ge 1),$$

where $P_j(x_0, \ldots, x_{j+1})$ is a universal homogeneous polynomial of degree $j + 1$ and weight $2j + 2$ (for example, $P_1(x_0, x_1, x_2) = x_0x_2 - x_1^2$) with

$$P_j(1, \xi, \ldots, \xi^{j+1}) = 0; \tag{3.10}$$

the latter follows from inserting $g(z) = e^{\xi z}$. In combination with

$$g^{(\ell)}(z_r)/g(z_r) = (\nu(r)/z_r)^\ell(1 + \epsilon_\ell(r)) = y(r)^\ell(1 + \epsilon_\ell(r)),$$

which holds whenever $|g(z_r)| = M(r, g)$ and $|z_r| = r \notin E$, (3.10) yields

$$w^{(j)}(z_r) = P_j(1, y(r)(1 + \epsilon_1(r)), \ldots, y(r)^{j+1}(1 + \epsilon_{j+1}(r))) = O(y(r)^{j+1}\epsilon(r))$$

with $\epsilon(r) = \max_{1\le\ell\le n+1} |\epsilon_\ell(r)| = o(1)$ outside E.

To proceed we need the following quantitative version: $\epsilon(r) = O(\nu(r)^{-\alpha})$ holds for every $\alpha < 1/4$ (see [202], p. 9). Since the coefficients in Ω do not grow faster than $|z|^s$ for some $s \ge 0$, we obtain $y(r)^k = O(r^s|y(r)|^{d_\Omega}\nu(r)^{-\alpha})$ if $y(r) \ge 1$, hence $\nu(r) = O(\max\{r, r^{s/\alpha}\})$.

3.2 Zeros of Differential Polynomials

In his seminal paper [77] Hayman proved several deep results on the value distribution of meromorphic and entire functions and their derivatives, and initiated a vast amount of research on the value distribution of differential polynomials, without using the name. A whole chapter in his famous treatise *Meromorphic functions* [78] is devoted to problems of this kind. In our presentation of the main progress since Hayman the focus will be on meromorphic functions, leaving aside the case of entire functions. For entire rather than meromorphic functions the results are stronger on one hand, and easier to accomplish on the other.

3.2.1 Hayman's Work on $f^{n+2} + af'$ and $f^nf' - a$

We will start with two major results in Hayman's paper [77] and discuss the progress on these and related problems since then.

Theorem 3.5 *For every transcendental meromorphic function f, $a \ne 0$, and $n \ge 3$,*

$$f^{n+2} + af' \quad \textit{assumes every finite value infinitely often} \qquad \textbf{(H1)}$$

and

$$f^n f' - a \quad \textit{has infinitely many zeros.} \tag{\textbf{H2}}$$

From $f^n f' - a = -ag^{-n-2}[g^{n+2} + g'/a]$ with $g = 1/f$ it follows that (**H1**) implies (**H2**). Mues [119] constructed counterexamples (see Exercise 3.9 below) to the cases $n = 1$ and $n = 2$ in (**H1**), showing that Hayman's hypothesis $n \geq 3$ in (**H1**) cannot be replaced with $n \geq 2$. We note, however, that (**H2**) remains true for $n = 1$ and $n = 2$, see Theorem 4.13.

Exercise 3.9 In (**H1**) and (**H2**) the value of $a \neq 0$ is irrelevant. Let f be any non-constant solution to the Riccati equation

$$w' = -(1 + 2\eta)(w + 1)(w + \eta) \ (\eta = e^{2\pi i/3}) \quad \text{resp.} \quad w' = 2(w^2 + 1).$$

Prove that $f^3 - f' + 1$ in the first case, and $f^4 + f' - 1$ in the second has no zeros. (**Hint.** f has two Picard values which may be read off the differential equation.)

3.2.2 *Generalisations of Hayman's Theorem*

Generalisations of Hayman's (**H2**) were proved by Sons [158] and Hennekemper [83], who examined the value distribution of $f^n(f')^{n_1} \cdots (f^{(k)})^{n_k} - a$ for $n \gg 2$ (depending on $n_1, \ldots, n_k$) and $(f^{n+k})^{(k)} - a$ for $n > 2$, respectively, with $k \geq 1$ in both cases. To proceed we will first prove a rather general theorem on the zeros of

$$\Psi = f^n P[f] + Q[f], \tag{3.11}$$

which covers the relevant results in [77, 83, 158]; P and Q denote not identically vanishing differential polynomials in f.

Theorem 3.6 ([32, 162]) *Let f be any transcendental meromorphic function and Ψ be given by* (3.11). *Then*

$$\overline{N}\Big(r, \frac{1}{\Psi}\Big) = S(r,f) \tag{3.12}$$

implies $n \leq \mathsf{d}_Q + 2$. Moreover, for $\mathsf{d}_Q = n - 2$,

$$m(r,f) + m(r, 1/f) + N_1(r,f) + N_1(r, 1/f) = S(r,f)$$

holds, where $N_1(r,f) = N(r,f) - \overline{N}(r,f)$ and $N_1(r, 1/f)$ 'counts' the multiple poles and zeros, respectively. The function f is a solution to the differential equation

$$wP[w]Q'[w] + n\chi w^n P[w] - (nw'P[w] - n\chi + wP'[w])Q[w] = 0, \tag{3.13}$$

where $\chi \not\equiv 0$ is a small meromorphic function with respect to f, and $P'[w]$ and $Q'[w]$ denote the total derivatives $\frac{d}{dz}P[w]$ and $\frac{d}{dz}Q[w]$.

Remark 3.2 In case of $\Psi = f^n f' - a$ and $\Psi = f^{n+2} + af' - c$ ($c \in \mathbb{C}$) as well as $\Psi = f^n (f')^{n_1} \cdots (f^{(k)})^{n_k} - a$ and $\Psi = (f^{n+k})^{(k)} - a$ (always $a \neq 0$) the inequality $n \leq 2$ is necessary for (3.12), hence the above mentioned results are corollaries of Theorem 3.6.

Proof We assume $n \geq \mathsf{d}_Q + 2$ and (3.12) to prove that Ψ does not vanish identically. Otherwise Clunie's Lemma (Exercise 3.4) applies to $f^n P[f] = -Q[f]$ and yields

$$m(r, P[f]) = S(r,f).$$

At any pole z_0 of f of order p which is not a pole of any of the coefficients, $Q[f]$ has a pole of order at most $p\mathsf{d}_Q \leq p(n-2)$, while f^n has a pole of order pn. Thus $P[f]$ vanishes at z_0, and this implies

$$N(r, P[f]) = S(r,f).$$

Finally, it follows from $f^n = -\frac{1}{P[f]}Q[f]$ and $T(r, P[f]) = S(r,f)$ that

$$nT(r,f) \leq T(r, Q[f]) + S(r,f) \leq \mathsf{d}_Q T(r,f) + S(r,f) \leq (n-2)T(r,f) + S(r,f),$$

which clearly is impossible. Differentiating (3.11) we obtain

$$f^{n-1} n\chi = \frac{\Psi'}{\Psi} Q[f] - Q'[f] \tag{3.14}$$

and

$$n\chi = nf' P[f] + fP'[f] - \frac{\Psi'}{\Psi} fP[f]. \tag{3.15}$$

Again Exercise 3.4, applied to (3.14), yields $m(r, \chi) = S(r,f)$. We note that Clunie's Lemma requires that the coefficients are small functions with respect to f. The 'coefficient' Ψ'/Ψ, however, is not necessarily small. Analysing the proof of Clunie's Lemma shows that it suffices to know that $m(r, \Psi'/\Psi) = S(r,f)$ holds. We also note that χ cannot vanish identically, since otherwise $\Psi \equiv cQ[f]$ and $f^n P[f] + (1-c)Q[f] \equiv 0$ holds for some $c \neq 0$. This, however, was excluded in the first part of the proof. Like there, it follows that

$$T(r, \chi) = S(r,f)$$

holds since χ is regular at almost every pole[2] of f. Clunie's Lemma applies to

$$f^{n-2}f = \frac{1}{n\chi}\Big(\frac{\Psi'}{\Psi}Q[f] - Q'[f]\Big) \tag{3.16}$$

and yields $m(r,f) = S(r,f)$, and in combination with

$$\frac{1}{f} = \frac{1}{\chi}\Big(\frac{f'}{f}P[f] + \frac{1}{n}P'[f] - \frac{\Psi'}{n\Psi}P[f]\Big) \tag{3.17}$$

we obtain $m(r, 1/f) = S(r,f)$. Since χ vanishes at almost every multiple zero and pole of f, it also follows that

$$N_1(r, 1/f) + N_1(r,f) = S(r,f).$$

Finally, the differential equation (3.13) is a combination of (3.14) and (3.15). Since for $n > \mathsf{d}_Q + 2$, χ vanishes at almost every pole of f, but does not vanish identically, $n \leq \mathsf{d}_Q + 2$ follows.

3.2.3 Limit Cases

We will now discuss the following differential polynomials which form various limit cases $n = \mathsf{d}_Q + 2$ with respect to Theorem 3.6.

$$\begin{aligned} &\textbf{1. } \Psi = f^2(f')^{n_1}\cdots(f^{(k)})^{n_k} - 1 \text{ with } n_k \geq 1,\\ &\textbf{2. } \Psi = f^2 f' - 1 \text{ as a special case,}\\ &\textbf{3. } \Psi = (f^{k+2})^{(k)} - 1 \text{ with } k \geq 1, \text{ and}\\ &\textbf{4. } \Psi = f^4 + f' - 1. \end{aligned} \tag{3.18}$$

The common idea will be to derive a Riccati differential equation from Eq. (3.13) and afterwards express the derivatives $f^{(j)}$ in terms of f, hence rewrite Ψ as an ordinary polynomial in f with small coefficients. To this end we will frequently make use of the following exercises.

Exercise 3.10 Let f be meromorphic with $m(r,f) + N_1(r,f) = S(r,f)$ such that

$$f(z) = \frac{\rho(z_0)}{z - z_0} + \sigma(z_0) + O(|z - z_0|) \quad (z \to z_0) \tag{3.19}$$

[2] Unless otherwise stated, 'almost every pole, zero, etc'. means 'up to a sequence of poles, zeros, etc. with counting function $S(r,f)$'.

holds at almost every simple pole, where $\rho \not\equiv 0$ and σ are small meromorphic functions w.r.t. f. Prove that $w = f(z)$ satisfies

$$w' = a_0(z) + a_1(z)w + a_2(z)w^2 \tag{3.20}$$

with $a_2 = -1/\rho$, $a_1 = (2\sigma - \rho')/\rho$, and $T(r, a_0) = S(r,f)$.
(**Hint.** It is convenient first to consider f/ρ instead of f, hence to assume $\rho \equiv 1$. In this case show that $f' + f^2 - 2\sigma f$ is regular at almost every pole of f.)

Exercise 3.11 Assuming (3.12) for $\Psi = f^2(f')^{n_1} \cdots (f^{(k)})^{n_k} - 1$, employ Eq. (3.13) to verify that (3.19) holds with coefficients $\rho = \frac{q+2}{2\chi}$, $\sigma = -\rho\frac{q+2}{q+4}\frac{\chi'}{\chi}$, and $q = 2n_1 + \cdots + (k+1)n_k \ge k+1$.

Exercise 3.12 Do the same for $\Psi = (f^{k+2})^{(k)} - 1$ with $\rho = \frac{2k+2}{k\chi}$ and $\sigma = 0$.
(**Hint.** Do not try to compute $P[f] = (f^{k+2})^{(k)}/f^2$ explicitly!)

Exercise 3.13 Let f be meromorphic with $m(r, 1/f) + N_1(r, 1/f) = S(r,f)$, and assume that, similar to Exercise 3.10, $f'(z_0) = \rho(z_0)$ and $f''(z_0) = 2\sigma(z_0)$ holds at almost every simple zero of f. Prove that $w = f(z)$ satisfies Eq. (3.20) with $a_0 = \rho$, $a_1 = -(2\sigma + \rho')/\rho$ and $T(r, a_2) = S(r, w)$.

Exercise 3.14 Assume that $\Psi = f^4 + f' - 1$ satisfies $\overline{N}(r, \Psi) = S(r,f)$, and let z_0 be any simple zero of f which is neither a zero of Ψ nor a zero or pole of χ. From (3.15) and (3.14) deduce that

- $f'(z_0) = \chi(z_0)$ and $f''(z_0) - \chi'(z_0) = \frac{1}{4}\chi(z_0)\psi(z_0)$ with $\psi = \Psi'/\Psi$;
- $f''(z_0) = \psi(z_0)(\dot\chi(z_0) - 1)$, hence $f''(z_0)(1 - \frac{3}{4}\chi(z_0)) = \chi'(z_0)(\chi(z_0) - 1)$;
- some Riccati equation (3.20) for $w = f(z)$ if $\chi(z) \not\equiv \frac{4}{3}$.

Theorem 3.7 *Let f be transcendental meromorphic. Then each of the differential polynomials* (3.18) **1.**, **2.** *and* **3.** *has infinitely many zeros in the stricter sense that*

$$\limsup_{r\to\infty} \frac{\overline{N}(r, 1/\Psi)}{T(r,f)} > 0.$$

Remark 3.3 The third case is due to Hennekemper [83]; Wang [195] settled the case $(f^m)^{(\ell)}$ for $m \ge 3$, $m > \ell$, which in view of Theorem 3.7 is relevant if $m = \ell+1 \ge 3$. Bergweiler and Eremenko [14] finished the case $m = \ell + 1 \ge 2$.

Proof We assume (3.12) for $\Psi = f^2(f')^{n_1} \cdots (f^{(k)})^{n_k} - 1$ (first case), and take from Exercise 3.10 and 3.11 that $w = f(z)$ solves

$$w' = a_0(z) + a_1(z)w + a_2(z)w^2 = P_2(z, w)$$

with $a_1 = -\frac{q}{q+4}\frac{\chi'}{\chi}$, $a_2 = \frac{2\chi}{q+2}$, and $T(r, a_0) = S(r,f)$. This implies

$$w'' = P_3(z, w) = 2!a_2(z)^2w^3 + \cdots, \; w''' = P_4(z, w) = 3!a_2(z)^3w^4 + \cdots \text{etc.},$$

hence $\Psi(z) = H(z, f(z))$ with

$$H(z, w) = w^2P_2(z, w)^{n_1} \cdots P_{k+1}(z, w)^{n_k} - 1 = A_{q+2}(z)w^{q+2} + \cdots + A_2(z)w^2 - 1$$

and $A_{q+2} = 2!^{n_2}3!^{n_3} \cdots k!^{n_k}a_2^{q-m}$, $q = \mathsf{d}_P = 2n_1 + \cdots + (k+1)n_k$, and $m = d_P = n_1 + \cdots + n_k$. Equation (3.14) with $n = 2$ and $Q[f] = -1$ takes the form

$$2\chi(z)wH(z, w) + H_z(z, w) + H_w(z, w)P_2(z, w) = 0 \tag{3.21}$$

for $w = f(z)$. Obviously this is an identity in (z, w). Now every (algebroid) solution $y = \eta(z)$ to $H(z, y) = 0$ solves

$$y' = P_2(z, y), \tag{3.22}$$

hence also

$$K[y] = y^2(y')^{n_1} \cdots (y^{(k)})^{n_k} - 1 = 0. \tag{3.23}$$

For a proof, write $H(z, w) = (w - \eta(z))^\ell \Delta(z, w)$ with $\Delta(z, \eta(z)) \not\equiv 0$ and employ

$$\begin{aligned} H_z(z, w) &= -\ell(w - \eta(z))^{\ell-1}\eta'(z)\Delta(z, \eta(z)) + O(|w - \eta(z)|^\ell), \text{ and} \\ H_w(z, w) &= \ell(w - \eta(z))^{\ell-1}\Delta(z, \eta(z)) + O(|w - \eta(z)|^\ell) \quad \text{as } w \to \eta(z) \end{aligned}$$

to obtain $-\eta'(z) + P_2(z, \eta(z)) = O(|w - \eta(z)|)$ from (3.21). Now (3.22) admits at most two solutions that are algebroid over the field of small functions w.r.t. f, since the cross-ratio $(f, \eta_1, \eta_2, \eta_3)$ of any four solutions to (3.22) is constant (see Exercise 1.31). Also the solutions to (3.23) cannot be single-valued. It follows that

$$\begin{aligned} H(z, w) &= A_{q+2}(z)(w - \eta_1(z))^{\lambda_1}(w - \eta_2(z))^{\lambda_2} \quad (\lambda_1 + \lambda_2 = q + 2), \\ &(-1)^{q+2}A_{q+2}\eta_1^{\lambda_1}\eta_2^{\lambda_2} = -1 \quad \text{and} \quad \lambda_1\eta_2 + \lambda_2\eta_1 \equiv 0 \end{aligned}$$

since H contains no linear term in w. The branches η_1 and η_2 are permuted under analytic continuation, and this yields $\lambda_1 = \lambda_2 = \lambda$ and $\eta_1^2 = \eta_2^2 = g$, where g is meromorphic, but $\sqrt{g}$ is not; g is even an entire function, since the solutions to $K[y] = 0$ cannot have poles. Now $(\sqrt{g})' = a_0 + a_1\sqrt{g} + a_2g$ is equivalent to $\frac{g'}{g} = 2a_1 = -\frac{2q}{q+4}\frac{\chi'}{\chi}$ and $a_0 + a_2g = 0$. This implies $g = \delta_1\chi^{-\frac{2q}{q+4}}$ $(\delta_1 \neq 0)$ on one hand, while $A_{q+2}g^\lambda = A_{q+2}(\eta_1\eta_2)^\lambda = -1$ and $A_{q+2} = \text{const} \cdot a_2^{q-m} = \text{const} \cdot \chi^{q-m}$ yield $g = \delta_2\chi^{-\frac{q-m}{\lambda}}$ on the other. From this and $2\lambda = q+2$ we obtain $m = 2\frac{\lambda-1}{\lambda+1} < 2$, hence $m = 1$, $\lambda = 3$, and $q = 4$. Now $m = 1$ means $K[y] = y^2y^{(k)} - 1$, so that

$K[\sqrt{g}]$ cannot vanish identically. This settles the first and, of course, also the second case. The proof of the third assertion runs along the same lines. This time $w = f(z)$ solves

$$w' = P_2(z,w) = a_0(z) - \tfrac{k}{2k+2}\chi(z)w^2 \quad (T(r,a_0) = S(r,f)),$$

and again $\Psi(z) = H(z,f(z))$ holds, where $H(z,w)$ is a polynomial with respect to w. The solutions $y = \eta(z)$ to $H(z,y) = 0$ now solve

$$y' = P_2(z,y) \quad \text{and} \quad (y^{k+2})^{(k)} = 1,$$

hence $\eta(z)^{k+2}$ is a polynomial $\Phi(z) = z^k/k! + \cdots$ of degree k. It is obvious that $\eta(z) = \sqrt[k+2]{\Phi(z)}$ cannot solve $\eta' = P_2(z,\eta) = a_0(z) + a_2(z)\eta^2$, which is equivalent to $\frac{1}{k+2}\Phi(z)^{\frac{1}{k+2}-1}\Phi'(z) = a_0(z) + a_2(z)\Phi(z)^{\frac{2}{k+2}}$. This settles the third case. ☕

While the cases **1.**, **2.**, and **3.** may be regarded as generalisations of Hayman's **(H2)**, the next theorem returns to case **(H1)** with $n = 2$.

Theorem 3.8 *For every transcendental meromorphic function, $f' + f^4$ has infinitely many zeros. If $f' + f^4$ omits the value $c = 1$, say, then f satisfies $f' = 2 + 2f^2$, hence $f(z) = \tan(2z + z_0)$ coincides with the second example in Exercise 3.9.*

Proof The first part (due to Mues [119]) is a corollary of Theorem 3.7: If $f' + f^4$ has only finitely many zeros, then so has $\Phi = -g^4(f' + f^4) = g^2g' - 1$ (set $f = 1/g$), since g and Φ have no common zeros. To prove (a slightly stronger version of) the second assertion (see [162]) we set $\Psi = f' + f^4 - 1$ and assume $\overline{N}(r, 1/\Psi) = S(r,f)$. According to Theorem 3.7 and Exercise 3.14, we have to consider two cases.

a) $\chi(z) \not\equiv 4/3$. Then $w = f(z)$ satisfies $w' = P_2(z,w) = a_0(z) + a_1(z)w + a_2(z)w^2$ by Exercise 3.14, hence $\Psi(z) = H(z,f(z))$ holds with

$$H(z,w) = w^4 + a_2(z)w^2 + a_1(z)w + a_0(z) - 1.$$

Again every solution $y = \eta(z)$ to $H(z,y) = 0$ satisfies $y' = P_2(z,y)$ and also $y^4 + y' - 1 = 0$. The first equation can have only two solutions $\eta_{1,2}$ of this kind. The second equation has four constant solutions $\sqrt[4]{1}$, while the non-constant solutions have algebraic singularities. In this (non-constant) case we have $\eta_1 = -\eta_2 = \eta$ since η_1 and η_2 are interchanged under analytic continuation, which, however, contradicts $\eta^4 \pm \eta' - 1 = 0$.

All in all we have to discuss the case $H(z,w) = (w^2 - \eta^2)^2 = (w^2 + 1)^2$ ($H(z,w) = (w - \eta_1)^4$ and $H(z,w) = (w - \eta_1)^3(w - \eta_2)$ cannot occur since H contains no term a_3w^3). This leads to the assertion $w' = 2(w^2 + 1)$ for $w = f(z)$.

b) $\chi(z) \equiv 4/3$. Here a non-trivial problem arises, namely to decide whether the differential equation $ww'' + \frac{16}{3}w^4 - 4(w' - \frac{4}{3})(w' - 1) = 0$ has a transcendental meromorphic solution such that $w'(z_0) = \frac{4}{3}$ holds at every zero of w. The equation passes the Painlevé test for zeros as well as for poles, that is, to every z_0 there exist formal solutions $\frac{4}{3}(z-z_0)+c_2(z-z_0)^2+\cdots$ and $\pm\frac{\sqrt{6}}{4}(z-z_0)^{-1}+c_0+c_1(z-z_0)+\cdots$. We will thus pursue a different track.

Exercise 3.15 From (3.15) with $\chi(z) \equiv 4/3$ deduce that

$$\frac{16}{f} = 12\frac{f'}{f} - 3\frac{\Psi'}{\Psi} = 3\frac{(f^4/\Psi)'}{f^4/\Psi}$$

and prove that Ψ is zero-free, hence $4/f = 3g'/g$ holds, where g is an entire function; g and g' have simple zeros, they coincide with the zeros and poles of f, respectively, and $\Psi = C/g'^4$ holds for some $C \neq 0$.

Exercise 3.16 (Continued) Deduce that

a) $256g^4 + 27g'^4 - 108gg'^2g'' - 81C = 0$,
b) $256g^2 - 54g''^2 - 27g'g''' = 0$,
c) $512gg' - 135g''g''' - 27g'g^{(4)} = 0$,

and prove that $\nu(r) \sim \frac{4}{3}r$ (central index) holds, hence $\varrho(g) = 1$.

To finish the proof of Theorem 3.8 we note that $g'(z_0) = 0$ implies $g'''(z_0) = 0$, hence $\kappa = g'''/g'$ is entire and has Nevanlinna characteristic $T(r,\kappa) = o(\log r)$. Thus $\kappa = \lambda^2$ is non-zero constant and $g(z) = c_0 + c_1e^{\lambda z} + c_2e^{-\lambda z}$ holds. This, however, is incompatible with b).

3.2.4 The Tumura–Clunie Theorem

The significance of the weight of a differential polynomial is restricted to the poles of the function f under consideration. If f is entire or 'almost' entire, that is, if $N(r,f) = S(r,f)$ holds, the role of the weight is taken by the degree. For example, $T(r,P[f]) \leq d_PT(r,f) + S(r,f)$ holds in place of $T(r,P[f]) \leq \mathsf{d}_PT(r,f) + S(r,f)$. It is self-evident that Theorem 3.6 has an analogue as follows.

Theorem 3.9 *Let f be an almost entire transcendental function and suppose that $\Psi = f^nP[f] + Q[f]$ with $P[f]Q[f] \not\equiv 0$ satisfies $\overline{N}(r,1/\Psi) = S(r,f)$. Then $d_Q \geq n-1$ holds. In the limit case $d_Q = n-1$, f satisfies the differential equation* (3.13) *and $m(r,1/f) + N_1(r,1/f) = S(r,f)$.*

Proof We assume $d_Q \leq n-1$ rather than $\mathsf{d}_Q \leq n-2$ in the proof of Theorem 3.6. Then (3.14) and (3.15) still hold and χ and $\psi = \Psi'/\Psi$ are small functions. To falsify the hypothesis $d_Q < n-1$ we remark that in this case $m(r,f) = S(r,f)$ would follow from (3.16) and Clunie's Lemma, in contrast to $N(r,f) = S(r,f)$. Again $m(r,1/f) = S(r,f)$ is obtained from (3.17), while $N_1(r,1/f) \leq N(r,1/\chi) = S(r,f)$ follows from the fact that χ has zeros of order at least $(p-1)$ at p-fold zeros of f. The differential equation (3.13) is just a combination of (3.14) and (3.15).

The classical *Tumura–Clunie Theorem* ([28], also [78], p. 69) deals with (almost) entire functions and differential polynomials

$$\begin{aligned} \Psi = f^n + Q[f] \quad (Q[f](z) \not\equiv 0,\ d_Q \le n-1) \\ \overline{N}(r, 1/\Psi) = S(r,f), \end{aligned} \tag{3.24}$$

and may be stated as follows.

Theorem 3.10 *Suppose f is (almost) entire and satisfies* (3.24). *Then*

$$\Psi(z) = \big(f(z) - a(z)\big)^n \quad \text{and} \quad \begin{cases} f'(z) = \chi(z) + \frac{1}{n}\psi(z)f(z) \\ a'(z) = \chi(z) + \frac{1}{n}\psi(z)a(z) \end{cases}$$

holds; a, χ, and $\psi = \Psi'/\Psi$ are small functions with respect to f.

We will now return to meromorphic functions and the zeros of ordinary polynomials $\Psi(z) = P(z, f(z))$. Without loss of generality we may assume that

$$P(z, w) = w^n + a_{n-2}(z)w^{n-2} + \cdots + a_0(z) \quad (n \ge 2)$$

is square-free. Theorem 2.6 yields $N(r, 1/\Psi) \ge (n-2)T(r,f) + S(r,f)$. The Tumura–Clunie Theorem referring to $\overline{N}(r, 1/\Psi)$ rather than $N(r, 1/\Psi)$ reads as follows.

Theorem 3.11 ([122]) *Under the hypotheses stated just now, $\overline{N}(r, 1/\Psi) = S(r,f)$ implies*

$$P(z, w) = w^2 + a_0(z).$$

Moreover, $w = f(z)$ satisfies the Riccati equation

$$w' = a_0(z)c(z) + \frac{a_0'(z)}{2a_0(z)}w + c(z)w^2 \quad (c \not\equiv 0 \text{ small}).$$

Remark 3.4 Since both branches of $\sqrt{-a_0(z)}$ also satisfy the above Riccati equation, $f(z)^2 + a_0(z) = (f(z) - \sqrt{-a_0(z)})(f(z) + \sqrt{-a_0(z)})$ has no zeros except possibly at poles of a_0'/a_0 and c.

Proof Theorem 3.6 applies to $\Psi(z) = f(z)^n + Q(z, f(z))$ with

$$Q(z, w) = a_{n-2}(z)w^{n-2} + \cdots + a_0(z).$$

In particular, $\deg_w Q = n-2$ and $a_{n-2} \not\equiv 0$ holds, and the differential equation (3.13) for $w = f(z)$ has the form

$$wQ_z(z, w) + n\chi(z)w^n + n\chi(z)Q(z, w) + (wQ_w(z, w) - nQ(z, w))w' = 0. \tag{3.25}$$

Since $wQ_w(z,w) - nQ(z,w) = -2a_{n-2}w^{n-2} + \cdots$ does not vanish identically, Malmquist's First Theorem yields

$$w' = b_0(z) + b_1(z)w + b_2(z)w^2 \quad (b_\nu \text{ small}). \tag{3.26}$$

Replacing w' in (3.25) with $b_0(z) + b_1(z)w + b_2(z)w^2$ we obtain an identity in (z, w), which in terms of P may be written as

$$wP_z(z,w) + n\chi(z)P(z,w) + (wP_w(z,w) - nP(z,w))(b_0(z) + b_1(z)w + b_2(z)w^2) = 0.$$

As in the proof of Theorem 3.7 it is shown that the solutions $w = \rho(z)$ to the algebraic equation $P(z,w) = 0$ also solve (3.26), and from $\deg_w P = n \geq 2$ it follows that there are two solutions ρ_1 and ρ_2. This implies $n = 2$ and $P(z,w) = w^2 + a_0(z)$, while $b_1 = \frac{a_0'}{2a_0}$ and $b_0 = a_0 b_2 = a_0 c$ follow from (3.26) for $w = \pm\sqrt{-a_0(z)}$.

Example 3.2 Pang and Ye [138] considered $\Psi = f^n + Q[f]$, where f is meromorphic and $Q[w] = w^{(k)} + a_1 w^{(k-1)} + \cdots + a_k w + a_{k+1}$ has constant coefficients, such that $Q[f](z) \not\equiv 0$. Assuming $\overline{N}(r, 1/\Psi) = S(r,f)$ and $n = k+3$, the authors derived a Riccati differential equation $f' = c(f-\alpha)(f-\beta)$ with constant coefficients (Theorem 1 in [138]) by a process similar to ours, but much more elaborate. We note that α and β are Picard values of f, and $n \leq k+3$ is a necessary condition for $\overline{N}(r, 1/\Psi) = S(r,f)$. Thus $\Psi = f^n + b_{n-2}f^{n-2} + \cdots + b_0$ is an ordinary polynomial in f, which necessarily has the form $\Psi = (f-\alpha)^{n_1}(f-\beta)^{n_2}$ with $n_1 + n_2 = n$ and $n_1\alpha + n_2\beta = 0$.

Exercise 3.17 Prove that, given n_1, n_2, α, β with $n_1\alpha + n_2\beta = 0$, $\alpha\beta \neq 0$, $n = n_1 + n_2$, and $k = n - 3 \geq 1$, there exist $c \neq 0$ and $a_1, \ldots, a_{k+1}$ such that

$$\Psi = f^n + f^{(k)} + a_1 f^{(k-1)} + \cdots + a_k f + a_{k+1} = (f-\alpha)^{n_1}(f-\beta)^{n_2}$$

holds for any non-constant solution to $f' = c(f-\alpha)(f-\beta)$.
(**Hint.** Prove that $f^{(\nu)} = c^\nu f^{\nu+1} + \cdots = c^\nu P_{\nu+1}(f)$ holds for $\nu = 1, 2, \ldots$.)
As an example we quote $\Psi = f^6 + f''' - \frac{1}{c}f' + 1$ with $c = \sqrt[3]{1/2}$, $f' = c(1+f^2)$, hence $f(z) = \tan(cz)$ and $\Psi(z) \equiv (1 + f(z)^2)^3 \neq 0$.

Remark 3.5 Theorems on the zeros of differential polynomials may have a different shape. They may be purely qualitative: 'Ψ has (infinitely many) zeros', 'semi'-quantitative: '$\limsup \overline{N}(r, 1/\Psi)/T(r,f) > 0$' like in the cases being discussed here, or quantitative like in Weissenborn [198], where the inequality

$$T(r,f) \leq \overline{N}(r, 1/\Psi) + \overline{N}(r,f) + S(r,f)$$

in the Tumura–Clunie context is proved. Perhaps the most prominent example is

Hayman's alternative, which is based on the inequality (see [77])

$$\ell T(r,f) \leq (2\ell+1)N\Big(r,\frac{1}{f}\Big) + (2\ell+2)N\Big(r,\frac{1}{f^{(\ell)}-1}\Big) + S(r,f) \quad (\ell \geq 1),$$

and says that f or $f^{(\ell)}-1$ (possibly both) have infinitely many zeros, with counting function a numerical portion of $T(r,f)$. Further examples of this kind are

$$(\ell-1)T(r,f) \leq N\big(r,1/\Psi\big) + o(T(r,f)) \quad (\Psi = f^{\ell}(f^{(k)})^n - 1,\ k,\ell,n \geq 1)$$

(Jiang and Huang [97]) and

$$(n-2)T(r,f) \leq N(r,1/\Psi) + S(r,f) \quad (\Psi = f^n + a_{n-1}(z)f^{n-1} + \cdots + a_0(z)).$$

An example for the qualitative case is Theorem 5 in [139], which says that if f is entire with zeros of multiplicity at least k, then $f^n f^{(k)}$ $(n,k \geq 1)$ assumes every value $a \neq 0$ infinitely often.

3.2.5 *Homogeneous Differential Polynomials*

Hayman [77] also considered homogeneous differential polynomials, thereby opening a second field of research concerning the zeros of differential polynomials. He proved that $f(z) = e^{az+b}$ $(a \neq 0)$ is the only transcendental meromorphic function of finite order such that $ff'f''$ has no zeros, and asked whether this will remain true if the additional hypotheses '$f' \neq 0$' and 'f has finite order' are dropped. This is the case for entire functions. Clunie [28] proved that $ff^{(k)} \neq 0$ for some $k \geq 2$ and f entire implies $f(z) = e^{az+b}$. For transcendental meromorphic functions and $k \geq 3$ this was proved by Frank [46], and by Mues [119] for $k = 2$ if f has finite (lower) order. Only much later and with quite different methods, Langley [104] was able to settle this problem in full generality. The corresponding result for $ff' \neq 0$ is not true: for example, $f(z) = e^{e^z}$ (entire of infinite order) and also the non-constant solutions to $w' = w(w+1)$ (meromorphic of finite order) have this property.

We will now describe a far-reaching method due to Frank [46] which applies to homogeneous differential polynomials $fL[f]$, where L is any differential operator with polynomial coefficients. It suffices to consider the canonical form

$$L[y] = y^{(k)} + a_{k-2}(z)y^{(k-2)} + \cdots + a_0(z)y \quad (k \geq 3). \tag{3.27}$$

Without any hypothesis on the zeros of $fL[f]$, Frank and Hellerstein [47] proved the *a priori estimate*

$$T(r,f'/f) = O\big(N(r,1/f) + N(r,1/L[f]) + r^{\varrho_{\max}}\big) + S(r,f'/f);$$

$\varrho_{\max}$ denotes the maximal order of growth of the solutions to $L[y] = 0$, and $r^{\varrho_{\max}}$ may be replaced with $\log r$ if $L[y] = y^{(k)}$. In particular, f'/f has finite order of growth if $fL[f]$ is zero-free. Assuming this, there exists an entire function satisfying $g^{-k} = L[f]/f$. The zeros of g are simple and coincide with the poles of f (which may have arbitrary multiplicities), hence $h = g(f'/f)$ is also an entire function. Both functions g and h have order of growth at most $\varrho_{\max}$, and even are polynomials if $L[y] = y^{(k)}$.

The main idea now is to consider the functions

$$w_\kappa = gy_\kappa' - hy_\kappa = gf\big(y_\kappa/f\big)' \quad (1 \le \kappa \le k),$$

where $(y_1, \ldots, y_k)$ denotes any fundamental set to the equation $L[y] = 0$, normalised by $W(y_1, \ldots, y_k) = 1$. From

$$\begin{aligned} f = g^k L[f] = g^k W(y_1, \ldots, y_k, f) &= g^k f^{k+1} W(y_1/f, \ldots, y_k/f, 1) \\ = (-1)^{k+1} f W\big(fg\big(y_1/f\big)', \ldots, fg\big(y_k/f\big)'\big) &= (-1)^{k+1} f W(w_1, \ldots, w_k) \end{aligned}$$

it follows that $W(w_1, \ldots, w_k) = (-1)^{k+1}$, hence these functions also annihilate some linear differential operator $M[w] = w^{(k)} + b_{k-2}(z)w^{(k-2)} + \cdots + b_0(z)w$; the coefficients are entire functions satisfying $T(r, b_\kappa) = m(r, b_\kappa) = S(r, f'/f)$, hence are polynomials, and even constants if L has constant coefficients. The fact that $M[gy' - hy] = 0$ is equivalent to $L[y] = 0$ then leads to the following relations:

$$\begin{aligned} -(ka_0 g' + a_0' g) &= (b_0 - a_0)h + \sum_{\kappa=1}^{k} b_\kappa h^{(\kappa)}, \\ (b_{\kappa-1} - a_{\kappa-1} - a_\kappa')g + (\kappa b_\kappa - k a_\kappa)g' &+ \sum_{j=\kappa+1}^{k} b_j \binom{j}{\kappa-1} g^{(j-\kappa+1)} \\ &= (b_\kappa - a_\kappa)h + \sum_{j=\kappa+1}^{k} b_j \binom{j}{\kappa} h^{(j-\kappa)} \qquad (1 \le \kappa \le k-2), \\ \binom{k}{2} g'' + (b_{k-2} - a_{k-2})g &= kh'. \end{aligned} \tag{3.28}$$

By an elimination process for linear operators we obtain some differential equation

$$K[v] = v^{(s)} + c_{s-1}(z)v^{(s-1)} + \cdots + c_0(z)v = 0$$

for $v = g(z)$, again with polynomial or even constant coefficients, and g, h, and $f'/f = h/g$ can be computed with considerable effort.

1. The case of $L[y] = y^{(k)}$ $(k \geq 3)$ leads to $w_1 = h$ and

$$w_\kappa = (\kappa - 1)z^{\kappa-2}g - z^{\kappa-1}h \quad (2 \leq \kappa \leq k),$$

and after intricate computations to $f'/f = h/g = a$ and $f(z) = e^{az+b}$ (Frank [46]).
2. If L has constant coefficients so have M and K. Then g and h are exponential polynomials, and tedious computations lead to three cases as follows [166].

a. $f(z) = \exp(az + b + e^{cz+d})$;
b. $f(z) = e^{az+b}/(z - z_0)^n$;
c. $f(z) = e^{az+b}/(e^{cz+d} - 1)^n$.

Apart from $f(z) = e^{az+b}$ (which just requires $a^k + a_{k-2}a^{k-2} + \cdots + a_0 = 0$), every case is realised by a unique operator L in the case of a. and b., and by $k+1$ uniquely determined operators L in the case of c.
3. Combining the arguments in [47] and [166] with the theory of asymptotic integration, Brüggemann [20] was able to prove the conjecture of Frank and Langley (see [17]) that if $fL[f]$ is zero-free (L with polynomial coefficients) and f has infinitely many poles, then

$$f = (H')^{-\frac{1}{2}(k-1)}H^{-n}, \ n \in \mathbb{N} \text{ and } H''/H' = p \tag{3.29}$$

holds for some non-constant polynomial p. We note that this occurs in 3.c. in a trivial way, and non-trivially for

$$L[y] = y''' - (z^2 + 2)y' - zy \quad \text{and} \quad 1/f(z) = e^{-z^2/2}\int_0^z e^{-t^2/2}\,dt;$$

f has infinitely many poles and $fL[f](z) = -6e^{-3z^2}f(z)^5$ is zero-free. Obviously (3.29) holds with $n = 1$, $k = 3$, and $H'(z) = e^{-z^2/2}$.
4. The case $k = 2$ resisted efforts for a long time and was finally settled by Langley [104], who also solved several related problems, see, for example, [105]. We note that for $k = 2$ the information in (3.28) is almost worthless, and the arguments differ substantially from those in the case of $k \geq 3$.

The following exercises will show that the results just stated are all sharp.

Exercise 3.18 Set $F(z) = e^z/(e^z - 1)$ and $f(z) = e^{az+b}F(z)^k$. Prove that $F' = F - F^2$ and $f^{(\kappa)}/f = Q_\kappa(F)$ hold, where Q_κ is a polynomial of degree κ. Then

$$L[f] = f^{(k)} + a_{k-1}f^{(k-1)} + \cdots + a_0 f \tag{3.30}$$

is zero-free if and only if $Q(w) = Q_k(w) + a_{k-1}Q_{k-1}(w) + \cdots + a_1Q_1(w) + a_0$ has the form const $\cdot\, w^m(w-1)^{k-m}$ for some $m \in \{0, 1, \ldots, k\}$; for each m this is possible in exactly one way.

Exercise 3.19 Set $f(z) = e^{az+e^z}$ and prove $f^{(\kappa)}/f = Q_\kappa(e^z)$, where Q_κ is a polynomial of degree κ. Then (3.30) is zero-free if and only if $Q(w) = \text{const} \cdot w^k$; this uniquely determines L. Similarly, for $f(z) = e^{az+bz^{-n}}$ $(n \geq 1)$ construct L in a unique way such that $L[f]$ has no zeros.

Exercise 3.20 ([103]) Suppose that p is a non-constant polynomial, H is entire and satisfies $H''/H' = p$, f (of finite order with infinitely many poles) and g (entire of infinite order) are given by $f = (H')^{-k}H^{-n}$ and $g = (H')^{-k}e^H$, respectively, and

$$L = (D + p(z))(D + 2p(z)) \cdots (D + kp(z)) \quad (D = \tfrac{d}{dz}).$$

Prove that $L[f] = cH^{-n-k}$ for some $c \neq 0$ and $L[g] = e^H$ hold.
(**Hint.** Prove $f' + kpf = -n(H')^{-(k-1)}H^{-(n+1)}$ and $g' + kpg = (H')^{-(k-1)}e^H$.)

3.3 Uniqueness of Meromorphic Functions

The very first application of Nevanlinna Theory was dedicated to the question of how many pre-images $f^{-1}(\{a\})$ determine a meromorphic function. Distinct meromorphic functions f and g are said to *share* the value a if $f^{-1}(\{a\}) = g^{-1}(\{a\})$. Moreover, if $f - a$ and $g - a$ (resp. $1/f$ and $1/g$ if $a = \infty$) have the same zeros counting multiplicities (the same divisor) we will speak of sharing *counting multiplicities*.

3.3.1 *The Five-Value Theorem*

Nevanlinna's first uniqueness result in [125] is known as the *Five-Value Theorem*; it initiated extensive research on sometimes obscure 'value sharing problems'. The serious investigations after Nevanlinna started in the late seventies and early eighties with the work of Gundersen [65, 66, 68]. Shortly afterwards Mues [120] introduced the 'method of auxiliary functions' with sustainable effect. The underlying idea is as follows: if f is meromorphic and ϕ is a small function (w.r.t. f), which vanishes on a sequence of poles, zeros, etc. of f, then either ϕ vanishes identically or else the sequence in question has counting function $S(r,f)$. Though the idea is simple, the main problem consists in constructing relevant auxiliary functions and requires a lot of experience. We set $T(r) = \max\{T(r,f), T(r,g)\}$ and denote by $S(r)$ the common error term.

Five-Value Theorem ([125]) *Distinct meromorphic functions f and g cannot share five values. In other words, five pre-images $f^{-1}(\{a\})$ determine f uniquely.*

Proof We assume that f and g share the finite values a_ν $(1 \le \nu \le q)$ and denote by $\overline{N}(r;a)$ the (integrated) counting function of common a-points, each point being counted once, ignoring multiplicities. The Second Main Theorem then yields

$$\begin{aligned}(q-2)T(r) &\le \sum_{\nu=1}^{q} \overline{N}(r;a_\nu) + S(r) \le \overline{N}\Big(r, \frac{1}{f-g}\Big) + S(r) \\ &\le T(r,f) + T(r,g) + S(r) \le 2T(r) + S(r),\end{aligned} \tag{3.31}$$

which obviously is a contradiction if $q \ge 5$.

3.3.2 Examples and Counterexamples

We will discuss several (counter-)examples to show that the Five-Value Theorem is the best possible and has no complete analogue for four values. To start with, the functions $f(z) = e^z$ and $g(z) = e^{-z}$ share four values: -1 and 1 counting multiplicities, and the Picard exceptional values 0 and ∞ (of course, also counting multiplicities). At the other end of the scale we have *Gundersen's example* [65]:

Exercise 3.21 Prove that the functions

$$f(z) = \frac{e^z+1}{(e^z-1)^2} \quad \text{and} \quad g(z) = \frac{(e^z+1)^2}{8(e^z-1)}$$

share the values $1, 0, \infty$ and $-1/8$ in the following manner: f has only simple zeros and 1-points, and only double poles and $(-1/8)$-points, while g has only simple poles and $(-1/8)$-points, and only double zeros and 1-points at the very same places. Prove also that $x = f(z)$ and $y = g(z)$ parametrise the rational curve

$$6xy + y + x^2 - 8xy^2 = 0$$

and satisfy the first-order differential equations

$$(x' + x + \tfrac{1}{8})^2 = 2(x + \tfrac{1}{8})(x - \tfrac{1}{4})^2 \quad \text{and} \quad (y' + y^2 - y)^2 = y(y-1)(y+\tfrac{1}{2})^2.$$

Gundersen's example has one more interesting property, namely $f(z) = -\frac{1}{2}$ implies $g(z) = \frac{1}{4}$, and *vice versa*. Reinders [142, 144] characterised this example three-fold:

Theorem 3.12 *Assume that F and G share four mutually distinct finite values a_ν, and one of the following conditions holds:*

- *there exist values $a, b \ne a_\nu$ $(1 \le \nu \le 4)$ such that $F(z_0) = a$ implies $G(z_0) = b$;*
- *for every ν the zeros of $(F - a_\nu)(G - a_\nu)$ have multiplicity three;*
- *for every ν either $F - a_\nu$ or else $G - a_\nu$ has only double zeros.*

Then F and G are either Möbius transformations of each other (this is possible in the first case) or else have the form $F = M \circ f \circ h$ and $G = M \circ g \circ h$, where f and g are the functions being discussed in Exercise 3.21; M is a Möbius transformation and h a non-constant entire function.

Exercise 3.22 (Reinders [143])   The elliptic functions defined by the differential equation $u'^2 = 12\,u(u+1)(u+4)$ have elliptic order two. Prove that the elliptic functions

$$\hat{f}(z) = \frac{1}{8\sqrt{3}}\frac{u(z)u'(z)}{u(z)+1} \quad \text{and} \quad \hat{g}(z) = \frac{1}{8\sqrt{3}}\frac{(u(z)+4)u'(z)}{(u(z)+1)^2}$$

of elliptic order four share the values $-1, 0, 1, \infty$ in the following manner: each of these values is assumed in an alternating way with multiplicity 1 by one of these functions, and with multiplicity 3 by the other, see Fig. 3.1. Prove also that $w = \hat{f}(z)$ solves

$$w'^4 - 4\sqrt{3}(w^2+1)w'^3 + 3888w^2(w^2-1)^2 = 0,$$

and $\hat{f}$ and $\hat{g}$ parametrise the algebraic curve

$$(x-y)^4 = 16xy(x^2-1)(y^2-1).$$

(Hint. From $u'^2 = 12u(u+1)(u+4)$ and $u'' = 18u^2+60u+24$ deduce $(u+1)\hat{f}^2 = P(u)$ and $(u+1)\hat{f}' = Q(u)$ with polynomials P and Q. To derive the differential equation for $\hat{f}$ compute the resultant $R(z, w, w_1)$ of the polynomials $(u+1)w^2 - P(u)$ and $(u+1)w_1 - Q(u)$ w.r.t. u. The algebraic curve is simply obtained by computing the resultant $R(x, y)$ of $yu(u+1)-x(u+4)$ and $16xy(u+1)^2-u^2(u+4)^2$ (corresponding to $y/x = \hat{g}/\hat{f}$ and $xy = \hat{f}\hat{g}$, respectively), again w.r.t. u; $u, x, y, w,$ and w_1 are just regarded as complex variables.)   ☕

The number five in the Five-Value Theorem is the best possible, as follows from our examples, which are of a quite different character otherwise. We note, however, that the proof of the Five-Value Theorem reveals important information common to all pairs of meromorphic functions f and g that share *four* values. This information has its origin in the equality sign in the Second Main Theorem.

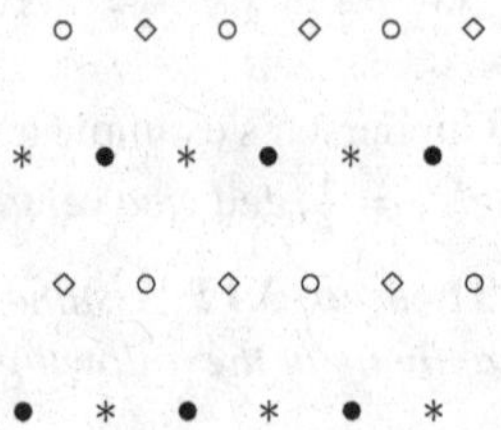

Fig. 3.1 The distribution of poles • (three-fold for f, simple for g, and two-fold for u) and ○ (three-fold for g and simple for f), and zeros ∗ (three-fold for f and simple for g) and ◇ (three-fold for g and simple for f)

Exercise 3.23 Suppose that f and g share four finite values a_ν. Use (3.31) for $q = 4$ to prove

- $T(r,f) = T(r) + S(r)$ and $T(r,g) = T(r) + S(r)$;
- $\sum_{\nu=1}^{4} \overline{N}(r;a_\nu) = 2T(r) + S(r)$;
- $\overline{N}\left(r, \frac{1}{f-g}\right) = 2T(r) + S(r)$;
- $\overline{N}\left(r, \frac{1}{f-b}\right) + \overline{N}\left(r, \frac{1}{g-b}\right) = 2T(r) + S(r)$ $(b \neq a_\nu)$, and also $\overline{N}(r,f) + \overline{N}(r,g) = 2T(r) + S(r)$.

Remark 3.6 We remind the reader of the special meaning of the term 'almost': by definition, almost every point in a given sequence (of poles, zeros, etc.) has some property P if the subsequence of points not having this property has counting function $S(r)$. For example, it follows from Exercise 3.23 that almost every zero of $f - g$ is simple and corresponds to some common a_ν-point of f and g, hence also almost every a_ν-point is simple for at least one of the functions f and g. Also almost all critical points of f and g correspond to the a_ν-points. Nevanlinna Theory tolerates sequences with counting function $S(r)$.

3.3.3 The Four-Value Theorem

The next result, known as the *Four-Value Theorem*, is also due to Nevanlinna [125]. It states that the first example $f(z) = e^z$, $g(z) = e^{-z}$ is unique if the pairs (f, g) and $(M \circ f \circ h, M \circ g \circ h)$ are identified (M any Möbius transformation, h any non-constant entire function).

Four-Value Theorem *Suppose f and g share four values a_ν counting multiplicities. Then relabelling the values if necessary, the cross-ratios of a_1, a_2, a_3, a_4 and f, g, a_3, a_4 satisfy*

$$(a_1, a_2, a_3, a_4) = (f, g, a_3, a_4) = -1;$$

a_1 and a_2 are Picard exceptional values of f and g, and $f = M \circ g$ holds with some Möbius transformation M that fixes a_3 and a_4 and permutes a_1 and a_2.

Proof (Mues [120]) It follows from Exercise 3.23 and the hypothesis 'counting multiplicities' that almost every a_ν-point is simple for f and g, f' and g' have almost no zeros, and at least two of the values a_ν satisfy $\overline{N}(r;a_\nu) \neq S(r)$; we may assume that $a_4 = \infty$ and $a_3 = 0$ have this property. The auxiliary function

$$\phi = \frac{f''}{f'} - \frac{g''}{g'}$$

satisfies $m(r, \phi) = S(r)$, and from the previous discussion of the critical points it follows that $N(r, \phi) = \overline{N}(r, \phi) = S(r)$ holds. Moreover, ϕ vanishes at almost every common pole of f and g. This yields a contradiction if $\phi \not\equiv 0$, namely $\overline{N}(r; \infty) = S(r)$. If, however, ϕ vanishes identically, $f = c_1 g + c_0$ follows at once by twofold integration, and $\overline{N}(r; 0) \neq S(r)$ yields $c_0 = 0$, while $c_1 \neq 1$ is part of the hypothesis. Since also the values a_ν ($\nu = 1, 2$) are shared by f and g, while $a_\nu = c_1 a_\nu$ is impossible, it follows that these values are Picard exceptional values for f and g, and from $f = c_1 g$, $a_1 = c_1 a_2$ and $a_2 = c_1 a_1$ it follows that $c_1 = -1$, $a_2 = -a_1$, $f = -g$, and $(a_1, a_2, 0, \infty) = (f, g, 0, \infty) = -1$.

Remark 3.7 Nevanlinna's proof was quite different and based on Theorem 2.10. Suppose that distinct meromorphic functions f and g share the values $0, 1, c$, and ∞ in the strong sense, namely assume that $h_1 = \frac{f}{g}$, $h_2 = \frac{f-1}{g-1}$, and $h_3 = \frac{f-c}{g-c}$ are zero-free entire functions. Then the Borel identity

$$(1-c)h_1 + ch_2 - h_3 - h_1 h_2 + ch_1 h_3 + (1-c)h_2 h_3 \equiv 0$$

holds, and by Theorem 2.10 at least two of the functions $h_1, h_2, h_3, h_1h_2, h_1h_3, h_2h_3$ are linearly dependent. No matter which are, this leads to the conclusion that g is a Möbius transformation of f. Since Möbius transformations may have only two fixed points, two of the shared values, 0 and ∞, say, must be Picard values of f and g, and this implies $g = a/f$. Then $f(z_0) = g(z_0) = 1$ implies $a = 1$ ($1 = a/1$) and $c = 1/c$, thus $g = 1/f$, $f = e^\phi$, $g = e^{-\phi}$, and $c = -1$.

Exercise 3.24 Discuss exemplarily the case $h_1 = \lambda h_2 h_3$, that is,

$$\frac{f}{(f-1)(f-c)} = \lambda \frac{g}{(g-1)(g-c)}.$$

Mues' idea to introduce auxiliary functions not only simplified the proof but pointed the way to further developments. For example, it follows from the proof that the hypothesis 'counting multiplicities' may be replaced with 'almost every a_ν-point is simple for f and g'. This is not just a trivial generalisation but will turn out to be crucial in what follows.

3.3.4 *Variations of the Four-Value Theorem*

The papers on 'value sharing problems' are legion. We will restrict ourselves to the fundamental question of whether and how far the hypothesis 'counting multiplicities' in the Four-Value Theorem may be relaxed. To this end we will prove two theorems due to Gundersen [65, 66, 68]. The proofs are based on suitably chosen auxiliary functions, the most important one,

$$\Psi = \frac{f'g'(f-g)^2}{P(f)P(g)} \quad \Big(P(x) = \prod_{\nu=1}^{4}(x-a_\nu)\Big) \tag{3.32}$$

was introduced by Mues in the early 1980s, and independently at almost the same time by Czubiak and Gundersen; if $a_4 = \infty$ the factor $x - a_4$ has to be omitted. Our general assumption is that f and g share four values a_ν. If appropriate, three of these values may be chosen at will. We note that Ψ carries the whole information on the shared values, but unfortunately is incapable of taking into account multiplicities.

Exercise 3.25 Prove that Ψ is a small entire function, that is, $m(r, \Psi) = T(r, \Psi) = S(r)$ holds. (**Hint.** $\Psi = \frac{f^2f'}{P(f)}\frac{g'}{P(g)} - 2\frac{ff'}{P(f)}\frac{gg'}{P(g)} + \frac{f'}{P(f)}\frac{g^2g'}{P(g)}$.)

We are now prepared to prove Theorem 3.14 below. Although this theorem is stronger than Theorem 3.13, it will be instructive to prove this result first. The proof, due to Rudolph [147], is 'auxiliary function-based'.

Theorem 3.13 *Suppose that f and g share four values a_ν, three of them counting multiplicities. Then the fourth value is also shared counting multiplicities.*

Proof Suppose f and g share the finite values a_ν $(1 \le \nu \le 3)$ counting multiplicities, and $a_4 = \infty$ without further hypothesis. Then

$$\phi = \frac{f'}{P(f)} - \frac{g'}{P(g)} \quad \Big(\text{now} P(x) = \prod_{\nu=1}^{3}(x-a_\nu)\Big)$$

satisfies $T(r, \phi) = S(r)$ and vanishes at the common poles of f and g. Thus we have either $\phi \not\equiv 0$ and $\overline{N}(r; \infty) = S(r)$ by the First Main Theorem, or else ϕ vanishes identically. In this case it follows from $P(f) = \kappa P(g)$ for some constant $\kappa \ne 0$ that $a_4 = \infty$ is also shared counting multiplicities. On the other hand it is not hard to show that the hypotheses '$\overline{N}(r; \infty) = S(r)$' and '$a_4 = \infty$ is shared counting multiplicities' are equally strong in the sense that in combination with the hypotheses on the other values a_ν they lead to the same conclusion.

The proof of Gundersen's Theorem 3.14 given here is due to Mues [120], see also [181]. It is even more 'auxiliary function-based' and requires repeated separation into cases. In part it will be organised in the form of exercises.

Theorem 3.14 *Suppose that f and g share four values a_ν, two of them counting multiplicities. Then all values a_ν are shared counting multiplicities.*

Proof We may assume that f and g share the values $a_3 = 0$ and $a_4 = \infty$ counting multiplicities and $a_1a_2 = 1$ holds; this may be achieved by considering cf and cg with $c^2a_1a_2 = 1$ in place of f and g, and will simplify matters significantly.

Exercise 3.26 Prove that under this hypothesis the auxiliary function

$$\phi = \frac{f''}{f'} + 2\frac{f'}{f} - \sum_{\nu=1}^{2} \frac{f'}{f - a_\nu} - \frac{g''}{g'} - 2\frac{g'}{g} + \sum_{\nu=1}^{2} \frac{g'}{g - a_\nu} \tag{3.33}$$

is regular at every zero and pole that is simple for f and g, and also at every a_ν-point ($\nu = 1, 2$), despite multiplicities. Prove also that $T(r, \phi) = S(r)$ and

$$\phi(z_\infty)^2 = (a_1 + a_2)^2 \Psi(z_\infty)$$

holds at every common simple pole z_∞. (**Hint.** Many computations are easily realised by using some computer algebra system like MAPLE, which is most suitable when operating with 'rational expressions' and finite Laurent series.)

One consequence is that either $\overline{N}(r; \infty) = S(r)$ or else $\phi^2 \equiv (a_1 + a_2)^2 \Psi$ holds. This argument may be repeated with f and g replaced by $F = 1/f$ and $G = 1/g$, which share the very same values $a_2, a_1, \infty, 0$.

Exercise 3.27 Prove that the pair $(F, G) = (1/f, 1/g)$ has the same Mues function (3.32), and the auxiliary function

$$\Phi = \frac{F''}{F'} + 2\frac{F'}{F} - \sum_{\nu=1}^{2} \frac{F'}{F - a_\nu} - \frac{G''}{G'} - 2\frac{G'}{G} + \sum_{\nu=1}^{2} \frac{G'}{G - a_\nu}$$

corresponding to (3.33) has the form

$$\Phi = \frac{f''}{f'} - 2\frac{f'}{f} - \sum_{\nu=1}^{2} \frac{f'}{f - a_\nu} - \frac{g''}{g'} + 2\frac{g'}{g} + \sum_{\nu=1}^{2} \frac{g'}{g - a_\nu}$$

when written in terms of f and g. This was the reason to insist on $a_2 = 1/a_1$!

Again either $\overline{N}(r; 0) = S(r)$ or else $\Phi^2 \equiv (a_1 + a_2)^2 \Psi$ holds. We thus have to discuss four cases as follows:

1. $\overline{N}(r; \infty) + \overline{N}(r; 0) = S(r)$, **2.** $\Phi \equiv \phi$, **3.** $\Phi \equiv -\phi$, and
4. $\overline{N}(r; \infty) = S(r)$ and $\Phi^2 \equiv (a_1 + a_2)^2 \Psi$ as one of two equivalent cases.

Exercise 3.28 In the *first case* prove that $\overline{N}(r; a_\nu) = T(r) + S(r)$ ($\nu = 1, 2$) holds (Exercise 3.23, second assertion). This is in some sense weaker than saying that f and g also share a_1 and a_2 counting multiplicities, but strong enough to ensure that the proof of the Four-Value Theorem works.

Exercise 3.29 Prove that in the *second case* $g = cf$ holds for some constant $c \neq 0$, this implying that a_1 and a_2 are Picard exceptional values of both functions, and f and g share all values counting multiplicities. In both cases the proof of Theorem 3.14 is finished.

Exercise 3.30 In the *third case* integrate $\Phi \equiv -\phi$ twice to obtain

$$\frac{f'}{(f-a_1)(f-a_2)} = \frac{\kappa g'}{(g-a_1)(g-a_2)} \quad \text{and} \quad \frac{f-a_1}{f-a_2} = C\Big(\frac{g-a_1}{g-a_2}\Big)^{\kappa}$$

for some integer κ and some constant $C \neq 0$. From $T(r,f) = T(r,g)+S(r)$ conclude that $\kappa = \pm 1$. Discuss the cases $\kappa = -1$ and $\kappa = 1$ separately to reduce the proof of Theorem 3.14 either-to the second case or else to the Four-Value Theorem.

In the *fourth case* we first assume $a_1 + a_2 \neq 0$ in addition, and consider the auxiliary function

$$\eta_1 = \Phi - (a_1 + a_2)\frac{f'(f-g)}{f(g-a_1)(f-a_2)}.$$

Exercise 3.31 Prove

$$N(r,\eta_1) \le N\Big(r, \frac{1}{g-a_1}\Big) - \overline{N}(r;a_1) \quad \text{and} \quad m(r,\eta_1) \le m\Big(r, \frac{1}{g-a_1}\Big) + S(r),$$

hence $T(r,\eta_1) \le T(r) - \overline{N}(r;a_1) + S(r)$, and similar estimates for

$$\theta_1 = \Phi - (a_1 + a_2)\frac{g'(f-g)}{g(f-a_1)(g-a_2)}$$

and the functions η_2 and θ_2 that are obtained by permuting a_1 and a_2 in the definition of η_1 and θ_1. (**Hint.** $\frac{f'(f-g)}{f(g-a_1)(f-a_2)} = -\frac{f'}{f(f-a_2)} + \frac{1}{g-a_1}\frac{f'(f-a_1)}{f(f-a_2)}$.)

Exercise 3.32 Prove that each of the functions $\eta_1, \theta_1, \eta_2, \theta_2$ vanishes at common simple zeros z_0 of f and g. (**Hint.** Use MAPLE and $1/a_1 + 1/a_2 = a_1 + a_2$.)

To proceed we have to discuss two subcases as follows.

a) $\eta_\nu \not\equiv 0$ or $\theta_\nu \not\equiv 0$ for $\nu = 1, 2$; **b)** $\eta_2 = \theta_2 \equiv 0$, say.

Exercise 3.33 Prove that in case **a)** $\overline{N}(r;0) \le T(r) - \overline{N}(r;a_\nu)$ $(\nu = 1,2)$ holds, and in combination with $\overline{N}(r;\infty) = S(r)$ and Exercise 3.23 also

$$\begin{aligned} 2\overline{N}(r;0) &\le 2T(r) - \overline{N}(r;a_1) - \overline{N}(r;a_2) + S(r) \\ &= \overline{N}(r;0) + \overline{N}(r;\infty) + S(r) = \overline{N}(r;0) + S(r), \end{aligned}$$

thus $\overline{N}(r;\infty) + \overline{N}(r;0) = S(r)$. This means that we are back in the first case. Prove also that in case **b)**

$$\frac{(f-a_2)f'}{f(f-a_1)} = \frac{(g-a_2)g'}{g(g-a_1)}$$

holds, hence f and g share four values counting multiplicities.

It remains to consider the fourth case with $a_1 + a_2 = 0$ and $a_1a_2 = 1$, hence $a_1 = i$, $a_2 = -i$, and $\Phi \equiv 0$. Integration yields

$$\frac{f'}{f^2(f-i)(f+i)} = \kappa \frac{g'}{g^2(g-i)(g+i)} \quad (\kappa \neq 0). \tag{3.34}$$

Exercise 3.34 Integrate (3.34) to obtain $\frac{1}{2i}\log\frac{(f+i)(g-i)^\kappa}{(f-i)(g+i)^\kappa} = \frac{1}{f} - \frac{\kappa}{g} + c.$

Since at least one of the values $\pm i$ is not a Picard exceptional value for f and g (otherwise we were already in the case of Theorem 3.13), κ (or $1/\kappa$) is a positive integer, hence f assumes the values $\pm i$ 'always' with multiplicity κ, while g has 'only' simple $\pm i$-points (up to a sequence of points with counting function $S(r)$). Now $\kappa = 1$ means that f and g also share the values $\pm i$ counting multiplicities, and we are done. From $\kappa \geq 2$ and Exercise 3.34, however, it follows that

$$\frac{(f+i)(g-i)^\kappa}{(f-i)(g+i)^\kappa} = e^H \quad (H = 2i(1/f - \kappa/g + c) \text{ an entire(!) function})$$

holds. At every common $\pm i$-point, H assumes the value $\pm 2(1-\kappa) + 2ic$, thus H is non-constant and $\overline{N}(r;i) + \overline{N}(r;-i) \leq 2T(r,H) + O(1)$ follows. To obtain a contradiction and to finishes the proof we refer to the following exercise.

Exercise 3.35 Use $\overline{N}(r;\pm i) \leq \frac{1}{\kappa}T(r) + O(1)$, $\overline{N}(r;\infty) = S(r)$ and Exercise 3.23 to deduce $\overline{N}(r;0) = T(r) + S(r)$ and

$$T(r,H) = m(r,H) \leq m(r,1/f) + m(r,1/g) + O(1) = S(r).$$

Remark 3.8 The diploma thesis [147] of Eva Rudolph was never published, so it is no wonder that several researchers [68, 93, 157, 190, 194, 196] independently rediscovered her results or results that may be deduced from [147]. To state these results we need to define the quantity (introduced by Mues [120])

$$\tau(a_\nu) = \liminf_{r\to\infty\ (r\notin E)} \frac{\overline{N}_s(r;a_\nu)}{\overline{N}(r;a_\nu)} \quad \text{if } \overline{N}(r;a_\nu) \neq S(r), \text{ and } \tau(a_\nu) = 1 \text{ otherwise;}$$

here $\overline{N}_s(r;a_\nu)$ denotes the counting function of those a_ν-points which are *simultaneously simple* for f and g, and E is the exceptional set for $S(r)$; we note that $\tau(a_\nu) = 1$ holds in particular if a_ν is shared counting multiplicities, and also if a_ν is a Picard value.

Theorem 3.15 *Suppose f and g share four values. Then these values are shared counting multiplicities if one of the following conditions is fulfilled in addition:*

- *one value satisfies $\tau(a_\nu) = 1$ and some other $\tau(a_\kappa) > \frac{2}{3}$; for $(a_1, a_2, a_3, a_4) = -1$ the constant $\frac{2}{3}$ may be replaced with $\frac{1}{2}$* (Mues [120]). This is the strongest result known so far.

- *one value is shared counting multiplicities, while the others satisfy* $\tau(a_\nu) > \frac{1}{2}$ (Wang [196], Huang [93]).
- *one of the values is shared counting multiplicities and satisfies* $\overline{N}(r; a_\nu) \geq \delta\, T(r)$ *for some* $\delta > \frac{4}{5}$ *on a set of infinite measure* (Gundersen [68]).
- *two of the values satisfy* $\tau(a_\nu) > \frac{4}{5}$*; for* $(a_1, a_2, a_3, a_4) = -1$ *the constant* $\frac{4}{5}$ *may be replaced with* $\frac{2}{3}$ (Wang [194], Huang [93]).
- $\tau(a_\nu) > \frac{3}{4}$ *holds for three of the values* (Song and Chang [157]).
- $\tau(a_\nu) > \frac{2}{3}$ *holds for* $1 \leq \nu \leq 4$ (Wang [196], conjectured in [157]).

Resume The reader will have noticed that the main problem in the context of the Four-Value Theorem was left aside. It is still an open question whether or not there exists some pair (f, g) sharing four values, exactly one of them counting multiplicities. There is some evidence that the answer to this question is 'No', but a proof is not in sight. Our examples indicate that meromorphic functions sharing four values (without any additional assumption) might be algebraically dependent and separately satisfy implicit first-order differential equations. So a step in the right direction could be to determine all pairs (f, g) of transcendental meromorphic functions that share four values and are algebraically dependent.

3.3.5 Reinders' Example Rediscovered

We will now show that the pair $(\hat{f}, \hat{g})$ in Reinder's Example 3.22 may be characterised as follows.

Theorem 3.16 ([143]) *Suppose f and g share the values* a_ν $(1 \leq \nu \leq 4)$ *such that* $(f - a_\nu)(g - a_\nu)$ *resp.* $1/(fg)$ *if* $a_\nu = \infty$ *has only zeros of order at least four. Then there exists some Möbius transformation M and some non-constant entire function h such that* $f = M \circ \hat{f} \circ h$ *and* $G = M \circ \hat{g} \circ h$ *holds.*

Proof Our goal is to derive an algebraic curve that is parametrised by (f, g). Choosing $a_1 = 0$, $a_2 = 1$, and $a_4 = \infty$ it will turn out that this curve has genus > 1 except when $a_3 \in \{-1, \frac{1}{2}, 2\}$. In case of $a_3 = -1$, say, the algebraic curve in question is

$$(y - x)^4 = 16xy(x^2 - 1)(y^2 - 1) \tag{3.35}$$

and has genus 1. We may assume that f and g are elliptic functions (actually $f = f_1 \circ h$, $g = g_1 \circ h$ holds, where f_1 and g_1 are elliptic functions and h is non-constant entire). Substituting $y = tx$, $x^2 = s$ transforms (3.35) into the algebraic curve

$$s(t-1)^4 - 16t(s-1)(st^2 - 1) = 0$$

of genus 0. The parametrisation $s = \frac{u^3(u+4)}{16(u+1)}$, $t = \frac{u+4}{u(u+1)}$ leads to

$$f(z)^2 = \frac{u(z)^3(u(z)+4)}{16(u(z)+1)} \quad \text{and} \quad g(z)^2 = \frac{u(z)(u(z)+4)^3}{16(u(z)+1)^3},$$

where now u is meromorphic and assumes the values $0, -1, -4$, and ∞ always with even multiplicities. Since $u^4 = 16f^3/g$ is an elliptic function, the just mentioned multiplicities are always equal to 2 and $u'^2 = \kappa^2 u(u+1)(u+4)$‡ holds. Choosing $4\kappa = 8\sqrt{3}$ just means a linear change of the independent variable.

Assuming $a_1 = 0$, $a_2 = 1$, and $a_4 = \infty$, we will now go into details to derive (3.35) as one of three equivalent equations. It is quite plausible that in 'almost' all cases f and g assume each value a_ν either in a (3:1) or (1:3) manner. To give a precise statement, let $\overline{N}_{(p:q)}(r; a_\nu)$ denote the counting functions of those a_ν-points which are p-fold for f and q-fold for g. Then $p + q \geq 4$ and $\min\{p, q\} = 1$ holds for almost all a_ν-points.

a) In the first step we will prove $\overline{N}(r; a_\nu) = \frac{1}{2}T(r) + S(r)$ and

$$\overline{N}_{(3:1)}(r; a_\nu) = \tfrac{1}{4}T(r) + S(r) \quad \text{and} \quad \overline{N}_{(1:3)}(r; a_\nu) = \tfrac{1}{4}T(r) + S(r).$$

The first assertion follows from Exercise 3.23 and $\overline{N}(r; a_\nu) \leq \frac{1}{2}T(r) + O(1)$. To prove the second for $a_4 = \infty$, say, observe that $\sum_{p,q>1} \overline{N}_{(p:q)}(r; \infty) = S(r)$ and

$$\begin{aligned} N(r,f) + N(r,g) &\geq \textstyle\sum_{\nu\geq 3}(\nu+1)(\overline{N}_{(\nu:1)}(r;\infty) + \overline{N}_{(1:\nu)}(r;\infty)) + S(r) \\ &\geq 4\overline{N}(r;\infty) + \textstyle\sum_{\nu\geq 4}(\nu-3)(\overline{N}_{(\nu:1)}(r;\infty) + \overline{N}_{(1:\nu)}(r;\infty)) + S(r) \\ &\geq \tfrac{1}{2}T(r) + \textstyle\sum_{\nu\geq 4}(\overline{N}_{(\nu:1)}(r;\infty) + \overline{N}_{(1:\nu)}(r;\infty)) + S(r), \end{aligned}$$

hence

$$\begin{aligned} \textstyle\sum_{\nu\geq 4}(\overline{N}_{(\nu:1)}(r;\infty) + \overline{N}_{(1:\nu)}(r;\infty)) &= S(r), \\ N(r,f) + N(r,g) &= 2T(r) + S(r), \\ \tfrac{1}{2}T(r) = \overline{N}(r;\infty) + S(r) &= \overline{N}_{(3:1)}(r;\infty) + \overline{N}_{(1:3)}(r;\infty) + S(r), \\ T(r) = N(r,f) + S(r) &\geq 3\overline{N}_{(3:1)}(r;\infty) + \overline{N}_{(1:3)}(r;\infty) + S(r) \\ &= 2\overline{N}_{(3:1)}(r;\infty) + \overline{N}(r;\infty) + S(r), \text{ and} \\ T(r) = N(r,g) + S(r) &\geq 2\overline{N}_{(1:3)}(r;\infty) + \overline{N}(r;\infty) + S(r) \end{aligned}$$

hold.

b) In the second step we will prove that the auxiliary functions

$$\Omega = 3\frac{f''}{f'} - 4\sum_{\nu=1}^{3}\frac{f'}{f-a_\nu} + 10\frac{f'}{f-g} - 3\frac{g''}{g'} + 6\frac{g'}{g-f}$$

‡ $\dfrac{u'^2}{u(u+1)(u+4)}$ is a zero-free polynomial!

and $\tilde{\Omega}$, which is obtained from Ω by permuting f and g, vanish identically. To prove that Ω is a small function w.r.t. f and g we observe that Ω is regular at every a_ν-point of type (1:3) and even vanishes at a_ν-points of type (3:1).[3] In particular, Ω has counting function of poles $N(r,\Omega) = S(r)$. From $10\frac{f'}{f-g} + 6\frac{g'}{g-f} = 6\frac{f'-g'}{f-g} + 4\frac{f'}{f-g}$ we obtain

$$m(r,\Omega) \le m\Big(\frac{f'}{f-g}\Big) + S(r).$$

To estimate the first term on the right-hand side we refer to Exercise 3.25, which states that the auxiliary function Ψ defined in (3.32) is small. From

$$\Psi\frac{f'}{f-g} = \frac{f'^2 g'(f-g)}{(f-a_1)(f-a_2)(f-a_3)(g-a_1)(g-a_2)(g-a_3)}$$

$m(r,\Omega) = S(r)$ and thus $T(r,\Omega) = S(r)$ follows. Since, however, Ω vanishes at every a_ν-point of type (3:1) $(1 \le \nu \le 4)$ with counting function $\frac{1}{4}T(r) + S(r)$, it must vanish identically. The same, of course, is true for $\tilde{\Omega}$.

c) To proceed we note that $\frac{1}{4}(\Omega + \tilde{\Omega})$ is the logarithmic derivative of

$$\lambda(z) = \frac{(f(z)-g(z))^4}{P(f(z))P(g(z))} \quad \Big(P(x) = \prod_{\nu=1}^{3}(x-a_\nu)\Big),$$

which thus is a non-zero constant, and f and g parametrise the algebraic curve

$$(x-y)^4 = \lambda P(x)P(y).$$

For $a_1 = 0$, $a_2 = 1$, $a_3 = c \ne 0, 1$, intricate computations involving poles of type (3:1) and (1:3) show that $2c^3 - 3c^2 - 3c + 2 = 0$ (with solutions $c = -1, 2, \frac{1}{2}$) and $\lambda = \frac{48}{c^2-c+1}$ holds, hence (3.35) in the case of $c = -1$; the interested reader is referred to [143]. We will give a 'semi-rigorous MAPLE-aided proof': When asked for the genus of the algebraic curve $C_{\lambda,c} : (x-y)^4 = \lambda xy(x-1)(y-1)(x-c)(y-c)$, MAPLE responds 'genus 5' if neither λ nor c is specified, 'genus 3' if $\lambda = 16$ but c is not specified, and 'genus 1' if $(\lambda, c) = (16, -1)$ (or $(16, 2)$ or $(64, \frac{1}{2})$). The 'proof' relies on the fact that the genus is > 1 if $(\lambda, c) \ne (16, -1)$, $(16, 2)$, and $(64, \frac{1}{2})$.

[3] Ω was constructed for that purpose; the verification is left to MAPLE.

3.3.6 Three Functions Sharing Four Values

The question whether there might be more than two meromorphic functions that share four values was answered in the negative by Cartan [21]. His argument, however, contained a gap which could not be bridged. Persistently pursuing his ideas instead leads to the following counterexample, and even more to a characterisation of the function triples sharing four values.

Example 3.3 Given any $a \neq 0, -1$, the non-constant solutions to the differential equation $u'^2 = 4u(u+1)(u-a)$ are elliptic functions of elliptic order two. For $a = \frac{1}{2}(-1 \pm i\sqrt{3})$ the algebraic equation

$$P(x,y) = y^3 - 3((\bar{a}-1)x^2 - 2x)y^2 - 3(2x^2 - (a-1)x)y - x^3 = 0 \tag{3.36}$$

has singular points $(x,y) = (0,0), (a,1), (-1,-a), (\infty,\infty)$, and no others. At each of these points two of the algebraic functions $y = \phi_\nu(x)$ defined by (3.36) have a square-root singularity, while the other behaves regularly (for example, $\phi_{1,2}(x) = \pm c_1\sqrt{x} + \cdots$ and $\phi_3(x) = c_2x^2 + \cdots$ hold at $x = 0$). The branches $f_\nu(z) = \phi_\nu(u(z))$ admit unrestricted analytic continuation in the complex plane, hence are meromorphic functions that share the values $0, 1, -a, \infty$; each value is assumed in turn with multiplicities $(1,1,4)$, $(1,4,1)$, and $(4,1,1)$, see Fig. 3.2.

The triple (f_1, f_2, f_3) not only represents a positive example but also the normalised solution to the above mentioned problem. Lack of space prevents us from presenting the elaborate and involved proof, which can be found in [168].

Theorem 3.17 *Any triple F_1, F_2, F_3 of meromorphic functions sharing four values is given by $F_j = M \circ f_j \circ h$, where h is non-constant entire, M is some Möbius transformation, and the functions f_j are defined in Example 3.3.*

Remark 3.9 The functions f_j have elliptic order six, hence they satisfy some Briot–Bouquet equation $Q(w,w') = 0$ of degree six with respect to w'. The problem of how to determine Q may be solved as follows: consider the polynomials $P(x,y)$ and $P_1(x,y,y_1) = 4x(x+1)(x-a)P_x(x,y)^2 - P_y(x,y)^2y_1^2$, which reflect the fact that $P(u(z), f(z)) \equiv 0$, $\frac{d}{dz}P(u(z), f(z)) \equiv 0$, and $u'^2 = 4u(u+1)(u-a)$, and compute the resultant $R(y, y_1)$ of P and P_1 with respect to the variable x to obtain

$$w'^6 - \tfrac{1}{2}(1-\bar{a})(3w-1+\bar{a})(w-1+\bar{a})(w-a)w'^4 - 256w^3(w-1)^3(w+a)^3 = 0.$$

In the same manner one can show that any two of these functions parametrise one and the same algebraic curve of genus one.

Exercise 3.36 (Continued) Prove that the algebraic curve $P(x,y) = 0$ has the parametrisation $x = r(t) = \frac{1}{2}(9 - 3\sqrt{3}i)\frac{t(1+t)^2}{(1+3t)^2}$, $y = s(t) = \frac{1}{2}(9 - 3\sqrt{3}i)\frac{t^2(1+t)}{(1+3t)}$. Determine meromorphic functions h_j $(1 \leq j \leq 3)$ such that $u(z) = r(h_j(z))$ and $f_j(z) = s(h_j(z))$ holds.

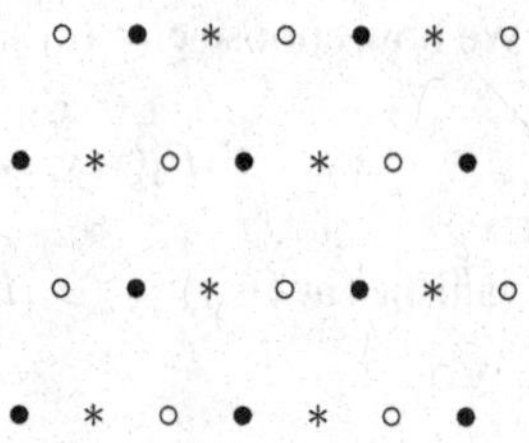

Fig. 3.2 u has double poles $* \circ \bullet$ and periods ϑ, ϑ'; f_1 has simple poles $* \circ$ and four-fold poles $\bullet$, and periods $3\vartheta, \vartheta + \vartheta'$. The four-fold poles of $f_2(z) = f_1(z + \vartheta)$ and $f_3(z) = f_1(z + 2\vartheta)$ are $*$ and $\circ$, respectively

3.3.7 *Pair Sharing*

A new direction of research was opened by Czubiak and Gundersen [31], who considered meromorphic functions f and g that 'share pairs' (a, b), this meaning $f^{-1}(\{a\}) = g^{-1}(\{b\})$, counting or disregarding multiplicities. For example, the functions f and g in Exercise 3.21 share five pairs $(0, 0)$, $(1, 1)$, $(-\frac{1}{8}, -\frac{1}{8})$, (∞, ∞), and $(-\frac{1}{2}, \frac{1}{4})$, the latter counting multiplicities. Czubiak and Gundersen introduced the useful technique of 'algebraic auxiliary functions' $P(f(z), g(z))$ to prove that the number 'five' is the best possible.

Theorem 3.18 *Meromorphic functions f and g that share more than five pairs are Möbius transformations of each other.*

Proof We will give a straightforward proof based on the main idea in [31]. We assume that f and g are not Möbius transformations of each other and share six finite pairs (a_ν, b_ν). By $\overline{N}(r; a_\nu, b_\nu)$ we denote the counting function of common (a_ν, b_ν)-points of (f, g), not counting multiplicities, and set $T(r) = \max\{T(r,f), T(r, g)\}$ with remainder term $S(r)$. Let L and L^* denote the Möbius transformations that map b_ν onto a_ν for $1 \le \nu \le 3$ and $4 \le \nu \le 6$, respectively. Then Nevanlinna's Second Main Theorem yields

$$\begin{aligned}
4T(r) &\le \sum_{\nu=1}^{6} \overline{N}(r; a_\nu, b_\nu) + S(r) \\
&\le \overline{N}\Big(r, \frac{1}{f - L(g)}\Big) + \overline{N}\Big(r, \frac{1}{f - L^*(g)}\Big) + S(r) \\
&\le 2T(r,f) + T(r, L(g)) + T(r, L^*(g)) + S(r) \\
&= 2T(r, g) + 2T(r, g) + S(r).
\end{aligned}$$

This implies $T(r,f) = T(r) + S(r)$, $T(r, g) = T(r) + S(r)$, and

$$\sum_{\nu=1}^{6} \overline{N}(r; a_\nu, b_\nu) = 4T(r) + S(r). \tag{3.37}$$

We now choose $\mathfrak{c} = (c_1, \ldots, c_6) \in \mathbb{C}^6$ non-trivially such that

$$F_6(x, y; \mathfrak{c}) = c_1xy^2 + c_2xy + c_3y^2 + c_4y + c_5x + c_6$$

vanishes at $(x, y) = (a_\nu, b_\nu)$, $1 \le \nu \le 5$. With any such $\mathfrak{c}$ consider

$$\Psi_6(z) = F_6(f(z), g(z); \mathfrak{c}),$$

and first assume $\Psi_6 \not\equiv 0$. Then $T(r, \Psi_6) \le T(r,f) + 2T(r, g) \le 3T(r) + S(r)$ holds, and Ψ_6 vanishes whenever $(f(z), g(z)) = (a_\nu, b_\nu)$, $1 \le \nu \le 5$. This implies

$$\sum_{\nu=1}^{5} \overline{N}(r; a_\nu, b_\nu) \le T(r, \Psi_6) \le 3T(r) + S(r)$$

hence $\overline{N}(r; a_6, b_6) = T(r) + S(r)$ follows from (3.37). Since, however, any pair (a_ν, b_ν) can take the role of (a_6, b_6), we obtain $\overline{N}(r; a_\nu, b_\nu) = T(r) + S(r)$ in contrast to (3.37). This contradiction shows that $\Psi_6 \equiv 0$ (representative for the other functions Ψ_ν) and

$$f(z) = -\frac{c_6 + c_4g(z) + c_3g(z)^2}{c_5 + c_2g(z) + c_1g(z)^2} = M(g(z))$$

holds. Since $T(r,f) = T(r, g) + S(r)$, the representation of M cannot be in lowest terms, hence f is a Möbius transformation of g in contrast to our hypothesis. ☕

Remark 3.10 It is obvious that the same proof allows us to replace the pairs of constants with pairs of small functions, if Möbius transformations over the field generated by the functions $a_1, \ldots, a_6, b_1, \ldots, b_6$ rather than $\mathbb{C}$ are considered. ☕

3.3.8 *Five Pairs*

We will first collect in the form of exercises some basic facts taken from [72, 92][4] about meromorphic functions f and g that share five pairs (a_ν, b_ν). It is assumed that f and g are not Möbius transformations of each other and that the values a_ν and b_ν are finite. If necessary, three of the pairs may be chosen at will. We remind the reader of the special meaning of the word 'almost': almost every zero, pole, etc. means 'up to a subsequence of zeros, poles, etc. with counting function $S(r)$'.

Exercise 3.37 Construct in a non-trivial manner

$$F(x, y; \mathfrak{c}) = c_1x^2y + c_2xy + c_3x^2 + c_4x + c_5y + c_6$$

[4] Cited in [72]. I had no access to [92].

such that $F(a_\nu, b_\nu; \mathfrak{c}) = 0$ $(1 \le \nu \le 5)$ holds. If $F(z) = F(f(z), g(z); \mathfrak{c}) \not\equiv 0$ prove

$$3T(r,f) \le \sum_{\nu=1}^{5} \overline{N}(r; a_\nu, b_\nu) + S(r) \le 2T(r,f) + T(r,g) + S(r).$$

Exercise 3.38 (Continued) $F(z) \equiv 0$ implies $g = M(f)$, where M is rational and $\deg M \le 2$. Prove that M is a Möbius transformation. (**Hint.** Prove that $M^{-1}(\{b_\nu\}) = \{a_\nu\}$ and $\deg M = 2$ is impossible for three pairs (a_ν, b_ν).)

In the same way, $\tilde{F}(x, y; \tilde{\mathfrak{c}}) = \tilde{c}_1 y^2 x + \tilde{c}_2 xy + \tilde{c}_3 y^2 + \tilde{c}_4 y + \tilde{c}_5 x + \tilde{c}_6$ can be constructed. From Exercise 3.38 it follows that neither F nor $\tilde{F}(z) = \tilde{F}(f(z), g(z); \tilde{\mathfrak{c}})$ can vanish identically. This will henceforth be assumed; it implies $T(r,f) = T(r,g) + S(r) = T(r) + S(r)$ and

$$\sum_{\nu=1}^{5} \overline{N}(r; a_\nu, b_\nu) = 3T(r) + S(r). \tag{3.38}$$

In particular, equality holds in the Second Main Theorem, and every value $a \ne a_\nu$ resp. $b \ne b_\nu$ is 'normal' for f resp. g (for example, $\overline{N}(r,f) + \overline{N}(r,g) = 2T(r) + S(r)$).

Exercise 3.39 (Continued) Prove $m(r, 1/F) = S(r)$ and

$$N(r, 1/F) = \overline{N}(r, 1/F) + S(r) = \sum_{\nu=1}^{5} \overline{N}(r; a_\nu, b_\nu) + S(r),$$

hence $T(r, F) = 3T(r) + S(r)$, and the same for $\tilde{F}$. Thus almost every zero of F and $\tilde{F}$ is simple and is also a zero of $f - a_\nu$ and $g - b_\nu$ for some ν, and *vice versa*.

Exercise 3.40 (Continued) Given any three pairs (a_κ, b_κ), (a_λ, b_λ), and (a_μ, b_μ), choose (d_1, d_2, d_3, d_4) non-trivially such that $H_{\kappa\lambda\mu}(x, y) = d_1 xy + d_2 x + d_2 y + d_4$ vanishes whenever (x, y) equals one of the distinguished pairs; $H_{\kappa\lambda\mu}(f(z), g(z)) \equiv 0$ is excluded since f and g are not Möbius transformations of each other. Prove that

$$H(z) = \left.\frac{H_{234}(x,y) H_{235}(x,y) H_{245}(x,y)}{F(x,y)\tilde{F}(x,y)}\right|_{x=f(z), y=g(z)}$$

vanishes at almost every zero of $f - a_2$ and $g - b_2$ and has poles of order at most two at almost every zero of $f - a_1$ and $g - b_1$. Prove also that there are almost no other poles since almost all poles of the numerator and zeros of the denominator, with the exception of the zeros of $(f - a_1)(g - b_1)$, are cancelled by the poles of the denominator and the zeros of the numerator, respectively. Also H satisfies $m(r, H) = S(r)$ by Exercise 3.40, hence $T(r, H) = N(r, H) + S(r) \le 2\overline{N}(r; a_1, b_1) + S(r)$.

Since $\mu = 1$ and $\nu = 2$ are only place markers,

$$\overline{N}(r; a_\nu, b_\nu) \leq 2\overline{N}(r; a_\mu, b_\mu) + S(r) \quad \text{and} \quad \overline{N}(r; a_\nu, b_\nu) \geq \tfrac{1}{3}T(r) + S(r) \tag{3.39}$$

follows from (3.38). In particular, none of the values a_ν and b_ν can be exceptional.

3.3.9 Gundersen's Example Rediscovered

The paper [183] contains yet another characterisation of Gundersen's example (Exercise 3.21). This characterisation uses Reinders' Theorem 3.12.

Theorem 3.19 *Suppose meromorphic functions f and g share four pairs (a_ν, b_ν), and a fifth pair (a_5, b_5) counting multiplicities such that*

$$m\big(r, 1/(f - a_5)\big) + m\big(r, 1/(g - b_5)\big) = S(r) \tag{3.40}$$

holds. Then there exists some Möbius transformation M such that either $f = M \circ g$ holds or else f and $M \circ g$ share four values and one pair.

Proof The proof is based on the technique developed in [31, 72]. We assume that f and g are not Möbius transformations of each other, and note that (3.40) is automatically fulfilled if $a_\nu = b_\nu$, $1 \leq \nu \leq 4$. Three of the pairs (a_ν, b_ν) may be prescribed, we will assume $(a_1, b_1) = (0, 0)$, $(a_2, b_2) = (2, -4)$, and, in particular, $a_5 = b_5 = \infty$. Then f and g have the same poles counting multiplicities, and (3.40) takes the form

$$m(r, f) + m(r, g) = S(r). \tag{3.41}$$

We note that almost all poles of f and g are simple since f and g have almost no critical points in common.

Exercise 3.41 Prove that there are at least two linearly independent vectors $\mathfrak{c} \in \mathbb{C}^5$ such that $P(x, y; \mathfrak{c}) = c_1x^2 + c_2xy + c_3y^2 + c_4x + c_5y$ satisfies

$$P(a_\nu, b_\nu; \mathfrak{c}) = 0 \quad (1 \leq \nu \leq 4). \tag{3.42}$$

Assuming $P(z; \mathfrak{c}) = P(f(z), g(z); \mathfrak{c}) \not\equiv 0$, deduce that

$$\sum_{\nu=1}^{4} \overline{N}(r; a_\nu, b_\nu) = 2T(r) + S(r) \quad \text{and} \quad T(r) = \overline{N}(r; \infty) + S(r).$$

(**Hint.** Apply the Second Main Theorem to $(f, a_1, \ldots, a_4)$ and $(g, b_1, \ldots, b_4)$.) ☕

Still assuming $P \not\equiv 0$ it follows from equality in the Second Main Theorem that $N(r, 1/P) = \overline{N}(r, 1/P) + S(r)$ and $m(r, 1/P) = S(r)$ holds. In particular, the ratio $\chi(z) = P(z; \tilde{\mathfrak{c}})/P(z; \mathfrak{c})$ has Nevanlinna characteristic $T(r, \chi) = S(r)$ and f and g parametrise the algebraic curve

$$F(x, y; z) = \chi_1 x^2 + \chi_2 xy + \chi_3 y^2 + \chi_4 x + \chi_5 y = 0 \quad (\chi_k = \chi c_k - \tilde{c}_k) \tag{3.43}$$

over the field $\mathbb{C}(\chi)$. This is also true if $P(z; \mathfrak{c})$ or $P(z; \tilde{\mathfrak{c}})$ vanishes identically.

Exercise 3.42 Prove that $\chi_1\chi_3 \not\equiv 0$. (**Hint.** For $\chi_3 \equiv 0$, say, g would be a Möbius transformation or a rational function of f of degree two over the field $\mathbb{C}(\chi)$, the latter contradicting $T(r, g) = T(r, f) + S(r)$. And the former?)

The algebraic curve (3.43) has the rational parametrisation (set $x = ty$)

$$x = \frac{p(z,t)}{s(z,t)} = -\frac{t(\chi_4 t + \chi_5)}{\chi_1 t^2 + \chi_2 t + \chi_3},\ y = \frac{q(z,t)}{s(z,t)} = -\frac{\chi_4 t + \chi_5}{\chi_1 t^2 + \chi_2 t + \chi_3} \quad (\chi_\nu = \chi_\nu(z))$$

over $\mathbb{C}(\chi)$. In terms of f and g this yields

$$\begin{aligned} f(z) &= \frac{p(z,\mathrm{t}(z))}{s(z,\mathrm{t}(z))} = -\frac{\mathrm{t}(z)(\chi_4(z)\mathrm{t}(z) + \chi_5(z))}{\chi_1(z)\mathrm{t}(z)^2 + \chi_2(z)\mathrm{t}(z) + \chi_3(z)} \\ g(z) &= \frac{q(z,\mathrm{t}(z))}{s(z,\mathrm{t}(z))} = -\frac{\chi_4(z)\mathrm{t}(z) + \chi_5(z)}{\chi_1(z)\mathrm{t}(z)^2 + \chi_2(z)\mathrm{t}(z) + \chi_3(z)} \end{aligned} \quad \text{with} \quad \mathrm{t}(z) = \frac{f(z)}{g(z)}.$$

From (3.39) with $a_1 = b_1 = 0$ it follows that f and g have 'many' common zeros. There are three possibilities to be discussed: the common zeros correspond to the

- *poles of* t, in which case $\chi_4 \equiv 0$ and almost all zeros of f are simple, while the zeros of g have order two. Moreover, t has almost no zeros ($N(r, 1/\mathrm{t}) = S(r)$).
- *zeros of* t, in which case $\chi_5 \equiv 0$ and almost all zeros of g are simple, while the zeros of f have order two. Moreover, t has almost no poles ($N(r, \mathrm{t}) = S(r)$).
- *zeros of* $\chi_4(z)\mathrm{t}(z) + \chi_5(z)$ with $\chi_4\chi_5 \not\equiv 0$. Then almost all zeros of f and g are simple, and t has almost no zeros and poles ($N(r, 1/\mathrm{t}) + N(r, \mathrm{t}) = S(r)$).

Taking all pairs (a_ν, b_ν) $(1 \le \nu \le 4)$ into account, the following holds: for every ν there exist $\phi_\nu, \psi_\nu, \alpha_\nu, \beta_\nu, \tilde{\beta}_\nu \in \mathbb{C}(\chi)$‖ such that

$$\begin{aligned} p(z,t) - a_\nu s(z,t) &= \phi_\nu(t - \alpha_\nu)(t - \beta_\nu) \\ q(z,t) - b_\nu s(z,t) &= \psi_\nu(t - \alpha_\nu)(t - \tilde{\beta}_\nu) \end{aligned} \qquad (\beta_\nu \not\equiv \tilde{\beta}_\nu)$$

‖At first glance one would expect that $\alpha_\nu, \beta_\nu, \tilde{\beta}_\nu$ are algebraic over $\mathbb{C}(\chi)$. But this is not the case, since an analytic continuation which permutes α_ν and β_ν would also permute α_ν and $\tilde{\beta}_\nu$, in contrast to $\beta_\nu \not\equiv \tilde{\beta}_\nu$.

holds; occasionally the factor $(t-\beta_\nu)$ and $(t-\tilde\beta_\nu)$ corresponding to $\beta_\nu\equiv\infty$ and $\tilde\beta_\nu\equiv\infty$, respectively, might be missing. The functions α_ν are mutually distinct, and the same is true for β_ν and $\tilde\beta_\nu$. Also β_ν and $\tilde\beta_\nu$ are exceptional for t, unless one of them coincides with α_ν. Since t has at most two exceptional functions, we obtain the following picture: for $\nu=1$ and $\nu=4$, say, we have $\beta_\nu\equiv\alpha_\nu$, that is, the pairs (a_ν,b_ν) are attained by (f,g) in a $(2:1)$ manner, while for $\nu=2$ and $\nu=3$ this happens the other way $(1:2)$. This means that, in addition to (3.43), we also have

$$F_y(a_\nu,b_\nu;z)\equiv 0\quad(\nu=1,4)\quad\text{and}\quad F_x(a_\nu,b_\nu;z)\equiv 0\quad(\nu=2,3).\qquad(3.44)$$

Exercise 3.43 Assume $\chi_3\equiv 1$ (this is possible since $\chi_3\not\equiv 0$ is already known). From (3.44), that is, $F_y(0,0)=F_x(2,-4)=F_x(a_3,b_3)=F_y(a_4,b_4)=0$ deduce

$$\chi_1=\frac{b_4(b_3+4)}{a_4(a_3-2)},\ \chi_2=-\frac{2b_4}{a_4},\ \chi_3=1,\ \chi_4=\frac{4b_4(b_3+2a_3)}{a_4(2-a_3)},\ \chi_5=0.\qquad(3.45)$$

In particular, the functions χ_k are constant, and f and g are ordinary rational functions of the meromorphic function $\mathsf{t}=f/g$.

Having determined the coefficients (3.45) we will now solve the nonlinear system

$$\frac{b_4(b_3+4)}{a_4(a_3-2)}a_\nu^2-\frac{2b_4}{a_4}a_\nu b_\nu+b_\nu^2+\frac{4b_4(b_3+2a_3)}{a_4(2-a_3)}b_\nu=0\quad(2\le\nu\le 4)\qquad(3.46)$$

obtained from $F(a_\nu,b_\nu;z)=0$ with $a_2=2$, $b_2=-4$. The first solution $b_3=-2a_3$, $b_4=-2a_4$ leads to $g=-2f$ against our hypothesis, while the second solution, again obtained with the help of MAPLE, may be expressed in terms of a_4 as follows:

$$a_1=b_1=0,\ a_2=2,\ b_2=-4,\ a_3=a_4-2,\ b_3=2a_4-4,\ b_4=2a_4-8.$$

Since $(0,2,a_4-2,a_4)=(0,-4,2a_4-4,2a_4-8)=\frac{(a_4-2)^2}{a_4(a_4-4)}$ (cross-ratio), there exists some Möbius transformation M such that $M(b_\nu)=a_\nu$ $(1\le\nu\le 4)$, hence f and $M\circ g$ share four *values* a_1,a_2,a_3,a_4 and the pair $(\infty,M(\infty))$, and this finishes the proof of Theorem 3.19.

Remark 3.11 It is open whether or not—and how—the hypothesis (3.40) may be relaxed. Is it sufficient to assume that the pair (a_5,b_5) is shared counting multiplicities? Is it even true that functions sharing five pairs are either Möbius transformations of each other or else are conjugate to the functions in Gundersen's Example? Recently [75] it was shown that there cannot exist meromorphic functions that share five pairs, two of them counting multiplicities, unless the functions are Möbius transformations of each other. Again it could be useful to consider algebraically dependent pairs (f,g) sharing five pairs (a_ν,b_ν).

Chapter 4
Normal Families

In this chapter the theory of Normal Families will be deepened and enlarged, and applied to various problems in the fields of entire and meromorphic functions, distribution of zeros of differential polynomials, ordinary differential equations and functional equations. The Yosida classes, which play an outstanding part in the theory of algebraic differential equations, will be introduced and discussed in the final part of the chapter.

4.1 Re-scaling

Re-scaling means any transformation $f_n \to \tilde{f}_n(z) = a_n f_n(b_n + c_n z)$ that converts a given non-normal sequence (f_n) into a normal sequence $(\tilde{f}_n)$. As an example, we consider the sequence $f_n(z) = z^n$, which is not normal at any point z_0 on the unit circle. Replacing z with $z_0 + \frac{1}{n}z$ and multiplying by z_0^{-n} yields a new sequence $\tilde{f}_n(z) = z_0^{-n} f_n\big(z_0 + \frac{z}{n}\big) = \big(1 + \frac{z/z_0}{n}\big)^n$, which converges to $\tilde{f}(z) = e^{z/z_0}$ with spherical derivative $\tilde{f}^\sharp(z) \le \tilde{f}^\sharp(0)$.

4.1.1 Zalcman's Lemma

What happened in our simple example happens *mutatis mutandis* in every family $\mathscr{F}$ of meromorphic functions that is *not normal* at some point. Written down formally this leads to the famous *Zalcman Re-scaling Lemma* [212, 213], which had and still has an enormous impact on many parts of Complex Analysis.

Theorem 4.1 *Suppose that the family $\mathscr{F}$ of meromorphic functions is not normal at some point z_0 of the common domain of definition. Then there exist sequences*

N. Steinmetz, *Nevanlinna Theory, Normal Families, and Algebraic Differential Equations*, Universitext, DOI 10.1007/978-3-319-59800-0_4

$f_k \in \mathscr{F}$, $z_k \to z_0$, and $\rho_k = 1/f_k^\sharp(z_k) \to 0$ such that the re-scaled sequence

$$g_k(z) = f_k(z_k + \rho_k z)$$

tends to some non-constant meromorphic function satisfying $g^\sharp(z) \le g^\sharp(0) = 1$.

Proof To simplify matters we assume that $z_0 = 0$ and all functions under consideration are meromorphic on $|z| \le 1$. Then $\mathscr{F}$ contains functions f_n satisfying $\sup_{|z|<1/(2n)} f_n^\sharp(z) \ge 2n^3$, hence there exists a ζ_n with $|\zeta_n| < 1/n$ such that

$$\max_{|z|\le 1/n} (1 - |nz|) f_n^\sharp(z) = (1 - |n\zeta_n|) f_n^\sharp(\zeta_n) \ge n^3.$$

Set $r_n = 1/f_n^\sharp(\zeta_n)$ and $h_n(z) = f_n(\zeta_n + r_n z)$. We claim that appropriately chosen subsequences $z_k = \zeta_{n_k}$, $\rho_k = r_{n_k}$, and $g_k = h_{n_k}$ will do. First of all, h_n is defined on $|z| < (1 - |\zeta_n|)/r_n = (1 - |\zeta_n|) f_n^\sharp(\zeta_n)$, hence on $|z| < n^3$, and

$$h_n^\sharp(z) = \frac{f_n^\sharp(\zeta_n + r_n z)}{f_n^\sharp(\zeta_n)} \le \frac{1 - n|\zeta_n|}{1 - n|\zeta_n| - nr_n|z|} \le \frac{1}{1 - 1/n} \quad (|z| < n)$$

follows from $nr_n \le \frac{1}{n^2}(1 - n|\zeta_n|)$. For every $m \in \mathbb{N}$ the sequence $(h_n)_{n>m}$ is normal on $|z| < m$, say, and the well-known Cantor diagonal process yields a subsequence $(g_k) = (h_{n_k})$ which converges, uniformly on every disc $|z| < R$. The limit function g satisfies $g^\sharp(z) \le \limsup_{n\to\infty} h_n^\sharp(z) \le 1 = g^\sharp(0)$.

4.1.2 Pang's Lemma

Zalcman's Lemma was generalised by Pang [136, 137] in a way which makes it more flexible, and, in particular, applicable to algebraic differential equations.

Theorem 4.2 *The statement of* Theorem 4.1 *remains valid if the sequence $g_k(z) = f_k(z_k + \rho_k z)$ is replaced with*

$$g_k(z) = \rho_k^\alpha f_k(z_k + \rho_k z) \quad (-1 < \alpha < 1 \text{ arbitrary}).$$

We note several important supplements to Theorem 4.2.

1. If the zeros and poles of the functions $f \in \mathscr{F}$ have order at least m and n, respectively, then α may vary in $-m < \alpha < n$.
2. In particular, if f has no zeros [$m = \infty$] resp. poles [$n = \infty$], then every $\alpha < n$ resp. $\alpha > -m$ is admitted.

3. If all zeros have multiplicity at least m, and $|f^{(m)}(\zeta)| \le K$ holds for every $f \in \mathscr{F}$ and every zero of f, then even $\alpha = -m$ is admitted.
4. $\alpha = n$ is admitted if $\lim_{z\to\zeta} |f(z)(z-\zeta)^n| \ge 1/K$ holds at every pole of every $f \in \mathscr{F}$.

Remark 4.1 The refinements 3. and 4. are due to Chen and Gu [23]. In the proof of Theorem 4.2 the expression

$$\frac{(1-|z|)^{1+\alpha}|f'(z)|}{(1-|z|)^{2\alpha}+|f(z)|^2}$$

takes the part of $(1-|z|)f^\sharp(z)$. The closely related expression

$$f^{\sharp_\alpha}(z) = \frac{|z|^\alpha |f'(z)|}{|z|^{2\alpha}+|f(z)|^2}$$

will play a crucial role in the context of Yosida functions, see Sect. 4.3.

Example 4.1 The solutions to Painlevé's first equation $w'' = z + 6w^2$ have double poles with principal part $(z-\zeta)^{-2}$, hence $-1 < \alpha \le 2$ is admitted. It will turn out that the right choice is $\alpha = 2$ in the following sense: suppose $z_k \to \infty$ and $w(z_k+z)$ is not normal at $z = 0$. Then some subsequence of $\rho_k^2 w(z_k + \rho_k \mathfrak{z})$ with $\rho_k = z_k^{-\frac{1}{4}}$ (note that ρ_k need not be positive) tends to some limit function $\mathfrak{w}$ which satisfies $\mathfrak{w}'' = 1 + 6\mathfrak{w}^2$. Choosing $\rho_k = z_k^{-\frac{1}{4}}$ is admitted since $w^{\sharp_{\frac{1}{2}}}(z) = O(|z|^{\frac{1}{4}})$ holds (a non-trivial fact). More on this kind of re-scaling in Sect. 4.3, Chaps. 5 and 6.

4.1.3 Functions of Poincaré, Abel, and Zalcman

Nothing can be said about the ratio $|z_k - z_0|/|\rho_k|$ in Zalcman's Lemma. The following examples deal with the extremal cases $z_k = z_0$ on one hand, and $\rho_k = o(|z_k - z_0|)$ on the other.

Poincaré Functions Suppose R is a rational function of degree $d > 1$ with *repelling fixed-point* at $z = z_0$, that is, we assume that

$$R(z) = z_0 + \lambda(z - z_0) + a_2(z - z_0)^2 + \cdots \quad (|\lambda| > 1)$$

holds on some neighbourhood of z_0. Then the sequence of *iterates* $R^n = R \circ R^{n-1}$ is not normal at z_0 since $R^n(z) = z_0 + \lambda^n(z - z_0) + \cdots$. By Zalcman's Lemma there exist sequences $\rho_k \to 0$, $z_k \to z_0$, and $n_k \uparrow \infty$ such that $R^{n_k}(z_k + \rho_k z)$ tends to some non-constant meromorphic function f satisfying $f^\sharp(z) \le f^\sharp(0) = 1$. If we dispense with the latter condition, it is possible to choose $n_k = k$, $z_k = z_0$, and $\rho_k = \lambda^{-k}$.

Assuming this for the moment we obtain

$$f(\lambda z) = R(f(z)) \quad \text{with } f(z) = z_0 + z + c_2 z^2 + \cdots \text{ at } z = 0 \tag{4.1}$$

from $R^{n+1}(z_0 + \lambda^{-(n+1)}(\lambda z)) = R(R^n(z_0 + \lambda^{-n}z))$; f is called a *Poincaré function* and solves *Schröder's functional equation* (4.1). Assuming that f exists on $|z| < r$, it may be analytically continued into the discs $|z| < |\lambda|r$, $|z| < |\lambda|^2 r$, etc., hence into the whole plane by applying $f(z) = R(f(z/\lambda))$ successively. Moreover,

$$f(z) = R^n(f(z/\lambda^n)) \sim R^n(z_0 + \lambda^{-n}z) \quad \text{and} \quad f(z) = \lim_{n\to\infty} R^n(z_0 + \lambda^{-n}z)$$

holds, locally uniformly on $\mathbb{C}$. To prove local existence it is convenient to consider the local inverse $\Phi(z) = R^{-1}(z) = \lambda^{-1}z + \cdots$ (we assume $z_0 = 0$), solve

$$F(\Phi(z)) = \lambda^{-1}F(z) \quad \text{with} \quad F(z) = z + b_2 z^2 + \cdots \text{ at } z = 0,$$

and set $f = F^{-1}$ locally.

Exercise 4.1 The local existence of F can be proved as follows ($\mu = \lambda^{-1}$).

- Choose $D = \{z : |z| < r\}$ such that $|\Phi(z)| \le |\mu|^{2/3}|z|$ holds on D (why is this possible?) and prove that $|\Phi(z) - \mu z| \le \frac{2}{r}|z|^2$.
  (**Hint.** Apply the Maximum Principle to $(\Phi(z) - \mu z)/z^2$.)
- Prove that the iterates Φ^n are well-defined and satisfy $|\Phi^n(z)| \le |\mu|^{2n/3}|z|$ on D, and deduce $|F_{n+1}(z) - F_n(z)| \le \frac{2}{r}\frac{|\Phi^n(z)|^2}{|\mu|^{n+1}} \le \frac{2r}{|\mu|}|\mu|^{n/3}$ ($F_n = \Phi^n/\mu^n$).
- Prove that $F_n \to F$, uniformly on D, and conclude that $F \circ \Phi = \mu F$.

Theorem 4.3 ([41, 177]) *The Poincaré functions are meromorphic on the plane with Nevanlinna characteristic*

$$T(r,f) \asymp r^{\varrho} \quad \Big(\varrho = \frac{\log \deg R}{\log |\lambda|}\Big).$$

Proof Set $T(r) = T(r,f)$, $q = |\lambda|$, and $d = \deg R$, and observe that Valiron's Lemma yields $T(qr) = dT(r) + \sigma(r)$, where σ is bounded. Iterating gives

$$T(q^n r) = d^n T(r) + d^{n-1}\sigma(r) + d^{n-2}\sigma(qr) + \cdots + \sigma(q^{n-1}r) = d^n(T(r) + O(1))$$

as $n \to \infty$, uniformly with respect to r. Set $s = q^n r$ and restrict r to $r_0 < r \le qr_0$ to obtain $0 < C_1 \le T(s)s^{-\varrho} \le C_2$ if r_0 is chosen sufficiently large.

Exercise 4.2 Determine the Poincaré function in the following cases

1. $R(z) = z^2$, $z_0 = 1$, $\lambda = 2$.
2. $R(z) = 2z^2 - 1$ (the second Chebychev polynomial), $z_0 = 1$, $\lambda = 4$.
3. $R(z) = \dfrac{2z}{1-z^2}$, $z_0 = 0$, $\lambda = 2$.

Solution 1. e^z, 2. $\cos\sqrt{-2z}$, 3. $\tan z$.

Example 4.2 Let f be any Poincaré function and suppose that $a \in \mathbb{C}$ has d^n mutually distinct pre-images a_ν under R^n. Then $f(\lambda^n z) - a = R^n(f(z)) - a$ implies

$$\begin{aligned} m\Big(|\lambda|^n r, \frac{1}{f-a}\Big) &= \sum_{\nu=1}^{d^n} m\Big(r, \frac{1}{f-a_\nu}\Big) + O(1) \\ &\le 2T(r,f) + O(\log r) = 2d^{-n}T(|\lambda|^n r, f) + O(\log r) \end{aligned}$$

by the Second Main Theorem. If this is true for every n, then

$$m\Big(r, \frac{1}{f-a}\Big) = o(T(r,f)) \tag{4.2}$$

holds (note that the $O(\log r)$-term depends on n). We note without proof that (4.2) even holds if $R^{-1}(\{a\}) \ne \{a\}$ (see [41, 177]). On the other hand, a is a Picard value for f if $R^{-1}(\{a\}) = \{a\}$. In particular, f is entire if R is a polynomial.

Example 4.3 The Weierstraß P-function has the duplication formula

$$\wp(2z) = \frac{(6\wp^2(z) - g_2/2)^2}{4(4\wp^3(z) - g_2\wp(z) - g_3)} - 2\wp(z) = R(\wp(z));$$

R has a repelling fixed-point at $z = \infty$ with 'multiplier' $\lambda = 4$ (this is defined by 'conjugation': $R_1(\zeta) = 1/R(1/\zeta)$ has a repelling fixed-point at $\zeta = 0$ with $\lambda = R_1'(0) = 4$). The corresponding Poincaré function, this time normalised by $f(z) = 1/z + \cdots$, is $f(z) = \wp(\sqrt{z})$:

$$f(4z) = \wp(\sqrt{4z}) = \wp(2\sqrt{z}) = R(\wp(\sqrt{z})) = R(f(z)).$$

Abel Functions The polynomial $P(z) = z + z^d$ of degree $d \ge 2$ has a *parabolic* fixed-point at the origin ($P(0) = 0$, $P'(0) = 1$), and the sequence P^n is not normal at $z = 0$ since $P^n(z) = z + nz^d + \cdots$. We assume $P^{n_k}(z_k + \rho_k z) \to f(z) \not\equiv \text{const}$, $z_k \to 0$, $\rho_k \to 0$, and $z_k\rho_k \ne 0$, and claim that $\rho_k = o(|z_k|^d)$. Otherwise we may assume $z_k^d = b_k\rho_k$ with $b_k \to b \in \mathbb{C}$. Then $P(z_k + \rho_k z) = z_k + \rho_k(z + b + o(1))$ and

$$P(f(z)) \sim P(P^{n_k}(z_k + \rho_k z)) = P^{n_k}(P(z_k + \rho_k z)) \sim P^{n_k}(z_k + \rho_k(z + b)) \sim f(z + b)$$

holds, uniformly on every disc $|z| < R$, and f satisfies $f(z+b) = P(f(z))$. Obviously $b = 0$ is impossible by Valiron's Lemma, hence we may assume $b = 1$, that is, f solves *Abel's functional equation*

$$f(z+1) = P(f(z)); \tag{4.3}$$

f has order of growth at most 1 since f is entire with bounded spherical derivative $f^\sharp$, see Theorem 4.8 below. This, however, is impossible, since any non-constant solution to Abel's equation has infinite order of growth (see Yanagihara [207] and the exercise below). This example also shows how sensitively the limit function may depend on the sequence (z_k): replacing z_k with $z_k + z_k^d$ while ρ_k remains unaltered leads to the limit function $P \circ f$.

Example 4.4 Let f be any non-constant solution to Abel's equation (4.3), where P is any nonlinear rational function with parabolic fixed-point at $z = 0$; global existence of f is assumed. From Cartan's Identity and Valiron's Lemma it easily follows that $T(r+1,f) \geq T(r, P \circ f) + O(1) = dT(r,f) + O(1)$, hence

$$T(r+n,f) \geq d^n[T(r,f) + O(1)] \quad (n \to \infty)$$

holds, uniformly w.r.t. r; this yields $T(r,f) \geq cd^r$ $(r \geq r_0)$ for some $c > 0$.

Zalcman Functions A rich class of meromorphic functions is obtained by a procedure similar to the construction of the Poincaré and Abel functions. Let R be any rational function of degree $d > 1$ and denote by (R^n) the sequence of its iterates. If (R^n) is not normal at z_0, that is, if z_0 is any point in the *Julia set* of R, Zalcman's Lemma yields meromorphic functions

$$f(z) = \lim_{k\to\infty} R^{n_k}(z_k + \rho_k z) \quad (z_k \to z_0,\ \rho_k \to 0), \tag{4.4}$$

which may be viewed as generalisations of the Poincaré and Abel functions; the term *Zalcman function* was coined in [177]. To describe the most interesting facts concerning the value distribution of these functions we need two concepts from iteration theory.

- A *critical point* is a zero of R' or a multiple pole of R. The critical points generate the *critical orbit* $C^+ = \{R^n(\zeta) : \zeta \text{ critical point},\ n \in \mathbb{N}\}$.
- The *exceptional set* $E = E(R)$ of R is the largest finite set such that $R^{-1}(E) \subset E$; it consists of at most two points and is 'usually' empty:

  card $E = 1$: R is conjugate to some polynomial;
  card $E = 2$: R is conjugate to $z \mapsto z^d$ or $z \mapsto z^{-d}$.

Theorem 4.4 *Zalcman functions for any rational function R and any z_0 in the Julia set of R have the following properties.*

- $m\Big(r, \dfrac{1}{f-a}\Big) = o(T(r,f))$ *if* $a \notin E$.
- $a \in E$ *is a Picard value of* f.
- $\vartheta(a,f)$ *is a positive rational number if* $a \in C^+ \setminus E$, *and* $\vartheta(a,f) = 0$ *otherwise.*
- $\sum_{a\in C^+\setminus E} \vartheta(a,f) + \sum_{a\in E} \delta(a,f) = \sum_{a\in\widehat{\mathbb{C}}} \Theta(a,f) = 2.$

For a proof and more results on Zalcman functions, see [177].

Example 4.5 Let f denote any Zalcman function for (R, z_0).

- $R(z) = z^2 + i$: f is entire and $\vartheta(i,f) = \frac{1}{2}$, $\vartheta(-1+i,f) = \frac{1}{3}$, $\vartheta(-i,f) = \frac{1}{6}$.
- $R(z) = z + z^2$: f is entire and $\vartheta(R^n(-\frac{1}{2}),f) = 2^{-n}$ $(n \in \mathbb{N})$.
- $R(z) = \frac{i}{2}(z - \frac{1}{z})$: $\vartheta(a,f) = \frac{1}{2}$ for $a = -1, 1, 0, \infty$,
  with critical orbit: $\pm i \mapsto \pm 1 \mapsto 0 \mapsto \infty = R(\infty)$.

4.1.4 The Theorems of Picard and Montel

The close relation between Picard- and Montel-type theorems, in other words between the qualitative theory of entire and meromorphic functions on one hand, and the theory of normal families of holomorphic and meromorphic functions on the other, has been known for a long time. Zalcman's Lemma perhaps provides the most elegant approach to both. We start with a proof of Picard's first or 'little' theorem.

Theorem 4.5 *Every transcendental meromorphic function assumes every value on the Riemann sphere with at most two exceptions.*

Proof (Ros [145]) Let f be any transcendental meromorphic function that omits three values, which may be assumed to be 0, 1, and ∞. For technical reasons it is required that $f'(0) \neq 0$, which is achieved by considering $f(z_0 + z)$ with $f'(z_0) \neq 0$ instead of f without changing notation. We consider the sequence (f_n) of entire functions $f_n(z) = \sqrt[2^n]{f(2^n z)}$, where $f_n(0)$ is any of the values $\sqrt[2^n]{f(0)}$. Then f_n omits the values 0 and $e^{2\nu\pi i/2^n}$ $(0 \le \nu < 2^n)$. If we assume that the sequence (f_n) is normal on the plane, each convergent subsequence (f_{n_k}) tends to some entire function g, which by Hurwitz' Theorem omits every value $e^{2\nu\pi i/2^n}$ $(0 \le \nu < 2^n,\ n \in \mathbb{N})$, hence omits every value on the circle $|w| = 1$. This implies that either $|g(z)| \le 1$ or else $|g(z)| \ge 1$ holds on $\mathbb{C}$, hence g is constant by Liouville's Theorem. This, however, is excluded by $g'(0)/g(0) = f'(0)/f(0) \neq 0$. On the other hand, if (f_n) is not normal at some point z_0, Zalcman's Lemma yields some non-constant entire limit function $g(z) = \lim_{k\to\infty} f_{n_k}(z_k + \rho_k z)$, which again omits the values $e^{2\nu\pi i/2^n}$, and the same contradiction is obtained.

In combination with Zalcman's Lemma, Picard's Theorem immediately leads to Montel's Second Normality Criterion.

Theorem 4.6 *The family $\mathscr{F}$ of meromorphic functions f omitting a fixed triple of values is normal on D.*

Proof If $\mathscr{F}$ is assumed to be non-normal at some point $z_0 \in D$, Zalcman's Lemma applies, yielding some non-constant meromorphic function that omits three values by Hurwitz' Theorem. This, however, contradicts Picard's Theorem.

Exercise 4.3 The omitted values a, b, c in Theorem 4.6 are assumed to be the same for every $f \in \mathscr{F}$. They may, however, also vary with f. Suppose that each $f \in \mathscr{F}$ omits three values a_f, b_f, c_f. Prove that $\mathscr{F}$ is normal, provided

$$\chi(a_f, b_f)\chi(b_f, c_f)\chi(c_f, a_f) \geq \delta > 0.$$

(**Hint.** Consider the family of cross-ratios (f, a_f, b_f, c_f) or use the Bolzano–Weierstraß Theorem.)

The step from Montel's Second Criterion to Picard's 'Great Theorem' is classical.

Theorem 4.7 *Let f be meromorphic on $0 < |z - z_0| < r$ with essential singularity at $z = z_0$. Then on every punctured neighbourhood of z_0, f assumes every value with at most two exceptions infinitely often.*

Proof Suppose that f takes on three values (0, 1, and ∞, say) only finitely often on $0 < |z| < \delta_0$ ($z_0 = 0$ is tacitly assumed), hence omits these values on some punctured disc $0 < |z| < \delta < \delta_0$. Then the sequence of functions $f_n(z) = f(2^{-n}z)$ $(2^{-n} < \delta)$ is normal on $0 < |z| < 1$, and we may assume that (f_{n_k}) converges to some limit g. If g is finite it is bounded by M on $|z| = 1/2$, thus f is bounded by $2M$, say, on the circles $|z| = 2^{-n_k-1}$. By the Maximum Principle, f is bounded by $2M$ on the annuli $2^{-n_{k+1}-1} < |z| < 2^{-n_k-1}$, hence on $0 < |z| < 2^{-n_1-1}$, which contradicts the fact that $z_0 = 0$ is an essential singularity. If, however, (f_{n_k}) converges to ∞, the same argument applies to the sequence $(1/f_{n_k})$ and limit function $g = 0$.

4.1.5 Normal Functions

Suppose f is meromorphic on the unit disc $\mathbb{D}$. Then the functions $f \circ M$, where $M(z) = \frac{ze^{i\alpha}+a}{1+\bar{a}ze^{i\alpha}}$ denotes any Möbius transformation mapping $\mathbb{D}$ onto itself, are also meromorphic on $\mathbb{D}$. They form the family $\mathscr{F}_f$, and f is called *normal* if the family $\mathscr{F}_f$ is normal. The term 'normal function' was coined by Lehto and Virtanen [108].[1]

[1]Normal functions (without the name) were considered much earlier by Noshiro [130].

An equivalent statement is

$$\sup_{|z|<1}(1-|z|^2)f^\sharp(z) < \infty.$$

Proof Normality of $\mathscr{F}_f$ at $z = 0$ implies that $(f \circ M)^\sharp(0) = (1 - |a|^2)f^\sharp(a)$ is bounded independent of a. On the other hand, $(1 - |\zeta|^2)f^\sharp(\zeta) \le \alpha$ implies

$$(f \circ M)^\sharp(z) = f^\sharp(M(z))|M'(z)| = f^\sharp(M(z))\frac{1 - |M(z)|^2}{1 - |z|^2} \le \frac{\alpha}{1 - |z|^2}$$

by the Schwarz–Pick Lemma. ☕

Exercise 4.4 The re-scaling method had already been used by Lohwater and Pommerenke [109] in the context of normal functions. Prove that if f is not normal there exist sequences $\rho_n \to 0$ and $z_n \to z_0$ on $\partial\mathbb{D}$ such that $f(z_n + \rho_n(1 - |z_n|^2)z)$ tends to some non-constant meromorphic function on the complex plane.
(**Hint.** Start with any sequence (ζ_n) such that $(1 - |\zeta_n|)f^\sharp(\zeta_n) \to \infty$ and consider $f_n(z) = f\Big(\dfrac{\zeta_n + z}{1 + \bar\zeta_n z}\Big)$; note that $\dfrac{\zeta_n + z}{1 + \bar\zeta_n z} = \zeta_n + (1 - |\zeta_n|^2)z + O(|z|^2)$ as $z \to 0$.) ☕

Meromorphic functions with bounded spherical derivative have order of growth at most two. For entire functions the true bound is one. This can be deduced from the inequality $(1 - |z|^2)|f'(z)| \le 2(\log^+|f(z)| + \alpha)\max\{|f(z)|, 1\}$ due to Pommerenke [141], which holds for α-normal holomorphic functions. Following Eremenko [36] we will prove

Theorem 4.8 *Entire functions satisfying $f^\sharp(z) \le \alpha$ also satisfy*

$$|f'(z)| \le 2\alpha \max\{1, |f(z)|\} \quad \text{and} \quad |f(z)| < (1 + |f(0)|)e^{2\alpha|z|}.$$

Proof Since $g(z) = f(z/\alpha)$ satisfies $g^\sharp(z) \le 1$, it suffices to handle the case $\alpha = 1$. Let E denote the set where $|f(z)| > 1$, and consider

$$u(z) = \frac{|f'(z)|}{|f(z)|(R + \log|f(z)|)} \quad (z \in E,\ R > 0 \text{ arbitrary}).$$

Then $|f'(z)| \le 2$ and $u(z) \le 2/R$ hold on $\mathbb{C} \cap \partial E$, while $\Delta \log u = u^2$ holds on $E \setminus \{\text{zeros of } f'\}$.[2] We fix $z_0 \in E$, set $E_{z_0} = \{z : |z - z_0| < R\}$,

$$v(z) = \frac{2R}{R^2 - |z - z_0|^2} \quad (\Delta \log v = v^2 \text{ on } E_{z_0}),$$

[2] $U(w) = \dfrac{1}{|w|(R + \log|w|)}$ is the Poincaré density of $e^{-R} < |w| < \infty$ and satisfies $\Delta \log U = U^2$. The chain-rule for the Laplacian yields $\Delta \log u = \Delta \log|f'| + (U^2 \circ f)|f'|^2 = u^2$ for $u = |f'|\, U \circ f$ whenever $f'(z) \ne 0$.

and $D = \{z \in E \cap E_{z_0} : u(z) > v(z)\}$. If D is empty, $u(z_0) \le 2/R$ holds. Otherwise $w = \log u - \log v$ satisfies $\Delta w = u^2 - v^2 > 0$ on each connected component C of D (note that $f'(z) \ne 0$), and has boundary values $w < 0$ on $\partial C \cap \partial E$ and $w = 0$ on $\partial C \cap E$. Since w is subharmonic on C, this contradicts the maximum principle. Thus $D = \emptyset$ and $u(z_0) \le 2/R$ holds. The inequality $u(z) \le 2/R$ can be written as

$$|f'(z)| \le 2|f(z)|\Big(1 + \frac{\log|f(z)|}{R}\Big).$$

Taking the limit $R \to \infty$ yields $|f'(z)| \le 2|f(z)|$ whenever $|f(z)| > 1$, while $|f'(z)| \le 2$ is trivially true if $|f(z)| \le 1$. The main assertion in Theorem 4.8 follows by integrating $|f'(z)| < 2\alpha(1 + |f(z)|)$.

Remark 4.2 Clunie and Hayman [30] proved $T(r, f) = O(r^{\beta+1})$ for entire functions with spherical derivative $f^\sharp(z) = O(|z|^\beta)$, $\beta > -1$.

4.2 Applications of the Zalcman–Pang Lemma

Bounded entire functions are constant, and locally bounded families of holomorphic functions are normal. Also meromorphic functions omitting a triple of values are constant, and families of meromorphic functions omitting a fixed triple of values are normal. These are two specifications of *Bloch's heuristic principle*, which may be vaguely verbalised as follows: Any 'property' P, which enforces meromorphic functions (on the plane) to be constant, imposed on a family of meromorphic functions on some domain D makes this family normal. Zalcman's Lemma was one by-product in a bid to formalise Bloch's principle. Much more on this principle, also called the *Robinson–Zalcman Heuristic Principle*, can be found in Schiff [148]. Zalcman's Lemma is, however, not restricted to verifying or falsifying Bloch's Principle in particular cases. For example, it may be employed to prove qualitative ('soft') versions of theorems on entire and meromorphic functions, or can be used to transfer results which are only known for meromorphic functions of finite order of growth to the general case. With regard to the extensive literature it would be presuming to aim at completeness. Our focus will be on topics that are in some sense representative and, in particular, closely related to the main issues of the present book.

4.2.1 Examples of Bloch's Principle

In accordance with the introductory words we will discuss some examples.

Theorem 4.9 *Let $\mathscr{F}$ be any family of meromorphic functions on some domain D such that every $f \in \mathscr{F}$ assumes each of the values a_ν ($1 \le \nu \le q$) always with multiplicity at least m_ν (omits a_ν if $m_\nu = \infty$). Then $\sum_{\nu=1}^{q}(1 - \frac{1}{m_\nu}) > 2$ implies normality of $\mathscr{F}$.*

Proof If $\mathscr{F}$ is not normal at some point we obtain in the usual way some non-constant meromorphic function $g(z) = \lim_{k\to\infty} f_k(z_k + \rho_k z)$ $(f_k \in \mathscr{F})$ with bounded spherical derivative. Whenever $g(\zeta_0) = a_\nu$ there exists some sequence $\zeta_k \to \zeta_0$ such that $f_k(z_k + \rho_k\zeta_k) = a_\nu$ $(k \ge k_0)$ with multiplicity at least m_ν, hence a_ν is totally ramified for g with $\Theta(a_\nu, g) \ge 1 - 1/m_\nu$ (or is a Picard value if $m_\nu = \infty$). This, however, contradicts Nevanlinna's inequality $\sum_{\nu=1}^{q} \Theta(a_\nu, g) \le 2$.

As a corollary we obtain the following normality criterion. The original result due to Lappan [107] corresponds to normal functions.

Corollary 4.1 *Let K be a positive number, a_ν $(1 \le \nu \le 5)$ distinct complex numbers, and $\mathscr{F}$ a family of meromorphic functions on some domain D, such that for every $f \in \mathscr{F}$, $f^\sharp(z) \le K$ holds on $f^{-1}(\{a_1, \ldots, a_5\})$. Then $\mathscr{F}$ is normal.*

Proof If $\mathscr{F}$ is not normal, the re-scaling process yields a non-constant meromorphic function $g(z) = \lim_{k\to\infty} f_k(z_k + \rho_k z)$ $(f_k \in \mathscr{F})$ with

$$g^\sharp(\zeta_0) = \lim_{k\to\infty} \rho_k f_k^\sharp(z_k + \rho_k\zeta_k) = 0 \quad (\zeta_0 = \lim_{k\to\infty} \zeta_k,\ f_k(z_k + \rho_k\zeta_k) = a_\nu)$$

whenever $g(\zeta_0) = a_\nu$. Thus each value a_ν is totally ramified or a Picard value for g, again in contrast to Nevanlinna's theorem.

Remark 4.3 If $\mathscr{F}$ consists of holomorphic functions only, then three finite values will suffice, since one can always take into consideration the Picard value $a_4 = \infty$ with $m_4 = \infty$ and $1 - 1/m_4 = 1$.

Exercise 4.5 To prove that the condition $\sum_{\nu=1}^{q}(1 - \frac{1}{m_\nu}) > 2$ as well as the number five are the best possible, consider the sequence $f_n(z) = \wp(nz)$ on $\mathbb{D}$, say; $\wp$ denotes the Weierstraß P-function for some lattice $(\wp'^2 = 4(\wp - e_1)(\wp - e_2)(\wp - e_3))$. Prove that (f_n) is not normal, although the values e_1, e_2, e_3, and $e_4 = \infty$ are totally ramified $(m_\nu = 2)$ and $f_n^\sharp(z) = 0$ whenever $f_n(z) \in \{e_1, e_2, e_3, e_4\}$. As a holomorphic example consider the non-normal sequence of functions $f_n(z) = \cos(nz)$ on $\mathbb{D}$, with $a_1 = 1$ and $a_2 = -1$ $(m_1 = m_2 = 2)$, and Picard value $a_3 = \infty$ $(m_3 = \infty)$.

For every transcendental meromorphic function f, $n \ge 1$, and $c \ne 0$, $f(f^{(n)} - c)$ has infinitely many zeros (Hayman's alternative [78]). This leads to the following normality criterion, see Gu [64].

Exercise 4.6 Prove that the family $\mathscr{F}$ of meromorphic functions such that f and $f^{(n)} + a_{n-1}f^{(n-1)} + \cdots + a_1 f' + a_0 f - 1$ is zero-free on D $(n \ge 1, a_0, \ldots, a_{n-1}$ fixed) is normal. We note that $a_0, \ldots, a_{n-1}$ may even depend on f provided $|a_\nu| < K$ $(0 \le \nu < n)$ holds for some $K > 0$ independent of f. (**Hint.** Since $f \in \mathscr{F}$ has no zeros, Theorem 4.2 applies with any $\alpha < 1$. Which α is appropriate?)

As we have seen, the Zalcman–Pang Lemma may be used to prove new normality criteria[3] inspired by Bloch's Principle and certain results in Nevanlinna theory.

[3] It seems that some criteria of this kind are easier to prove than to apply to non-artificial problems.

Note, however, that neither Bloch's Principle nor its reversal is a theorem. There are 'properties' P which lead to normality criteria though there exist non-constant meromorphic functions having these properties.

Theorem 4.10 ([13, 136, 137]) *Let $a \neq 0$ and b be complex constants and $n \geq 3$ an integer. Then the family $\mathcal{F}$ of meromorphic functions f such that $f' + af^n - b$ has no zeros in the common domain of definition, is normal.*

Proof We may assume $a = 1$. If $\mathcal{F}$ is not normal at z_0, we choose $f_k \in \mathcal{F}$, $z_k \to z_0$, $\rho_k \to 0$, and $\alpha = \frac{1}{n-1}$ such that $g_k(z) = \rho_k^\alpha f_k(z_k + \rho_k z)$ tends to some non-constant meromorphic function g. From $n\alpha = \alpha + 1$ it then follows that

$$g_k'(z) + g_k(z)^n = \rho_k^{\alpha+1}[f_k'(z_k + \rho_k z) + f_k(z_k + \rho_k z)^n]$$

never assumes the value $\rho_k^{\alpha+1} b$. Since $g' + g^n$ cannot be zero-free by Corollary 4.2 below ($n = 3$) and Theorem 3.8 ($n \geq 4$) if g is transcendental, and *a fortiori* if g is rational, there exists some zero ξ of $g' + g^n$. The image of $D : |z - \xi| < \delta$ under $g' + g^n$ covers some disc $|w| < \epsilon$, and so $(g_k' + g_k^n)(D)$ covers $|w| < \epsilon/2$ for $k \geq k_0$, say, by Hurwitz' Theorem. This, however, contradicts $\rho_k^{\alpha+1} b \to 0$. ☕

Remark 4.4 It is obvious that $a = a_f$ and $b = b_f$ may depend on f, provided $1/K \leq |a_f| \leq K$ and $|b_f| \leq K$. ☕

Exercise 4.7 In [138] Pang and Ye proved the following Criterion.
Let $k \geq 1$ and $n \geq k + 3$ be integers, and $a_1, \ldots, a_{k+1}$ holomorphic functions on some domain D. Then the family $\mathcal{F}$ of meromorphic functions f such that $f(z)^n + f^{(k)}(z) + a_1(z)f^{(k-1)}(z) + \cdots + a_k(z)f(z) + a_{k+1}(z) \neq 0$ on D, is normal.

For a proof assume non-normality at some point and apply Zalcman–Pang Rescaling with $\alpha = \frac{k}{n-1} \in (0, 1)$ to obtain a non-constant meromorphic function g such that $\Psi = g^n + g^{(k)}$ has no zeros. By Exercise 3.2, $n = k + 3$ is necessary, and g satisfies $g' = c(g - \beta_1)(g - \beta_2)$. Deduce that $g^{(k)} = P_k(g)$, where P_k is a polynomial of degree $k+1$ with simple zeros at β_1 and β_2. Prove that $g^n + P_k(g) = (g - \beta_1)^{n_1}(g - \beta_2)^{n_2}$ with $n_1 + n_2 = n$ and $n_1\beta_1 + n_2\beta_2 = 0$ is impossible. ☕

Remark 4.5 By Exercise 3.9 there exist transcendental meromorphic functions such that $f' - f^3 - 1$ resp. $f' + f^4 - 1$ is zero-free. It is obvious how to modify these functions such that this holds for $f' + af^n + b$ ($n = 3$ resp. $n = 4$ and $ab \neq 0$). More importantly, for $n = 4$ these functions satisfy some fixed Riccati equation $w' = c_0 + c_1 w + c_2 w^2$ with constant coefficients (see Theorem 3.8 for $a = 1$ and $b = -1$), hence the meromorphic functions with the property $f' + af^4 + b \neq 0$ form a normal family ($f^\sharp(z) \leq |c_0| + |c_1| + |c_2|$). This matches the following modification of Bloch's principle, which was suggested in [13], namely to replace the above condition 'any property P, which enforces *meromorphic functions to be constant*' with the less restrictive 'any property P, which enforces *the family of meromorphic functions with this property to be normal on* $\mathbb{C}$'. ☕

4.2.2 From Finite to Infinite Order

The Zalcman–Pang Lemma can also be used to prove theorems on entire and meromorphic function in two steps. In the first part the proof is given for functions of finite order or even functions having bounded spherical derivative (so-called Yosida functions), while the general case is reduced to the special one by applying the Zalcman–Pang Lemma. Of course, the results obtained this way can only be qualitative. As an example, we quote the following results due to Pang and Zalcman [139].

Theorem 4.11 *Let f be a transcendental entire function all of whose zeros have multiplicity at least $k \geq 1$. Then $f^{(k)}f^n$ $(n \geq 1)$ assumes every non-zero value infinitely often.*

Proof We first suppose that f has bounded spherical derivative, hence f is of exponential type ($|f(z)| \leq Ae^{B|z|}$ holds). If $f^{(k)}f^n$ assumes some value $a \neq 0$ only finitely often, then f must have infinitely many zeros. For otherwise, $f(z) = P_1(z)e^{\alpha z}$ and $f^{(k)}(z) = P_2(z)e^{\alpha z}$ hold with polynomials P_1 and P_2, hence also $f^{(k)}(z)f(z)^n = P_3(z)e^{(n+1)\alpha z}$. Entire functions of this type, however, assume every non-zero value infinitely often. We now assume that f has infinitely many zeros, each of order at least k. If $f^{(k)}f^n$ assumes $a \neq 0$ only finitely often, then $f^{(k)}(z)f(z)^n = P_4(z)e^{\gamma z} + a$ if $k + n \geq 3$ resp. $\frac{1}{2}f(z)^2 = P_5(z)e^{\gamma z} + az + a_0$ if $k = n = 1$ has infinitely many multiple zeros, which clearly is impossible. To settle the general case consider any sequence $z'_\nu \to \infty$ such that $f^\sharp(z'_\nu) \to \infty$, and apply the Zalcman–Pang Lemma to $f_\nu(z) = f(z'_\nu + z)$. Then (some subsequence of) $g_\nu(z) = \rho_\nu^\alpha f_\nu(z'_\nu + z_\nu + \rho_\nu z)$ with $z_\nu \to 0$, $\rho_\nu \to 0$, and $\alpha = -\frac{k}{n+1} > -k$ tends to some non-constant entire function g of exponential type (this includes polynomials), while $h_\nu = g_\nu^{(k)}g_\nu^n$ tends to $h = g^{(k)}g^n$. By Hurwitz' Theorem g has zeros of order at least k, hence $g \not\equiv 0$ cannot be a polynomial of degree less than k. In other words, h is non-constant and assumes every non-zero value (even infinitely often if g is transcendental). Suppose $h(\zeta_0) = a$. By Hurwitz' Theorem there exists some sequence $\zeta_\nu \to \zeta_0$ such that $h_\nu(\zeta_\nu) = a$ $(\nu \geq \nu_0)$, hence $f^{(k)}f^n$ assumes a at $z'_\nu + z_\nu + \rho_\nu\zeta_\nu \to \infty$. ☕

Theorem 4.12 *Let $k \geq 1$ and $n \geq 1$ be integers, and let $\mathscr{F}$ be the family of holomorphic functions f on some domain D, with zeros having multiplicity at least $k \geq 1$ and such that $f^{(k)}f^n$ omits some fixed value $a \neq 0$. Then $\mathscr{F}$ is normal.*

Proof If $\mathscr{F}$ is not normal we obtain, similar to the proof of Theorem 4.11, some non-constant entire function g of exponential type, such that either $g^{(k)}g^n \neq a$ or else $g^{(k)}g^n \equiv a$. The latter is clearly impossible for transcendental functions as well as for non-constant polynomials, while the other contradicts Theorem 4.11. ☕

One more result with this kind of proof is

Theorem 4.13 ([14]) *Let f be non-constant meromorphic and $m > \ell \geq 1$. Then $(f^m)^{(\ell)}$ assumes every finite non-zero value infinitely often.*

Remark 4.6 For $m > \ell + 1$, see the stronger (semi-quantitative) Theorem 3.7. ☕

Proof If f has only finitely many zeros, then the assertion follows from Theorem 3.5 in [78] (Hayman's alternative for f^m). We thus may assume that f has infinitely many zeros, and will start by explaining the main idea which leads to a proof if f has finite order of growth. Let h be any transcendental meromorphic function with infinitely many multiple zeros z_j, and set $g(z) = z - h(z)/c$ ($c \neq 0$ arbitrary). The zeros z_j are parabolic fixed-points of g, that is, $g(z_j) = z_j$ and $g'(z_j) = 1$ hold. Fixed points z_j of this kind attract some invariant domain L_j under iteration (known as a *Leau domain* or *parabolic basin*), that is, the sequence of iterates $g^n = g \circ g^{n-1}$ tends to $z_j \in \partial L_j$ as $n \to \infty$, locally uniformly on L_j, and L_j contains some critical or asymptotic value of g ($\zeta_j = g(\xi_j)$ with $g'(\xi_j) = 0$, $\xi_j \in L_j$, or $g(z) \to \zeta_j$ as $z \to \infty$ on some path γ_j tending to infinity in L_j).[4] Since the domains L_j are mutually disjoint, the critical and asymptotic values ζ_j are mutually distinct and g has infinitely many critical or asymptotic values. If in the present case g has only finitely many asymptotic values, then g has infinitely many critical values. If, however, the number of asymptotic values is infinite, then also the number of critical points is infinite by a deep theorem due to Bergweiler and Eremenko [14].[5] If g and so h has finite order, h has infinitely many critical points ξ and $g'(\xi_j) = c$ holds. Since the zeros of f are multiple zeros of $h = (f^m)^{(\ell-1)}$, this argument applies if f has finite order of growth.

We now assume that f has infinite order and $(f^m)^{(\ell)}$ assumes the value $c = 1$, say, only finitely often. Consider the sequence of functions $f_n(z) \doteq 2^{-kn} f(2^n z)$ with $k = \ell/m$ on $\frac{1}{4} < |z| < 2$, say. This sequence cannot be normal, since otherwise $f_n^\sharp(z) \leq M$ holds on $\frac{1}{2} < |z| < 1$, say, which implies $f^\sharp(2^n z) \leq 2^{(k-1)n} M < M$ on $\frac{1}{2} < |z| < 1$ and $f^\sharp(\zeta) < M$ on $|\zeta| > 1$ in contrast to our hypothesis that f has infinite order. The Zalcman–Pang Lemma with $\alpha = -k \in (-1, 0)$ yields a non-constant meromorphic function

$$F(z) = \lim_{j\to\infty} \rho_j^{-k} f_{n_j}(z_j + \rho_j z) = \lim_{j\to\infty} \rho_j^{-k} 2^{-kn_j} f(2^{n_j}(z_j + \rho_j z)) \quad (z_j \to z_0, \rho_j \to 0)$$

of finite order. Since $(f_n^m)^{(\ell)}(z) - 1 = (f^m)^{(\ell)}(2^n z) - 1$ has only a fixed number of zeros, namely $z = 2^{-n}\zeta$ with $(f^m)^{(\ell)}(\zeta) = 1$, and z_0 is non-zero, $(F^m)^{(\ell)} - 1$ is zero-free on $\mathbb{C}$ (note that $(f_{n_j}^m)^{(\ell)}(z_j + \rho_j z)$ tends to $(F^m)^{(\ell)}(z)$ on $\mathbb{C} \setminus \{\text{poles of } F\}$). This contradicts the first part of the proof, provided F is transcendental. Now suppose that F is rational. Then $(F^m)^{(\ell)} - 1 \neq 0$ on $\mathbb{C}$ implies

$$F(z)^m = z^\ell/\ell! + c_{\ell-1} z^{\ell-1} + \cdots \quad (z \to \infty),$$

which is clearly impossible since $1 \leq \ell < m$. ☕

---

[4]For rational functions g this is a basic fact in iteration theory, see [174], Chap. 3, §5, or any other textbook in this field. Abel's functional equation $\phi \circ g = \phi + 1$ has a local solution (the inverse to Abel's function being discussed in Sect. 4.1.3) on some subdomain P_j of L_j (called a *petal*), and analytic continuation of ϕ leads to singularities arising from critical values of g in L_j. In the transcendental case the singularities may also come from asymptotic values in L_j.

[5] **Corollary 3** ([14]) *Meromorphic functions of finite order having only finitely many critical points cannot have infinitely many asymptotic values.*

As a corollary we obtain the following completion of Theorem 3.8.

Corollary 4.2 *For every transcendental meromorphic function g, $g' + g^3$ has infinitely many zeros.*

Proof Theorem 4.13 applies to $f = 1/g$, $m = 2$, and $\ell = 1$. Hence $(f^2)' - 2$, say, has infinitely many zeros, and so has $g' + g^3$.

4.2.3 From Normal Families to Differential Equations

Meromorphic solutions to (implicit) first-order algebraic differential equations have finite order of growth. This was proved by Gol'dberg [56] (see also Exercise 3.2), and generalised by Bergweiler [11] as follows.

Theorem 4.14 *Let $P[w]$ be a differential polynomial, that is, a finite sum of terms $q_M(z, w)M[w]$, where $q_M \not\equiv 0$ may be any polynomial in w with rational coefficients, and each monomial M has the form $M[w] = (w')^{r_1}(w'')^{r_2}\cdots(w^{(m)})^{r_m}$ for some $r_1, \ldots, r_m$,* with reduced weight *$\delta_M = r_1 + 2r_2 + \cdots + mr_m < n$. Then every meromorphic solution to the differential equation $w'^n = P[w]$ has finite order of growth.*

Proof Assuming that w has infinite order, the spherical derivative tends to infinity faster than any power of z on some sequence $z = z_k' \to \infty$. Then the sequence of functions $w(z_k' + z)$ is not normal at $z = 0$, and by Zalcman's Lemma there exist z_k with $z_k' - z_k \to 0$ and $\rho_k = 1/w^\sharp(z_k) \to 0$, such that (some subsequence of) $\mathfrak{w}_k(\mathfrak{z}) = w(z_k + \rho_k\mathfrak{z})$ tends to some non-constant meromorphic function $\mathfrak{w}$, and $\mathfrak{w}_k$ satisfies

$$\mathfrak{w}_k'^n = \sum q_M\big(z_k + \rho_k\mathfrak{z}, \mathfrak{w}_k\big)\rho_k^{n-\delta_M}M[\mathfrak{w}_k].$$

Now $z_k^s\rho_k \to 0$ for every integer s, $n > \delta_M$, and $\mathfrak{w}_k^{(\nu)} \to \mathfrak{w}^{(\nu)}$, locally uniformly on $\mathbb{C} \setminus \{\text{poles of } \mathfrak{w}\}$, yields $\mathfrak{w}'^n \equiv 0$ in contrast to the fact that $\mathfrak{w}$ is non-constant.

Exercise 4.8 Prove a quantitative version of Theorem 4.14 as follows. Suppose that the coefficients in $q_M(z, w)$ do not grow faster than $|z|^t$ for some $t \geq 0$. Then $w^\sharp(z) = O(|z|^t)$ holds, hence w has order of growth at most $2t + 2$.

Not every attempt is successful. The most general second-order differential equations without 'movable singularities' other than poles have the form

$$w'' = L(z, w)w'^2 + M(z, w)w' + N(z, w), \tag{4.5}$$

the most prominent examples are Painlevé's equations I, II, and IV, which will be discussed in Chap. 6. We assume that Eq. (4.5), where L, M, and N are rational functions in both variables, possesses a transcendental meromorphic solution.

Unfortunately, Theorem 4.14 does not apply to Eq. (4.5). If w has infinite order, the method of proof of Theorem 4.14 leads to some limit function $\mathfrak{w}$ that solves $\mathfrak{w}'' = \Lambda(\mathfrak{w})\mathfrak{w}'^2$ with $\ell\Lambda = P'/P$ for some polynomial P of degree $\ell = \deg_w L$ ($\Lambda \equiv 0$ and $P = 1$ if $L \equiv 0$), hence $\mathfrak{w}'^{\ell} = \text{const} \cdot P(\mathfrak{w})$. What can be shown is $\ell \leq 6$ (Bergweiler [11]), but nothing else; actually $\deg L \leq 4$ is known.

Example 4.6 In the case of equation $w'' = z + 6w^2$ we try again. Since w has poles of order two with principal part $(z-p)^{-2}$ we may consider any sequence $\mathfrak{w}_k(\mathfrak{z}) = \rho_k^{\alpha} w(z_k + \rho_k \mathfrak{z})$ with $-1 < \alpha \leq 2$ and $\rho_k = 1/w^\sharp(z_k)$. Assuming $z_k \rho_k^{2+\alpha} \to 0$ on some sequence $z_k \to \infty$ and $\mathfrak{w}_k \to \mathfrak{w}$, we obtain $\mathfrak{w}'' = 6\mathfrak{w}^2$ if $\alpha = 2$ and $\mathfrak{w}'' = 0$ if $\alpha < 2$. In this case $\mathfrak{w} = a\mathfrak{z} + b$ and $\frac{|a|}{1+|a\mathfrak{z}+b|^2} = \mathfrak{w}^\sharp(\mathfrak{z}) \leq \mathfrak{w}^\sharp(0) = \frac{|a|}{1+|b|^2}$ hold for every $\mathfrak{z}$. This is only possible if $b = 0$ and $\mathfrak{w} = a\mathfrak{z}$, and may be interpreted as follows: for every $\epsilon > 0$, $w^\sharp(z) = O(|z|^{\frac{1}{4-\epsilon}})$ holds outside the discs $|z - q| < \delta|q|^{-\frac{1}{4-\epsilon}}$ ($\delta > 0$ arbitrary) about the zeros of w. We note that $w^\sharp(z) = O(|z|^{\frac{1}{4}})$ without exception would lead to the sharp estimate $T(r, w) = O(r^{\frac{5}{2}})$. Though the latter is true, the former is not. Actually, the best one can prove is $w^\sharp(z) = o(|z|^{\frac{3}{4}})$ (Chap. 6).

Exercise 4.9 State and prove an analogous result for $w'' = \alpha + zw + 2w^3$.

4.2.4 *From Differential Equations to Normal Families*

Normality of any family $\mathscr{F}$ of meromorphic functions on some domain D is not only equivalent to local boundedness of the family $\mathscr{F}^\sharp$ (Marty's Criterion), but also follows from the fact that the family $1/\mathscr{F}^\sharp$ of reciprocals $1/f^\sharp$ is locally bounded. It suffices to consider the unit disc $\mathbb{D}$.

Theorem 4.15 ([59, 178]) *Let $\mathscr{F}$ be a family of meromorphic functions on $\mathbb{D}$ and assume that $f^\sharp(z) \geq \epsilon > 0$ holds for every $f \in \mathscr{F}$. Then $\mathscr{F}$ is normal.*

Proof It is, of course, not always the case (although one might believe) that the Zalcman–Pang method provides the fastest way to prove normality criteria. Actually the first authors [59] successfully used this method. We will prove the explicit estimate

$$f^\sharp(z) \leq \frac{2/\epsilon}{(1-|z|)^2}.$$

Since $f^\sharp$ is non-zero, f is locally univalent and its Schwarzian derivative (see Sect. 1.5.4) is holomorphic on $\mathbb{D}$. This implies the representation $f = w_1/w_2$, where w_1 and w_2 form a fundamental set of the linear differential equation

$$w'' + \tfrac{1}{2}S_f(z)w = 0,$$

normalised by $W(w_1, w_2) = 1$. From $f' = -1/w_2^2$ and $f^\sharp = (|w_1|^2 + |w_2|^2)^{-1}$ it then follows that $|w_1(z)|^2 + |w_2(z)|^2 \le 1/\epsilon$ holds. To prove the assertion we just remark that the Cauchy–Schwarz inequality, applied to $1 = |w_1 w_2' - w_1' w_2|$, yields

$$f^\sharp(z) = \frac{1}{|w_1(z)|^2 + |w_2(z)|^2} \le |w_1'(z)|^2 + |w_2'(z)|^2.$$

The standard Cauchy estimate: $|w(z)| \le 1/\sqrt{\epsilon}$ *on* $\mathbb{D}$ *implies* $(1-|z|)|w'(z)| \le 1/\sqrt{\epsilon}$ then gives the desired estimate.

Exercise 4.10 Prove that $f(z) = \left(\frac{1+z}{1-z}\right)^{i\lambda}$ $(\lambda > 0, f(0) = 1)$ has spherical and Schwarzian derivative $f^\sharp(z) = \frac{\lambda}{|1-z^2|} \frac{2}{|f(z)| + |f(z)|^{-1}}$ and $S_f(z) = \frac{2+2\lambda^2}{(1-z^2)^2}$, with $f^\sharp(z) > f^\sharp(\pm i) = \frac{\lambda}{2\cosh\frac{\pi}{2}\lambda}$ on $\mathbb{D}$ and $f^\sharp(x) = \frac{\lambda}{1-x^2}$ on $-1 < x < 1$.[6]

(**Hint.** $\log f^\sharp$ obeys the Minimum Principle; $|f(z)| = e^{\mp\lambda\pi/2}$ holds on $|z| = 1$, $\pm\operatorname{Im} z > 0$.)

4.3 The Yosida Classes

Entire and meromorphic functions having bounded spherical derivative are known as *Yosida functions*, in higher dimensions also called *Brody curves*. They were introduced by Yosida [210] (classes A and A_0). In this section Yosida functions and their generalisations will be discussed. They occur in a natural way in the context of algebraic differential equations. Unless otherwise stated, all functions are assumed to be non-constant meromorphic on $\mathbb{C}$.

4.3.1 Yosida Functions

Yosida functions may be equivalently defined by the fact that the translates

$$f_h(\mathfrak{z}) = f(z + h)$$

[6]For f to be univalent in the unit it is necessary that $|S_f(z)| \le \frac{6}{(1-|z|^2)^2}$ (W. Krauss, Über den Zusammenhang einiger Charakteristiken eines einfach zusammenhängenden Bereiches mit der Kreisabbildung, *Mitt. Math. Sem. Giessen* **21** (1932), 1–28), and sufficient that $|S_f(z)| \le \frac{2}{(1-|z|^2)^2}$ (Z. Nehari, The Schwarzian derivative and schlicht functions, *Bull. Amer. Math. Soc.* **55** (1949), 545–551). The present example (E. Hille, Remarks on a paper by Zeev Nehari, *Bull. Amer. Math. Soc.* **55** (1949), 552–553) shows that Nehari's criterion is sharp.

form a normal family on $\mathbb{C}$. Of particular interest are the Yosida functions such that every limit function

$$\mathfrak{f}(\mathfrak{z}) = \lim_{h_k\to\infty} f_{h_k}(\mathfrak{z}) \tag{4.6}$$

is non-constant (class A_0 in [210]). Then these limits are also Yosida functions. Yosida's definition was extended in [179] to a wider class of meromorphic functions as follows: f belongs to the class $\mathfrak{Y}_{\alpha,\beta}$ with $\alpha \in \mathbb{R}$ and $\beta > -1$, if the family $(f_h)_{|h|>1}$ of functions

$$f_h(\mathfrak{z}) = h^{-\alpha} f(h + h^{-\beta}\mathfrak{z}) \tag{4.7}$$

is normal on the plane and *at least one* of the limit functions (4.7) is non-constant, and to the Yosida class $\mathfrak{Y}^0_{\alpha,\beta}$, if *all* limit functions are non-constant; $\mathfrak{Y}^0_{0,0}$ coincides with Yosida's class A_0; it contains the elliptic functions.

4.3.2 A Modified Spherical Derivative

The expression

$$f^{\sharp_\alpha}(z) = \frac{|z|^\alpha |f'(z)|}{|z|^{2\alpha} + |f(z)|^2} \tag{4.8}$$

occurs quite naturally in the context of nonlinear algebraic differential equations.

Exercise 4.11 Prove that $|z|^{-|\alpha|} f^\sharp(z) \le f^{\sharp_\alpha}(z) \le |z|^{|\alpha|} f^\sharp(z)$ holds for $|z| \ge 1$. (**Hint.** First assume $\alpha \ge 0$, afterwards consider $1/f$.)

The next theorem will show how normality of the family of functions (4.7) is related to the *modified spherical derivative* (4.8).

Theorem 4.16 *The family of functions* (4.7) *is normal on* $\mathbb{C}$ *if and only if*

$$f^{\sharp_\alpha}(z) = O(|z|^\beta) \tag{4.9}$$

holds. Moreover, (4.9) *implies* $f^\sharp(z) = O(|z|^{|\alpha|+\beta})$.

Proof Normality of the family (f_h) at $\mathfrak{z} = 0$ implies that the family $(f_h^\sharp(0))_{|h|>1}$ is bounded, that is,

$$f_h^\sharp(0) = \frac{|h|^{-\alpha-\beta}|f'(h)|}{1 + |h|^{-2\alpha}|f(h)|^2} = \frac{|h|^{\alpha-\beta}|f'(h)|}{|h|^{2\alpha} + |f(h)|^2} = |h|^{-\beta} f^{\sharp_\alpha}(h) \le C \quad (|h| > 1),$$

and thus (4.9) holds. To prove the converse we note that

$$\begin{aligned}\limsup_{h\to\infty} f_h^\sharp(\mathfrak{z}) &= \limsup_{h\to\infty} \frac{|h|^{\alpha-\beta}|f'(h+h^{-\beta}\mathfrak{z})|}{|h+h^{-\beta}\mathfrak{z}|^{2\alpha}+|f(h+h^{-\beta}\mathfrak{z})|^2} \\ &= \limsup_{z\to\infty} |z|^{-\beta} f^{\sharp_\alpha}(z) < \infty\end{aligned} \tag{4.10}$$

($z = h + h^{-\beta}\mathfrak{z}$) holds, uniformly with respect to $|\mathfrak{z}| < R$. This shows that the continuous function $(h,\mathfrak{z}) \mapsto f_h^\sharp(\mathfrak{z})$ is bounded on $|h| > 1, |\mathfrak{z}| < R$, and the family $(f_h)_{|h|>1}$ is normal on $\mathbb{C}$ by Marty's Criterion.

Remark 4.7 For $f \in \mathfrak{Y}_{\alpha,\beta}$ any finite and non-constant limit function (4.6) is a Yosida function ($\mathfrak{f} \in \mathfrak{Y}_{0,0}$) with $\mathfrak{f}^\sharp(\mathfrak{z}) \le \limsup_{z\to\infty} |z|^{-\beta} f^{\sharp_\alpha}(z)$.

Exercise 4.12 In order that at least one limit function is non-constant it is necessary and sufficient that the right-hand side limsup in (4.10) is positive. Prove that the condition $\limsup_{z\to\infty} |z|^{-\beta} f^{\sharp_\alpha}(z) < \infty$ combined with

$$\liminf_{z\to\infty} \sup_{|\zeta - z|<\epsilon} |\zeta|^{-\beta} f^{\sharp_\alpha}(\zeta) > 0 \quad (\text{for some } \epsilon > 0)$$

implies $f \in \mathfrak{Y}^0_{\alpha,\beta}$, and *vice versa*.

Exercise 4.13 Given $f \in \mathfrak{Y}_{\alpha,\beta}$, prove that $F(\zeta) = \zeta^m f(\zeta^n)$ ($n \in \mathbb{N}, m \in \mathbb{Z}$) belongs to $\mathfrak{Y}_{m+\alpha n,(\beta+1)n-1}$. Discuss the case $\alpha \in \mathbb{Q}$.

Exercise 4.14 It is known [118] that the Weierstraß elliptic functions

$$\wp(z;g_2,g_3),\ \wp'(z;0,g_3),\ \wp^2(z;g_2,0),\ \text{and}\ \wp^3(z;0,g_3)\ \text{resp.}\ \wp'^2(z;0,g_3)$$

may be written as $f_n(z^n)$ with $n = 2,3,4,6$, respectively. Prove that the corresponding meromorphic functions f_n belong to $\mathfrak{Y}^0_{0,\frac{1}{n}-1}$.

4.3.3 A Modified Ahlfors–Shimizu Formula

Although $w = z^{-\alpha} f$ need not be single-valued, $w^\sharp$ is well-defined, and an elementary calculation shows that $w^\sharp(z) = f^{\sharp_\alpha}(z) + O(|z|^{-1})$. Now a closer inspection of the proof of the Ahlfors–Shimizu Formula yields

$$N(r,f) + \frac{1}{4\pi}\int_0^{2\pi} \log(1 + r^{-2\alpha}|f(re^{i\theta})|^2)\,d\theta = \frac{1}{\pi}\int_1^r A_\alpha(t,f)\frac{dt}{t}, \tag{4.11}$$

with $A_\alpha(t,f) = \int_{|z|<t} [f^{\sharp_\alpha}(z) + O(|z|^{-1})]^2 \, d(x,y) \sim \int_{|z|<t} f^{\sharp_\alpha}(z)^2 \, d(x,y)$ as $t \to \infty$. The left-hand side of (4.11) differs from $T(r,f)$ by a term $O(\log r)$, so that

$$T(r,f) \sim \frac{1}{\pi} \int_1^r \int_{|z|<t} f^{\sharp_\alpha}(z)^2 \, d(x,y) \, \frac{dt}{t} \quad (r \to \infty)$$

holds for any transcendental meromorphic function. Hence functions $f \in \mathfrak{Y}_{\alpha,\beta}$ have Nevanlinna characteristic

$$T(r,f) = O(r^{2+2\beta}). \tag{4.12}$$

4.3.4 The Distribution of Zeros and Poles

Let $\mathfrak{P}$ denote the set of non-zero poles of $f \in \mathfrak{Y}_{\alpha,\beta}$, and set

$$\Delta_\delta(p) = \{z : |z - p| < \delta |p|^{-\beta}\} \quad \text{and} \quad \mathfrak{P}_\delta = \bigcup\nolimits_{p \in \mathfrak{P}} \Delta_\delta(p).$$

Re-scaling about any sequence $z_n \to \infty$ with $f(z_n)z_n^{-\alpha} \to \infty$ then yields some limit function $\mathfrak{f} = \lim_{z_n \to \infty} f_{z_n}$, which either is $\equiv \infty$ or else has a pole at $\mathfrak{z} = 0$. If the former case is excluded we obtain

Theorem 4.17 *Suppose $f \in \mathfrak{Y}_{\alpha,\beta}$ is such that all limit functions $\lim_{h_k \to \infty} f_{h_k}$ are meromorphic ($\not\equiv \infty$). Then $f(z) = O(|z|^\alpha)$ holds outside $\mathfrak{P}_\delta$, $\delta > 0$ arbitrary, and $m(r,f) \le \alpha^+ \log r + O(1)$.*

Proof The first part follows from Hurwitz' Theorem: $f(z_n)z_n^{-\alpha} \to \infty$ implies that (for every convergent subsequence) $\lim_{z_{n_k} \to \infty} f_{z_{n_k}} = \mathfrak{f} \not\equiv \infty$ has a pole at $\mathfrak{z} = 0$, hence $|z_n|^\beta \operatorname{dist}(z_n, \mathfrak{P}) \to 0$ $(n \to \infty)$ and $z_n \in \Delta_\delta(p_n)$ for some p_n (and $n \ge n_0$) holds. To prove the second assertion, let (r_n) be any sequence of radii that tends to infinity. The interval $[-\pi, \pi]$ is divided into $K_n = [r_n^{1+\beta}]$ subintervals of length $2\pi/K_n \sim 2\pi r_n^{-\beta-1}$. Then for one of these intervals, $[-\pi/K_n, \pi/K_n] = I_n$, say,

$$m(r_n,f) \le \frac{K_n}{2\pi} \int_{I_n} \log^+ |f(r_n e^{i\theta})| \, d\theta$$

holds. We may assume that $f_{r_n} \to \mathfrak{f} \not\equiv \infty$. From $z = r_n e^{i\theta} = r_n + r_n^{-\beta}\mathfrak{z}$, $d\theta = r_n^{-\beta-1}|d\mathfrak{z}|$, $K_n r_n^{-\beta-1} \to 1$, and $\log^+ |f(z)| \le \alpha^+ \log r_n + \log^+ |f_{r_n}(\mathfrak{z})|$ it follows that

$$\limsup_{r_n \to \infty} \, [m(r_n,f) - \alpha^+ \log r_n] \le \frac{1}{2\pi} \int_{-\pi i}^{\pi i} \log^+ |\mathfrak{f}(\mathfrak{z})| \, |d\mathfrak{z}|$$

holds (we integrate along the segment from $-\pi i$ to πi). ☕

Remark 4.8 Since $f \in \mathfrak{Y}_{\alpha,\beta}$ implies $1/f \in \mathfrak{Y}_{-\alpha,\beta}$, Theorem 4.17 applies to $1/f$ and $\mathfrak{P}$ replaced with the set $\mathfrak{Q}$ of non-zero zeros of f, provided none of the limit functions (4.6) vanishes identically. If all limit functions are $\not\equiv 0$ and $\not\equiv \infty$, then

$$f(z) \asymp z^\alpha \quad \text{holds outside } \mathfrak{P}_\delta \cup \mathfrak{Q}_\delta.$$

The growth of f is measured by α, while β determines the size of the 'local unit discs' $\Delta_1(z_0)$. We also note that for $f \in \mathfrak{Y}_{\alpha,\beta}, f^{(k)} \in \mathfrak{Y}_{\alpha+k\beta,\beta}$ holds if and only if the poles of f are β- *separated* from each other, that is, if for $\delta > 0$ sufficiently small, the discs $\Delta_\delta(p)$ are mutually disjoint.

For $f \in \mathfrak{Y}_{\alpha,\beta}$ it is necessary that the zeros of f are β-separated from the poles, that is, $\inf_{q\in\mathfrak{Q}} |q|^\beta \operatorname{dist}(q, \mathfrak{P}) > 0$ holds; equivalently, $\Delta_\delta(p) \cap \Delta_\delta(q) = \emptyset$ holds for $p \in \mathfrak{P}$, $q \in \mathfrak{Q}$, and $\delta > 0$ sufficiently small. In many cases the poles of solutions to algebraic differential equations are simple with residues $\sigma = \sigma(p)$ and are β-separated from each other. Re-scaling along any sequence of poles yields

$$w_p(\mathfrak{z}) = p^{-\alpha} w(p + p^{-\beta}\mathfrak{z}) = \sigma(p)p^{\beta-\alpha}/\mathfrak{z} + \cdots,$$

hence for $w \in \mathfrak{Y}_{\alpha,\beta}$ it is necessary that $\sigma(p) \asymp |p|^{\alpha-\beta}$ as $p \to \infty$, and $w = O(|z|^\alpha)$ outside $\mathfrak{P}_\delta$ hold. In combination, these conditions are also sufficient for $f \in \mathfrak{Y}_{\alpha,\beta}$.

Theorem 4.18 *Let f be meromorphic and suppose that the poles p of f are β-separated and simple with residues $\sigma = \sigma(p)$ satisfying $|\sigma(p)| \asymp |p|^{\alpha-\beta}$ as $p \to \infty$. Assume also that $f(z) = O(|z|^\alpha)$ holds outside $\mathfrak{P}_\delta$. Then $f \in \mathfrak{Y}_{\alpha,\beta}$.*

Proof We first assume $|\sigma(p)| = 1$, hence $\alpha = \beta$. The well-known Cauchy estimate

$$r|f'(z)| \le \max_{|\zeta - z| = r} |f(\zeta)| \quad (z \notin \mathfrak{P}_\delta,\ r = \tfrac{\delta}{2}|z|^{-\beta})$$

yields $f'(z) = O(|z|^{2\beta})$, hence $f^{\sharp_\beta}(z) = O(|z|^\beta)$ outside $\mathfrak{P}_\delta$. The Cauchy estimate applied to

$$f(z) = \frac{\sigma}{z-p} + \sigma_0 + \cdots \quad \text{on } \Delta_\delta(p)$$

gives $\sigma_0 = O(|p|^\beta)$, and the Maximum Principle applied to the regular function

$$F = \sigma f' + f^2 - 2\sigma_0 f$$

yields $|F(z)| = O(|z|^{2\beta})$, $|f'(z)| = O(|z|^{2\beta} + |f(z)|^2)$ and

$$f^{\sharp_\beta}(z) = O(|z|^\beta) \quad \text{on } \Delta_\delta(p) \text{ as } p \to \infty.$$

In the general case, $g(z) = f(z)/\sigma(p)$ satisfies $g(z) = O(|z|^{\alpha-(\alpha-\beta)}) = O(|z|^\beta)$ on $\partial\Delta_\delta(p)$, hence $|g'(z)| = O(|z|^{2\beta} + |g(z)|^2)$ holds on $\Delta_\delta(p)$. In terms of f this means

$$\begin{aligned}|f'(z)| &= O(|\sigma(p)||z|^{2\beta} + |f(z)|^2|\sigma(p)|^{-1})\\ &= |\sigma(p)|^{-1}O(|z|^{2\beta+2(\alpha-\beta)} + |f(z)|^2)\\ &= (|z|^{2\alpha} + |f(z)|^2)\,O(|z|^{\beta-\alpha}) \qquad \text{on } \Delta_\delta(p),\end{aligned}$$

hence $f^{\sharp_\alpha}(z) = O(|z|^\beta)$ holds on $\mathfrak{P}_\delta$ in any case, while $f(z) = O(|z|^\alpha)$ and the Cauchy estimate imply $f'(z) = O(|z|^{\alpha+\beta})$ and $f^{\sharp_\alpha}(z) = O(|z|^\beta)$ outside $\mathfrak{P}_\delta$. ☕

Exercise 4.15 Suppose that the poles and zeros of $f \in \mathfrak{Y}_{\alpha,\beta}$ are β-separated from the zeros and poles of $\tilde f \in \mathfrak{Y}_{\tilde\alpha,\beta}$, respectively. Prove that $f\tilde f \in \mathfrak{Y}_{\alpha+\tilde\alpha,\beta}$. ☕

4.3.5 The Class $\mathfrak{Y}^0_{\alpha,\beta}$

Much more is known about the functions $f \in \mathfrak{Y}^0_{\alpha,\beta}$, $\alpha \in \mathbb{R}$, $\beta > -1$. We just list the most important properties. For detailed proofs, see [179].

1. $\sup_{|\zeta-z|<\delta|z|^{-\beta}} f^{\sharp_\alpha}(\zeta) \asymp |z|^\beta$.
2. $T(r,f) \asymp r^{2+2\beta}$.
3. $m(r,f) \sim \alpha^+ \log r$, $m(r,1/f) \sim \alpha^- \log r$, and $m(r,1/f') = O(\log r)$.
4. Every sufficiently large disc $\Delta_R(z_0)$ ($|z_0| > 1$ arbitrary) contains poles and zeros.
5. There exists an $M > 0$ such that each 'local unit disc' $\Delta_1(z_0)$ contains at most M poles and zeros; in particular, the multiplicities are bounded.
6. The poles of f are β-separated from the zeros.
7. $f(z) \asymp |z|^\alpha$ holds on $\mathbb{C} \setminus (\mathfrak{P}_\delta \cup \mathfrak{Q}_\delta)$.
8. $f^\sharp(z) = O(|z|^{\beta-|\alpha|})$ and $f'(z)/f(z) = O(|z|^\beta)$ hold on $\mathbb{C} \setminus (\mathfrak{P}_\delta \cup \mathfrak{Q}_\delta)$.
9. $f' \in \mathfrak{Y}^0_{\alpha+\beta,\beta}$ if and only if the poles are β-separated.
10. The value distribution of f takes place in $\mathfrak{P}_\delta$ if $\alpha < 0$, and in $\mathfrak{Q}_\delta$ if $\alpha > 0$.
11. $\mathfrak{Y}^0_{0,0}$ contains every limit $\mathfrak{f} = \lim_{h_k\to\infty} f_{h_k}$ for $f \in \mathfrak{Y}^0_{\alpha,\beta}$: $\mathfrak{Y}^0_{0,0}$ is universal.

Most of the results also hold in $\mathfrak{Y}^0_{\alpha,-1}$; 2. has to be replaced with $T(r,f) \asymp \log^2 r$. The interested reader is also referred to [43, 53–55, 112, 206].

Example 4.7 ([7]) The differential equation $w'^2 = \frac{1}{1-z^2}w(4w^2 - g_2)$, where g_2 is chosen such that π is one of the periods of the corresponding P-function (with $\wp'^2 = 4\wp^3 - g_2\wp$), has a solution such that $w(\sin\zeta) = \wp(\zeta)$; w and $w^{(k)}$ belong to $\mathfrak{Y}^0_{0,-1}$ and $\mathfrak{Y}^0_{-k,-1}$, respectively. The limit functions $\mathfrak{w} = \lim_{h_k\to\infty} w_{h_k}$ are non-constant and satisfy $\mathfrak{w}'^2 = -\frac{1}{(\mathfrak{z}+1)^2}\mathfrak{w}(4\mathfrak{w}^2 - g_2)$, hence are given by $\mathfrak{w}(\mathfrak{z}) = \wp(i\log(\mathfrak{z}+1) + c)$ with essential singularity at $\mathfrak{z} = -1$. ☕

Remark 4.9 One may also consider meromorphic functions f on sectors $S : |\arg z - \theta_0| < \eta$. Restricting the sequences (h_k) to $|\arg z - \theta_0| < \eta - \delta$, $\delta > 0$ arbitrary, leads to the classes $\mathfrak{Y}_{\alpha,\beta}(S)$ and $\mathfrak{Y}^0_{\alpha,\beta}(S)$. The solutions to several important classes of differential equations belong to appropriate classes $\mathfrak{Y}_{\alpha,\beta}$. It is an open problem to determine the solutions w that belong to some hybrid class between $\mathfrak{Y}^0_{\alpha,\beta}$ and $\mathfrak{Y}_{\alpha,\beta}$ as follows: the plane is divided into finitely many alternating sectors S_ν and Σ_ν, such that $w \in \mathfrak{Y}^0_{\alpha,\beta}(S_\nu)$ while $w(z)z^{-\alpha}$ has a constant limit as $z \to \infty$ on every closed sub-sector of Σ_ν, and thus has an asymptotic expansion. Sometimes the sectors S_ν degenerate to smaller domains like $|\arg z - \theta_\nu| < \phi(|z|)$ where $\phi(r) \to 0$ as $r \to \infty$, and even to rays $\arg z = \theta_\nu$; more in Chaps. 5 and 6.

4.3.6 Yosida Classes and Riccati Differential Equations

The theory of generalised Yosida functions is most effective in the context of algebraic differential equations. This will be demonstrated in the case of the Riccati differential equations

$$w' = P(z) - w^2,$$

where, for the present, $P(z) = z^n + a_{n-1}z^{n-1} + \cdots + a_0$ is any polynomial. From

$$|w'| \le |P(z)| + |w|^2 = O(|z|^n + |w|^2)$$

it follows that w is a candidate for $\mathfrak{Y}_{\frac{n}{2},\frac{n}{2}}$. Suppose $w_{h_k} \to \mathfrak{w} \not\equiv \infty$. Then $w'_{h_k} \to \mathfrak{w}'$, locally uniformly on $\mathbb{C} \setminus \{\text{poles of } \mathfrak{w}\}$, and $\mathfrak{w}$ satisfies $\mathfrak{w}' = 1 - \mathfrak{w}^2$, hence the possible finite limit functions are $\mathfrak{w} \equiv \pm 1$ and $\mathfrak{w} = \coth(\mathfrak{z} + \mathfrak{z}_0)$; the limit $\mathfrak{w} \equiv \infty$ does not occur since $\mathfrak{v}_{h_k} = 1/w_{h_k}$ satisfies $\mathfrak{v}'_{h_k} = 1 - h_k^{-n}P(h_k + h_k^{-\frac{n}{2}}\mathfrak{z})\mathfrak{v}^2_{h_k}$, which excludes $\mathfrak{v}_{h_k} \to 0$. Thus $w \in \mathfrak{Y}_{\frac{n}{2},\frac{n}{2}}$ and $1/w \in \mathfrak{Y}_{-\frac{n}{2},\frac{n}{2}}$.

Exercise 4.16 Prove that to $\epsilon > 0$ there exists an $r_0 > 0$ such that each disc $\Delta_4(p)$: $|z - p| < 4|p|^{-\frac{n}{2}}$ about some pole p of w with $|p| > r_0$ contains exactly three poles $p, p + (\pi i + \epsilon_1)p^{-\frac{n}{2}}$, and $p - (\pi i + \epsilon_2)p^{-\frac{n}{2}}$, and two zeros $p + (\frac{\pi}{2}i + \epsilon_3)p^{-\frac{n}{2}}$ and $p - (\frac{\pi}{2}i + \epsilon_4)p^{-\frac{n}{2}}$, with $|\epsilon_j| < \epsilon$. The same is true if the terms 'zero' and 'pole' are permuted. In particular, the poles and zeros of w are $\frac{n}{2}$-separated from each other. (**Hint.** Re-scale about any sequence (p_n) of poles and consider the poles and zeros of $\coth \mathfrak{z}$; note that $\pi < 4 < \frac{3}{2}\pi$.)

Henceforth we assume that $\delta > 0$ is chosen so small that the discs

$$\Delta_\delta(p) : |z - p| < \delta|p|^{-\frac{n}{2}} \quad (p \in \mathfrak{P}) \quad \text{and} \quad \Delta_\delta(q) \quad (q \in \mathfrak{Q})$$

are mutually disjoint; again $\mathfrak{P}$ and $\mathfrak{Q}$ denote the set of non-zero poles and zeros of w, respectively.

Exercise 4.17 Let the closed simple curve Γ_r be obtained by modifying the circle $C_r : |z| = r$ as follows: $\Gamma_r = C_r$, if C_r does not intersect any disc $\Delta_\delta(p)$. If, however, $C_r \cap \Delta_\delta(p) \neq \emptyset$ for some pole p, replace this sub-arc of C_r with the part of $\partial\Delta_\delta(p)$ outside C_r if $|p| \le r$, and inside C_r otherwise. Prove that Γ_r has length $O(r)$ and deduce from $\operatorname{res}_p w = 1$ and $|w| = O(|z|^{\frac{n}{2}})$ on Γ_r that

$$n(r, w) = O(r^{\frac{n}{2}+1}) \quad \text{and} \quad T(r, w) = O(r^{\frac{n}{2}+1}).$$

Note that for $w \in \mathfrak{Y}_{\frac{n}{2},\frac{n}{2}}$, $T(r, w) = O(r^{n+2})$ had to be expected!

Remark 4.10 The construction of Γ_r works in a much more general context: Let (Δ_j) be any sequence of mutually disjoint discs with centres $z_j \to \infty$. Then a simple closed curve Γ_r with winding number 1 about points z_j with $|z_j| \le r$ and 0 if $|z_j| > r$ can be constructed which has length $O(r)$ and avoids the discs Δ_j.

Exercise 4.18 Deduce from Exercise 4.16 that each pole of w of sufficiently large modulus belongs to a unique sequence (p_k) of poles such that

$$p_{k+1} = p_k \pm (\pi i + o(1))p_k^{-\frac{n}{2}} \tag{4.13}$$

holds with fixed sign $\pm$. Sequences of this kind are called *strings* or, more precisely, $(\pm\pi i, \frac{n}{2})$-strings.

The recursion (4.13) can easily be resolved. Set $q_k = p_k^{\frac{n}{2}+1}$ to obtain

$$q_{k+1} = q_k(1 \pm (\pi i + o(1))q_k^{-1})^{\frac{n}{2}+1} = q_k \pm ((\tfrac{n}{2} + 1)\pi i + o(1)),$$

hence $q_k = \pm[(\frac{n}{2} + 1)\pi i + o(1)]k$ and $p_k = [\pm((\frac{n}{2} + 1)\pi i + o(1))k]^{\frac{2}{n+2}}$.

Exercise 4.19 (Continued) Prove that $\arg p_k \to \hat\theta_\nu = \frac{2\nu+1}{n+2}\pi$ holds for some ν. The rays $\arg z = \hat\theta_\nu$ are called *Stokes rays;* asymptotically they attract all poles.

Exercise 4.20 Prove that each string has counting function

$$n(r, (p_k)) = \frac{r^{\frac{n}{2}+1}}{(\frac{n}{2} + 1)\pi} + o(r^{\frac{n}{2}+1}).$$

In combination with Exercise 4.17 deduce that there are only finitely many strings, and show that on each sector $S_\delta : \theta_{\nu-1} + \delta < \arg z < \theta_\nu - \delta$, $w(z)z^{-\frac{n}{2}}$ tends to some $\epsilon_\nu = \epsilon_\nu(w) \in \{-1, 1\}$ as $z \to \infty$. (**Hint.** Re-scale along any sequence in S_δ and look for poles of the limit functions $\mathfrak{w} = \lim_{h_k\to\infty} w_{h_k}$.)

Actually the number of strings is $\sum_{\nu=0}^{n+1}(-1)^\nu\epsilon_\nu$ if n is even, and $\sum_{\nu=-m}^{m}(-1)^\nu\epsilon_\nu$ if $n = 2m - 1$ is odd (due to the discontinuity of $z^{\frac{n}{2}}$ along $\arg z = \pi$); more on Riccati equations and the Yosida class $\mathfrak{Y}_{\frac{n}{2},\frac{n}{2}}$ in Chap. 5.

Chapter 5
Algebraic Differential Equations

One of the most difficult problems in the theory of Algebraic Differential Equations is to decide whether or not the solutions are meromorphic in the plane. In case this question has been answered satisfactorily, which by experience requires particular strategies adapted to the equations under consideration, there remain several major problems to be solved: to determine the Nevanlinna functions, the distribution of zeros and poles, zero- and pole-free regions, asymptotic expansions on pole-free regions, and solutions deviating from the 'generic' case. This program will be pursued in the subsequent sections on linear, Riccati, and implicit first-order differential equations.

5.1 Linear Differential Equations

Throughout this section we will consider linear homogeneous differential equations

$$L[w] = w^{(n)} + a_{n-1}(z)w^{(n-1)} + \cdots + a_1(z)w' + a_0(z)w = 0 \tag{5.1}$$

with entire coefficients. The solutions are entire functions and the Wronskian determinant $W(w_1, \ldots, w_n)$ of any fundamental set $w_1, \ldots, w_n$ is an exponential function, $W(z) = e^{-\int a_{n-1}(z)\,dz}$. The operator L may be rediscovered as a formal determinant,

$$L[w] = \frac{W(w_1, \ldots, w_n, w)}{W(w_1, \ldots, w_n)}. \tag{5.2}$$

N. Steinmetz, *Nevanlinna Theory, Normal Families, and Algebraic Differential Equations*, Universitext, DOI 10.1007/978-3-319-59800-0_5

5.1.1 The Order of Growth

By Theorem 1.7 all solutions have finite order of growth if the coefficients in L are polynomials. In the particular case of constant coefficients the order of growth is at most one, this following from the well-known form of the distinguished fundamental set, which consists of functions $z^j e^{\lambda z}$. The converse is also true.

Theorem 5.1 *Suppose every solution to equation (5.1) has finite order resp. order of growth at most one. Then the coefficients in L are polynomials resp. constants.*

Proof We will prove by induction on n that if (5.1) has *meromorphic* coefficients and n linearly independent *meromorphic* solutions of finite order of growth resp. order at most one, then $m(r, a_\nu) = O(\log r)$ resp. $m(r, a_\nu) = o(\log r)$ $(0 \le \nu < n)$ holds. This is obvious if $n = 1$, where $a_0 = -w'/w$. To reduce the order of the equation we set $w = w_1 y$. The new equation for $u = y'$ is then

$$u^{(n-1)} + b_{n-2}(z)u^{(n-2)} + \cdots + b_1(z)u' + b_0(z)u = 0,$$
$$b_{\nu-1} = a_\nu + \sum_{j=\nu+1}^{n} \tbinom{j}{\nu} a_j w_1^{(j-\nu)}/w_1 \quad (1 \le \nu \le n-1).$$

Then $m(r, b_{\nu-1}) = O(\log r)$ holds by hypothesis, and assuming $m(r, a_j) = O(\log r)$ for $\nu < j \le n$ yields

$$m(r, a_\nu) \le m(r, b_{\nu-1}) + \sum_{j=\nu+1}^{n} \left[m(r, a_j) + m\big(r, w_1^{(j-\nu)}/w_1\big)\right] + O(1) = O(\log r).$$

In the second case (order ≤ 1) one has just to replace $O(\log r)$ with $o(\log r)$. ☕

Theorem 5.1 is a special case of the following Theorem of Frei [51].

Theorem 5.2 *Suppose that the coefficients $a_{j+1}, \ldots, a_{n-1}$ are polynomials, while a_j is transcendental. Then there are at most j linearly independent solutions of finite order.*

Exercise 5.1 Frei's Theorem may be proved by induction on n: adapt the above proof to show that if $m(r, a_\nu) = O(\log r)$ holds for $\nu = k, \ldots, n-1$, and if $k+1$ linearly independent solutions $w_1, \ldots, w_{k+1}$ have finite order of growth, then also $m(r, a_{k-1}) = O(\log r)$ holds. In any case prove that

$$m(r, a_\nu) = O(\log \max_{1 \le \mu \le n} rT(r, w_\mu))$$

holds outside some exceptional set of finite measure. ☕

To determine the possible orders of growth there are two methods in common. The first one employs the Wiman–Valiron Theory, see Sect. 1.5.3. Suppose $a_j(z) =$

$A_j z^{\alpha_j} + \cdots$ as $z \to \infty$, and let w be any entire transcendental solution. Then the central index $\nu(r)$ of w approximately solves the reduced equation

$$y^n + \sum_{j=0}^{n-1} A_j x^{\alpha_j + n - j} y^j = 0 \quad (y \sim \nu(r),\ x = z_r) \tag{5.3}$$

with z_r any point on $|z| = r$ such that $M(r, w) = |w(z_r)|$, and $r \notin F$, where F has finite logarithmic measure. The order of growth of w coincides with one of the non-positive (actually negative) slopes of the associated Newton–Puiseux polygon.

Exercise 5.2 Prove that the solutions to $w''' + zw' - (z+1)w = 0$ have order of growth either 1 or $\frac{3}{2}$. One solution is $w_1 = e^z$ of order 1. Derive the differential equation for $y = u'$ with $w = ue^z$ to prove that any solution $w \not\equiv ce^z$ has order of growth $\frac{3}{2}$.

5.1.2 *Asymptotic Expansions*

The second far-reaching method is based on asymptotic integration. Linear equations (5.1) with polynomial (or rational) coefficients may be transformed into linear systems

$$z^{-q}\mathfrak{y}' = A(z)\mathfrak{y}$$

which were discussed in Sect. 1.4. The procedure is not unique, in particular, the *rank* $q + 1$ depends on the transformation, and it may be difficult to determine the smallest possible integer q. If Theorem 1.9 is translated into the context of equation (5.1) we obtain:

Theorem 5.3 *Equation (5.1) has a* formal *fundamental system*

$$\mathring{w}_\nu(z) = e^{P_\nu(z^{1/p})} z^{\lambda_\nu} H_\nu(z, \log z) \quad (1 \le \nu \le n), \tag{5.4}$$

where

- *p is a positive integer,*
- *P_ν is a polynomial with $\deg P_\nu = d_\nu$ and $P_\nu(0) = 0$,*
- *λ_ν is some complex number, and*
- *$H_\nu(z, t)$ is a polynomial in t, its coefficients have asymptotic expansions in $z^{1/p}$.*

Given any sector $S : |\arg z - \theta| < h$ with sufficiently small central angle, there exists some fundamental system $w_1, \ldots, w_n$ such that w_ν is represented by $\mathring{w}_\nu$ on S.

Remark 5.1 In this form Theorem 5.3 goes back to Sternberg [187]. The first approximation $\mathring{w}(z) \approx e^{P(z^{1/p})} z^\lambda (\log z)^\mu$ with $P(z^{1/p}) = bz^\varrho + \cdots$ (we omit the index ν) may be differentiated arbitrarily often:

$$\frac{d^k}{dz^k}\mathring{w}(z) \approx (b\rho)^k e^{P(z^{1/p})} z^{\lambda+k\rho-k}(\log z)^\mu.$$

The triples $(P_\nu, \lambda_\nu, \mu_\nu)$ are mutually distinct if different branches of $z^{1/p}$ are well taken into account; in particular, logarithmic terms may occur only if $(P_\nu, \lambda_\nu) = (P_\kappa, \lambda_\kappa)$ for some $\kappa \neq \nu$. The similarity to linear differential equations with constant coefficients is apparent: solutions $z^j e^{\omega z}$ $(j \geq 1)$ occur if and only if ω is a multiple characteristic root.

It is obvious that the orders of growth of transcendental solutions w are among the positive numbers d_ν/p, that is $T(r, w) \asymp r^{d_\nu/p}$ holds for some ν. In particular, there always exist solutions of maximal order of growth $\varrho_{\max} = \max_{1\leq\nu\leq n} d_\nu/p$.

Exercise 5.3 Prove that the leading terms $b_\nu \zeta^{d_\nu}$ of the polynomials P_ν may be determined as follows: the simplified algebraic equation (5.3) has n branches $y_\nu(x) = c_\nu x^{\sigma_\nu} + \cdots$ as $x \to \infty$. Then $d_\nu = p\sigma_\nu$ and $b_\nu = pc_\nu/d_\nu$ if $\sigma_\nu > 0$.

Example 5.1 To determine the essential terms, namely the number p and the leading terms of the polynomials P_ν (or more) it is sometimes more convenient to set $y = w'/w$, derive a differential equation for y, and determine the first terms in the asymptotic expansions. For $w''' + z^2 w = 0$ this leads to $y'' + 3yy' + y^3 + z^2 = 0$. To determine the leading term we plug in $y \sim kz^\alpha$, $y' \sim \alpha k z^{\alpha-1}$, $y'' \sim \alpha(\alpha-1)kz^{\alpha-2}$ to obtain $k^3 z^{3\alpha} + z^2 = O(|z|^{\max\{2\alpha-1,\alpha-2\}})$, hence $3\alpha = 2$, $p = 3$, and $k = -1$, say. Given any sector of central angle $\frac{2}{3}\pi$ we obtain three mutually distinct solutions $y \sim -z^{2/3} - \frac{2}{3}z^{-1} + \cdots$, hence three linearly independent solutions $\mathring{w}(z) = e^{-\frac{3}{5}z^{5/3}} z^{-2/3}(1 + O(|z|^{-1}))$, one for each branch of $z^{1/3}$; logarithmic terms do not occur, and every non-trivial solution has maximal order of growth $\varrho_{\max} = \frac{5}{3}$.

Exercise 5.4 ([3]) Employ Wiman–Valiron theory to compute the possible orders of the solutions to $w''' + z^2 w'' + zw' + w = 0$, and determine the principal terms of the solutions $\mathring{w}_\nu$ (see Theorem 5.3). Prove that the solution represented by $\mathring{w}_2 - \mathring{w}_3$ on $|\arg z| < \delta$ has zeros $\approx e^{k\pi}$ $(k \to \infty)$.

Solution $\mathring{w}_1 \sim e^{-z^3/3} z^{-3}(1 + \frac{10}{3}z^{-3} + \cdots)$, $\mathring{w}_{2,3} \sim z^{\pm i}(1 - \frac{7\pm 9i}{39}z^{-3} + \cdots)$.

5.1.3 Sub-normal Solutions

Asymptotic integration and Wiman–Valiron theory applied to Eq. (5.7) provide a finite set of possible orders of growth; $\varrho = 0$ is reserved for polynomial solutions

($\varrho = 0$ means $\nu(r) = O(1)$). Let S be any sufficiently small sector and let w_ν, $\nu \in J$, denote the solution that is represented on S by $\mathring{w}_\nu$ with maximal degree $\deg P_\nu = p\varrho_{\max}$. Obviously, these functions are linearly independent, hence every fundamental system contains at least as many solutions of maximal order of growth as are formal solutions with maximal degree of P_ν. Nothing more can be said in general. The non-trivial solutions of order of growth less than $\varrho_{\max}$ are called *sub-normal*, in combination with the trivial solution they form a linear subspace. Necessary for the existence of sub-normal solutions, but not sufficient, is that the Newton–Puiseux diagram has at least two non-positive slopes. For example, this is the case in Exercise 5.2.

Example 5.2 The existence of sub-normal solutions is in some sense exceptional. The Newton–Puiseux method applied to equation $w''' + zw' - (z + 1 + \lambda)w = 0$ indicates linearly independent solutions having orders of growth $1, \frac{3}{2}, \frac{3}{2}$. The method of Example 5.1 yields $y'' + 3yy' + y^3 + zy - (z + 1 + \lambda) = 0$ having solutions $y_1 \sim 1 + \lambda z^{-1} - 3\lambda z^{-2} + \cdots$ and $y_{2,3} \sim iz^{\frac{1}{2}} - \frac{1}{2} + \frac{3}{8}iz^{-\frac{1}{2}} - (\frac{\lambda}{2} + \frac{3}{4})z^{-1} + \cdots$, one for each branch of $z^{\frac{1}{2}}$, hence $\mathring{w}_1 = e^z z^\lambda(1 + O(|z|^{-1}))$ and

$$\mathring{w}_{2,3} = \exp\left(\tfrac{2}{3}iz^{\frac{3}{2}} - \tfrac{1}{2}z + \tfrac{3}{4}iz^{\frac{1}{2}}\right) z^{-\frac{\lambda}{2}-\frac{3}{4}}(1 + O(|z|^{-\frac{1}{2}})).$$

A simple compactness argument shows that solutions of order of growth *one* (if any) are given by $w = Q_\lambda(z)e^z$ with some polynomial Q_λ of degree λ. An elementary computation shows that this happens for every $\lambda \in \mathbb{N}_0$ with $Q_0(z) = 1$, $Q_1(z) = z + 3$, $Q_2(z) = z^2 + 6z + 12$, etc. In any other case the non-trivial solutions have maximal order of growth $\varrho_{\max} = \frac{3}{2}$.

5.1.4 The Phragmén–Lindelöf Indicator

Let w be any solution to (5.1) of order of growth $\varrho > 0$. Then

$$h_w(\theta) = \limsup_{r\to\infty} \frac{\log|w(re^{i\theta})|}{r^\varrho} \tag{5.5}$$

is called the *Phragmén–Lindelöf indicator* or just *indicator* of w. Given any angle $\hat{\theta}$, w is represented on $S : |\arg z - \hat{\theta}| < \delta$ by some linear combination $\sum_{\nu=1}^n c_\nu \mathring{w}_\nu$ with coefficients $c_\nu = c_\nu(w, \hat{\theta})$. If the polynomials P_ν of degree $p\varrho$ have mutually distinct leading coefficients b_ν, an elementary computation yields

$$h_w(\theta) = \max\{\operatorname{Re} b_\nu e^{i\varrho\theta} : c_\nu \neq 0,\ \deg P_\nu = p\varrho\}$$

on $|\theta - \hat{\theta}| < \delta$ (with the convention that the maximum of the empty set is zero). Then $h_w(\theta)$ is continuous and piecewise continuously differentiable. The discontinuities

$\hat{\theta}_j$ of $h_w'(\theta)$ are called *Stokes directions* (of w of order ϱ); they occur if the maximum is attained for $\nu = \kappa$ on $0 < \theta - \hat{\theta} < \delta' \leq \delta$ and for $\nu = \lambda \neq \kappa$ on $0 < \hat{\theta} - \theta < \delta' \leq \delta$. If $\hat{\theta}$ is not a Stokes direction,

$$\log|w(z)| = h_w(\theta)r^\varrho + O(r^{\varrho-\epsilon}) \quad (0 < \epsilon < 1/p) \tag{5.6}$$

holds as $z = re^{i\theta} \to \infty$ on S. We mention without proof that this remains true if $\hat{\theta}$ is a Stokes direction and S is replaced with $S \setminus \bigcup_j \{z : |z - z_j| < r_j\}$, such that $\frac{1}{r}\sum_{|z_j|<r} r_j$ tends to zero as $r \to \infty$, and also (in arbitrary directions) if the leading coefficients b_ν are not necessarily mutually distinct.

Remark 5.2 Entire functions of finite order of growth $\varrho > 0$ satisfying (5.6) with error term $o(r^\varrho)$ outside some so-called C^0-set $\bigcup_j \{z : |z - z_j| < r_j\}$ with $\frac{1}{r}\sum_{|z_j|<r} r_j \to 0$ as $r \to \infty$ are called of *completely regular growth* in the sense of Levin–Pfluger (see, e.g. [110, 111]). In the general case the proof that (5.6) holds on S outside some C^0-set is quite elaborate and much more delicate. The interested reader is referred to [3, 5, 9, 18, 171–173].

5.1.5 *The Distribution of Zeros*

From now on the coefficients in (5.1) are assumed to be polynomials. It will be convenient just to consider the canonical form

$$L[w] = w^{(n)} + a_{n-2}(z)w^{(n-2)} + \cdots + a_1(z)w' + a_0(z)w = 0 \tag{5.7}$$

with constant Wronskian for any fundamental system. This is easily obtained by the transformation $w \mapsto w\exp(-\frac{1}{n}\int a_{n-1}(z)\,dz)$ (without changing notation), and does not affect the distribution of zeros.

Exercise 5.5 Let w be any solution to (5.7) of order ϱ. Assuming (5.6), prove that

$$m(r, w^{\pm 1}) = \frac{r^\varrho}{2\pi}\int_0^{2\pi} h_w^{\pm}(\theta)\,d\theta + O(r^{\varrho-\epsilon})$$

holds for some $\epsilon > 0$. In particular, the deficiency $\delta(0, w)$ is either less than one or else 0 is a Borel or even Picard exceptional value of w. Prove also that the counting function of zeros outside the union of sectors $|\arg z - \hat{\theta}_j| < \delta$ is $O(r^{\varrho-\epsilon})$.

Remark 5.3 If the polynomials P_ν are not mutually distinct one has to estimate from below functions f represented by finite sums $\sum e^{Q_\nu(z^{1/p})}z^{\lambda_\nu}H_\nu(z, \log z)$ on sectors S, but outside small neighbourhoods of the zeros; f behaves like an entire function of order of growth at most $\varrho - \epsilon$ and may have infinitely many zeros with counting function $O(r^{\varrho-\epsilon})$. Nevertheless, the statements in Exercise 5.5 also remain valid in this case. Bank [3] seems to have been the first to notice that there exist solutions having infinitely many zeros outside arbitrarily small sectors $|\arg z - \hat{\theta}_j| < \epsilon$ about the Stokes directions, a phenomenon that does not occur if $n = 2$.

Example 5.3 Exponential polynomials

$$w(z) = \sum_{\mu=1}^{m} p_\mu(z) e^{\omega_\mu z}$$

($p_\mu \not\equiv 0$ polynomials, ω_μ mutually distinct) satisfy some Eq. (5.1) with constant coefficients. Prove that w has indicator $h_w(\theta) = \max_{1\le\mu\le m} \operatorname{Re} \omega_\mu e^{i\theta}$. We mention without proof that $\frac{1}{2\pi}\int_0^{2\pi} h_w(\theta)\,d\theta$ is the circumference of the so-called *indicator diagram* I, the convex hull of the points $\bar{\omega}_1, \ldots, \bar{\omega}_m$ (see [140], Vol. II, p.163). Prove that $\delta(0, w) = 0$ if and only if $0 \in I$.

5.1.6 Exceptional Fundamental Sets

We will now discuss the question of whether there may exist fundamental systems $w_1, \ldots, w_n$ that are exceptional in the sense that $E = w_1 \cdots w_n$ has fewer zeros than expected ($N(r, 1/E) = o(r^{\varrho_{\max}})$). The case $n = 2$ is very easy to deal with on one hand, but is not very instructive on the other. This is true in many respects, most phenomena occur in the case of $n \ge 3$ only.

Exercise 5.6 For $w'' + a_0(z)w = 0$ ($a_0 \not\equiv 0$), $E = w_1w_2$, and $W(w_1, w_2) = c \ne 0$ derive the differential equation $2EE'' - E'^2 + 4a_0(z)E^2 = c^2$, and deduce $m(r, 1/E) = O(\log r)$. To prove that either $N(r, 1/E) \asymp \max\{T(r, w_1), T(r, w_2)\}$ or else $a_0(z) \equiv -\lambda^2$ and $w_{1,2}(z) = c_{1,2}e^{\pm\lambda z}$ holds, compare the central index $\nu_w(r)$ of any solution $w \not\equiv 0$ with the central index $\nu_E(r)$ if $E \not\equiv$ const.: prove $\nu_E(r) \sim 2\nu_w(r)$. If, however, E is a non-zero constant, then $a_0(z)$ is also a non-zero constant.

The goal of this section is to transfer this alternative to the case $n > 2$.

Theorem 5.4 *Let $w_1, \ldots, w_n$ be some fundamental set to Eq. (5.7) such that $\delta(0, w_\nu) = 1$ holds for every w_ν of maximal order of growth $\varrho_{\max}$. Then L has constant coefficients.*

Exercise 5.7 The differential equation $w''' - 4z^2w' - 12zw = 0$ is solved by $w_1 = e^{z^2}$ and $w_2 = e^{-z^2}$. Expand $W(w_1, w_2, w_3) = c$ to obtain the linear inhomogeneous equation $-4zw_3'' + 4w_3' + 16z^3w_3 = c$; for $c \ne 0$ deduce $m(r, 1/w_3) = O(\log r)$ and $\nu(r) \sim 2r^2$ (central index of w_3). By Theorem 5.4 the zeros of w_3 have exponent of convergence 2.

The proof of Theorem 5.4 requires some preparation. We first note that by Exercise 5.5, zero is a Borel (or Picard) exceptional value for every w_ν of maximal order $q = \varrho_{\max}$; q is a positive integer, and every w_ν has the form $w_\nu(z) = e^{b_\nu z^q} g_\nu(z)$, where g_ν is an entire function of order of growth less than q (or a polynomial), with $b_\nu = 0$ if $\varrho(w_\nu) < q$. The functions w_ν with fixed $b_\nu = b$ span the subspace $\mathbf{V}_b$.

Lemma 5.1 *Let $w(z) = g(z)e^{bz^q}$ with $b \neq 0$ and $\varrho(g) < q$ be any solution to Eq. (5.1). Given any sufficiently small sector S, let $\hat{w}_1, \ldots, \hat{w}_n$ denote the solutions that are represented on S by the formal solutions $\mathring{w}_1, \ldots, \mathring{w}_n$, and assume $P_\mu(z^{1/p}) = bz^q + \cdots$ exactly for $\mu = 1, \ldots, m$. Then $w, \hat{w}_1, \ldots, \hat{w}_m$ are linearly dependent.*

Proof Suppose that $w(z) - \sum_{\mu=1}^{m} c_\mu \hat{w}_\mu(z) = \sum_{\nu=m+1}^{n} c_\nu \hat{w}_\nu(z) \not\equiv 0$ holds on S; then the left-hand side has indicator $h(\theta) = \operatorname{Re} be^{iq\theta}$ (of order q), while the right-hand side has not. This proves $c_\nu = 0$ for $m < \nu \leq n$.

Lemma 5.2 *Suppose that $\mathbf{V}_b$ is spanned by $w_1, \ldots, w_m$. Then $W(w_1, \ldots, w_m)$ has only finitely many zeros, and the operator*

$$M[w] = \frac{W(w_1, \ldots, w_m, w)}{W(w_1, \ldots, w_m)} = w^{(m)} + s_{m-1}(z)w^{(m-1)} + \cdots + s_0(z)w$$

has rational coefficients $s_\mu(z) = \binom{m}{\mu}(-bqz^{q-1})^{m-\mu} + \cdots$ at $z = \infty$.

Proof The coefficients s_μ of M have only finitely many poles if and only if the Wronskian $W(w_1, \ldots, w_m)$ has only finitely many zeros, and then, indeed, are rational functions, this following from the proof of Theorem 5.1 and Exercise 5.1, which show that $m(r, s_\mu) = O(\log r)$. Since the algebraic equation

$$y^m + s_{m-1}(x)y^{m-1} + \cdots + s_0(x) = 0$$

has m solutions $y_\mu(x) = bqx^{q-1} + \cdots$ with the same leading term, hence has the simplified equation $(y - bqx^{q-1})^m = 0$, the shape of the rational functions s_μ follows from the binomial theorem. To prove that $W(w_1, \ldots, w_m)$ has only finitely many zeros we consider any sufficiently small sector S and denote by $\hat{w}_1, \ldots, \hat{w}_m$ the solutions that occur in Lemma 5.1. Then $w_1, \ldots, w_m$ and $\hat{w}_1, \ldots, \hat{w}_m$ span the same linear space, and this implies $W(w_1, \ldots, w_m) = cW(\hat{w}_1, \ldots, \hat{w}_m)$ on S for some $c = c(S) \neq 0$. To compute $W(\hat{w}_1, \ldots, \hat{w}_m)$ we will have a closer look at a single solution $\hat{w}_\mu$; it is represented on S by $e^{P_\mu(z^{1/p})} z^{\lambda_\mu} H_\mu^{[0]}(z, \log z)$ with $P_\mu(z^{1/p}) = bz^q + Q_\mu(z^{1/p})$ ($\deg Q_\mu < \deg P_\mu$), and the kth derivative $\hat{w}_\mu^{(k)}$ has the representation $e^{P_\mu(z^{1/p})} z^{\lambda_\mu + k(q-1)} H_\mu^{[k]}(z, \log z)$ (at each step the factor qbz^{q-1} occurs in addition; the constant qb is incorporated into $H_\mu^{[k]}$). Thus $W(\hat{w}_1, \ldots, \hat{w}_m)$ has the representation $e^{P(z^{1/p})} z^\lambda H(z, \log z)$ with $\lambda = \sum_{\mu=1}^m \lambda_\mu + \frac{1}{2}(q-1)m(m-1)$ and $P = \sum_{\mu=1}^m P_\mu$; $H(z, t)$ is some polynomial in t with coefficients having asymptotic expansions in $z^{1/p}$. On S, $H(z, \log z)$ represents some analytic function $h(z) = c(1 + o(1))(\log z)^K z^{-N/p}$ (K, N integers, $z \to \infty$ on S), which obviously has only finitely many zeros. A simple compactness argument then shows that $W(w_1, \ldots, w_m)$ has only finitely many zeros on the plane. ☕

Proof of Theorem 5.4 Let $\mathbf{V}_b$ be the vector space spanned by the solutions $w_\mu(z) = e^{bz^q} g_\mu(z)$ $(1 \leq \mu \leq m)$ with $q = \varrho_{\max}$, $b \neq 0$, and $\varrho(g_\mu) < q$. The Wronskian $W(w_1, \ldots, w_n)$ is constant and

$$W(M[w_{m+1}], \ldots, M[w_n]) = \frac{W(w_1, \ldots, w_n)}{W(w_1, \ldots, w_m)} \tag{5.8}$$

holds by Exercise 1.13. The right-hand side of (5.8) has no zeros and at most finitely many poles by Lemma 5.2. We will show that the left-hand Wronskian has at least $m(n-m)(q-1)$ zeros, which leads to the following alternative:

- $m = n$, hence $W(w_1, \ldots, w_n) = e^{nbz^q + \cdots}$ is non-constant.
- $q = \varrho_{\max} = 1$, hence L has constant coefficients.

The solution space $\mathbf{V}$ of $L[w] = 0$ is the direct sum $\mathbf{V}_b \oplus \mathbf{V}_{\beta_1} \oplus \cdots \oplus \mathbf{V}_{\beta_k}$, where $\mathbf{V}_{\beta_\kappa}$ is spanned by the functions $w_\nu(z) = e^{b_\nu z^q} g_\nu(z)$ with $b_\nu = \beta_\kappa \neq b$ ($\beta_\kappa \neq b$ mutually distinct; $\beta_\kappa = 0$ is admitted, $\mathbf{V}_0$ collects the sub-normal solutions). Given any sufficiently small sector S, $\mathbf{V}_{\beta_\kappa}$ is also spanned by the functions represented on S by $\hat{w}_\ell$ with $b_\ell = \beta_\kappa$. Let $u(z) = e^{\beta_\kappa z^q + Q(z^{1/p})} z^\lambda H(z, \log z)$ denote any solution with $\beta_\kappa \neq 0$ and $\frac{1}{p} \deg Q < q$. From $u^{(j)}(z) = (\beta_\kappa q z^{q-1} + o(|z|^{q-1}))^j u(z)$ and $s_j(z) = \binom{m}{j}(-bqz^{q-1} + o(|z|^{q-1}))^{m-j}$ it easily follows that

$$\begin{aligned} M[u] &= \sum_{j=0}^{m} \tbinom{m}{j}(-bqz^{q-1} + o(|z|^{q-1}))^{m-j}(\beta_\kappa q z^{q-1} + o(|z|^{q-1}))^j u(z) \\ &= (\beta_\kappa - b + o(1))^m q^m z^{(q-1)m} u(z). \end{aligned}$$

Similarly we obtain $\frac{d}{dz} M[u] = (\beta_\kappa - b + o(1))^m q^m z^{(q-1)m} u'(z)$, etc., hence

$$W(M[w_{m+1}], \ldots, M[w_n]) = (C + o(1)) z^{m(n-m)(q-1)} W(u_{m+1}, \ldots, u_n)$$

with $C = \prod_{j=1}^k q^m (\beta_\kappa - b)^m \neq 0$. Since $W(M[w_{m+1}], \ldots, M[w_n])$ has at most finitely many poles and $W(u_{m+1}, \ldots, u_n)$ has no poles and only finitely many zeros on S, a compactness argument shows that

$$W(M[w_{m+1}], \ldots, M[w_n]) = r(z) e^{Q(z)}$$

holds, where Q is some polynomial and r is a rational function which has a pole of order at least $m(n-m)(q-1)$ at infinity, hence also has at least $m(n-m)(q-1)$ finite zeros. ☕

5.1.7 Fundamental Sets with Zeros Along the Real Axis

Let g be an entire function with Nevanlinna characteristic $T(r,g) \asymp r^{\varrho}$. The zeros of g are said to be asymptotically distributed along the rays

$$\arg z = \hat{\theta}_j \quad (\hat{\theta}_1 < \hat{\theta}_2 < \cdots < \hat{\theta}_k < \hat{\theta}_{k+1} = \hat{\theta}_1 + 2\pi)$$

if for every $\epsilon > 0$ the counting function of zeros on the union of sectors $\hat{\theta}_j + \epsilon < \arg z < \hat{\theta}_{j+1} - \epsilon$ is $o(r^{\varrho})$. For entire functions of completely regular growth the distribution of zeros has an effect on the order of growth as follows.

Theorem 5.5 *Let g be an entire function of completely regular growth of order ϱ, and suppose that its zeros are asymptotically distributed along the rays* $\arg z = \hat{\theta}_j$. *Then $\delta(0,g) = 0$ implies $\min_{1\le j\le k}(\hat{\theta}_{j+1} - \hat{\theta}_j) \le \pi/\varrho$.*

Proof The indicator of g is piecewise trigonometric, that is, $h_g(\theta) = \operatorname{Re} c_j e^{i\varrho\theta}$ holds on $\hat{\theta}_j < \arg z < \hat{\theta}_{j+1}$, and $\delta(0,g) = 0$ yields $\int_0^{2\pi} h_g^+(\theta)\,d\theta > \int_0^{2\pi} h_g^-(\theta)\,d\theta = 0$. Thus h_g is never negative, and there is at least one interval $(\hat{\theta}_j, \hat{\theta}_{j+1})$ where $h_g(\theta)$, hence $\cos(\varrho\theta + \gamma_j)$ is positive. This implies $\varrho(\hat{\theta}_{j+1} - \hat{\theta}_j) \le \pi$.

Remark 5.4 In [33] a similar result is proved: *Suppose the zeros of some entire function g* (not necessarily of completely regular growth) *are distributed along the rays* $\arg z = \hat{\theta}_j$ *and ϱ is sufficiently large* (depending on the geometry of the rays). *Then* $\delta(0,g) > 0$. In other words, ϱ cannot be arbitrarily large if $\delta(0,g) = 0$.

Theorem 5.6 *Let $w_1, \ldots, w_n$ be any fundamental set of Eq. (5.7). Then each of the conditions below implies that the coefficients of L are constants.*

- *$T(r, w_1 \cdots w_n) = o(r^{\varrho_{\max}})$;*
- *the zeros of the solutions w_ν of order $\varrho_{\max}$ are asymptotically distributed along the real axis.*

Proof Set $E = w_1 \cdots w_n$. In the first case, $\delta(0, w_\nu) = 1$ holds for every solution w_ν of maximal order, this following from $N(r, 1/w_\nu) \le N(r, 1/E) = o(r^{\varrho_{\max}})$. By Theorem 5.4 the coefficients of L are constants. In the second case the entire function E satisfies

$$m\Big(r, \frac{1}{E}\Big) = m\Big(r, \frac{W}{w_1 \cdots w_n}\Big) + O(1) = O(\log r)$$

(the Wronskian $W = W(w_1, \ldots, w_n)$ is constant), and E is either a polynomial and Theorem 5.4 applies, or else is a transcendental function of order of growth

$\varrho(E) \leq 1$ by Theorem 5.5. In case of $\varrho_{\max} \leq 1$, Theorem 5.1 applies. If, however, $\varrho_{\max} > 1$ holds, then Theorem 5.4 applies since $N(r, 1/w_\nu) \leq N(r, 1/E) = O(r) = o(r^{\varrho_{\max}})$ holds for every solution w_ν.

Exercise 5.8 We note that Theorem 5.6 fails to hold if in (5.1) arbitrary rational coefficients are admitted. To give an example, let $\omega_1, \ldots, \omega_n$ be non-zero and mutually distinct complex constants, and set $w_\nu(z) = e^{\omega_\nu z^2}$. Prove that the Wronskian $W(w_1, \ldots, w_n)$ has only finitely many zeros, hence the functions w_ν form a zero-free fundamental set of some Eq. (5.1) with non-constant rational coefficients.

Remark 5.5 Theorem 5.4 was proved by Frank [45] under the hypothesis that $E = w_1 \cdots w_n$ has only finitely many zeros, that is, if 0 is a Picard exceptional value for every w_ν, and by Brüggemann [18] under the weaker assumption that 0 is a Borel exceptional value for each w_ν, thereby solving a problem posed by Frank and Wittich. Brüggemann [19] also settled a conjecture of Hellerstein and Rossi (see [17]) by proving the second part of Theorem 5.6, assuming that the non-real zeros of E have exponent of convergence less than $\varrho_{\max}$. In the present form, Theorem 5.6 is due to the author [171–173]. There is an extensive literature on second-order equations, where naturally much more is known; we mention [3–5, 38–40, 45, 50, 67, 70, 74, 81, 82]. In the next section we will consider second-order equations via a new approach to Riccati equations.

5.2 Riccati Equations

The basic facts concerning the value distribution of the solutions to Riccati differential equations

$$w' = a(z) + b(z)w + c(z)w^2 \quad (c \not\equiv 0) \tag{5.9}$$

with polynomial coefficients are well understood due to the pioneering work of Wittich (see his book [202], Chapter V, p. 73–80). The solutions are meromorphic in the complex plane, and every non-rational solution has Nevanlinna characteristic $T(r, w) \asymp r^{\frac{n}{2}+1}$; the non-negative integer n can easily be computed from the coefficients a, b, and c. Equation (5.9) is closely related to some second-order linear differential equation (transformation $c(z)w = -u'/u$), hence it would be possible to take the detour via this also well-understood linear equation. We will, however, use a new approach which was initiated in Sect. 4.3 and also applies to higher-order equations and systems.

5.2.1 A Canonical Form

The change of variables $w_1 = cw + \frac{1}{2}b + \frac{1}{2}(c'/c)$ transforms (5.9) into

$$\begin{aligned} w_1' &= A(z) - w_1^2, \\ A(z) &= \tfrac{1}{4}b^2 - ac - \tfrac{1}{2}b' + \tfrac{3}{4}(c'/c)^2 + \tfrac{1}{2}b(c'/c) - \tfrac{1}{2}(c''/c) \\ &= a_n z^n + a_{n-1}z^{n-1} + \cdots \text{ as } z \to \infty. \end{aligned}$$

Replacing the variables z and w_1 with $\tilde{z} = z/\lambda$ and $\tilde{w}(\tilde{z}) = \lambda w_1(z)$ ($\lambda^{n+2}a_n = 1$), we eventually obtain the *canonical form*, which, of course, will be written $\tilde{}$ -free as

$$\begin{aligned} w' &= P(z) - w^2, \\ P(z) &= z^n + c_{n-1}z^{n-1} + \cdots \text{ as } z \to \infty. \end{aligned} \tag{5.10}$$

This equation was the object in Exercises 4.16–4.20, where P was assumed to be a polynomial, while now P is a rational function. This, however, is not really restrictive since we are interested in asymptotic results. Of course, we will only consider solutions that are meromorphic in the plane. This is the case if the coefficients in (5.9) are polynomials. The method applies if $n \geq -1$, though $n = -1$ never occurs in the polynomial case.

Exercise 5.9 Prove that $w' = \frac{1}{2z}(z + w + w^2)$ has the meromorphic solution $w = \sqrt{z}\tan\sqrt{z}$ of order of growth $\frac{1}{2}$. Deduce the canonical form $w' = z^{-1} - \frac{3}{16}z^{-2} - w^2$. Note, however, that just one solution is single-valued.

Exercise 5.10 The degrees of the coefficients a, b, c in (5.9) may be arbitrarily large compared with n. Prove that the Riccati equation

$$w' = z^{2m-1} + 2z^{m-1} - (m-1)z^{m-2} + 2(z^m + 1)w + zw^2$$

has the canonical form $w' = 1 + z^{-1} + \frac{3}{4}z^{-2} - w^2$.

The main result on the solutions to the canonical equations (5.10) may be stated as follows.

Theorem 5.7 *Let w be any transcendental meromorphic solution to the canonical Riccati equation* (5.10). *Then*

- *w belongs to the Yosida class $\mathfrak{Y}_{\frac{n}{2},\frac{n}{2}}$;*
- *the poles of w are distributed in $(\pm\pi i, \frac{n}{2})$-strings; each string (p_k)*
  - *satisfies the iterative scheme $p_{k+1} = p_k \pm (\pi i + o(1))p_k^{-\frac{n}{2}}$;*
  - *is asymptotic to some ray $\arg z = \hat{\theta}_\nu = \frac{2\nu+1}{n+2}\pi$, at most one for each ν;*
  - *has counting function $n(r, (p_k)) \sim \dfrac{r^{\frac{n}{2}+1}}{(\frac{n}{2}+1)\pi}$;*

– *on* $\Sigma_\nu : \hat{\theta}_{\nu-1} < \arg z < \hat{\theta}_\nu$, *w has an asymptotic expansion* $w \sim \epsilon z^{\frac{n}{2}} + \sum_{j=1}^{\infty} c_j(\epsilon) z^{\frac{1}{2}(n-j)} \quad (\epsilon = \epsilon(w, \nu) \in \{-1, 1\})$;

– *w has Nevanlinna characteristic* $T(r, w) = \ell(w)\dfrac{r^{\frac{n}{2}+1}}{(\frac{n}{2}+1)^2\pi} + o(r^{\frac{n}{2}+1})$, *where* $\ell(w)$ *denotes the number of strings* $(1 \le \ell(w) \le n+2)$.

Remark 5.6 The rays $\sigma_\nu : \arg z = \hat{\theta}_\nu$ and sectors $\Sigma_\nu : \hat{\theta}_{\nu-1} < \arg z < \hat{\theta}_\nu$ are called *Stokes rays* and *sectors*, respectively. The coefficients c_j in the asymptotic expansions depend only on ϵ, and c_{2j-1} vanishes if n is even. The reader who is acquainted with the field of Complex Dynamics (see, for example, [174]) will notice the close relationship of the iterative scheme with the iteration of the rational function $R_\pm(z) = z \pm \pi i z^{-m}$ (assuming $\frac{n}{2} = m$); $R_\pm$ has a *parabolic* fixed-point at ∞ and $m+1$ *attracting petals* around the rays $\arg(\pm\pi i z^{m+1}) = 0 \bmod 2\pi$; convergence to infinity takes place along the rays $\arg z = \hat{\theta}_\nu$.

Proof Most of the above statements may be deduced from Exercises 4.16–4.20; they are based solely on the elementary estimate $|w'(z)| = O(|z|^n + |w(z)|^2)$. It remains to prove the existence of asymptotic expansions and the fact that there is at most one string along each Stokes ray $\arg z = \hat{\theta}_\nu$.

Re-scaling along any sequence $h_k \to \infty$ with $\hat{\theta}_\nu + \delta < \arg z < \hat{\theta}_{\nu+1} - \delta$ yields solutions $\mathfrak{w} = \lim_{h_k \to \infty} w_{h_k}$ $(w_h = h^{-\frac{n}{2}} w(h + h^{-\frac{n}{2}}\mathfrak{z}))$ to $\mathfrak{w}' = 1 - \mathfrak{w}^2$ without poles, hence $\mathfrak{w} = \pm 1$. Since the *cluster set*[1] of $z^{-\frac{n}{2}} w(z)$ as $z \to \infty$ on each sector $\hat{\theta}_\nu + \delta < \arg z < \hat{\theta}_{\nu+1} - \delta$ is connected, it follows that $z^{-\frac{n}{2}} w(z) \to \epsilon_\nu \in \{-1, 1\}$. In other words,

$$w(z) = \epsilon_\nu z^{\frac{n}{2}} + o(|z|^{\frac{n}{2}}) = \psi_0(z) + o(|z|^{\frac{n}{2}})$$

holds as $z \to \infty$, uniformly on $\hat{\theta}_\nu + \delta < \arg z < \hat{\theta}_{\nu+1} - \delta$. To prove that w has an asymptotic expansion we could apply Theorem 1.11. It is, however, more instructive to give an *ad hoc* proof. Assuming

$$w(z) = \epsilon_\nu z^{\frac{n}{2}} + \sum_{j=1}^{k} c_j z^{\frac{n-j}{2}} + o(|z|^{\frac{n-k}{2}}) = \psi_k(z) + o(|z|^{\frac{n-k}{2}}) \tag{5.11}$$

for some $k \ge 0$, hence $-w^2 + P(z) = \psi_k'(z) + o(|z|^{\frac{n-k}{2}-1})$, we may compare w with the solution $y = \epsilon_\nu z^{\frac{n}{2}} + \sum_{j=1}^{\infty} c_j^{[k]} z^{\frac{n-j}{2}}$ to the algebraic equation $-y^2 + P(z) = \psi_k'(z)$.

---

[1] Let f be a bounded holomorphic function on some unbounded domain D. By definition, the cluster set of f as $z \to \infty$ on D is the set of all limits $\lim_{n\to\infty} f(z_n)$ with $z_n \to \infty$ in D. The cluster set is always compact, and connected if D is *locally connected at infinity.* A sufficient condition for the latter is that any two points $a, b \in D$ may be joined by a curve in $D \cap \{z : |z| \ge \min\{|a|, |b|\}\}$. For example, this is true for sectors, and also for sectors with mutually disjoint 'holes' $|z - z_j| \le r_j$, $z_j \to \infty$.

This leads to $(w-y)(w+y) = o(|z|^{\frac{n-k}{2}-1})$ and $w+y \sim 2\epsilon_\nu z^{\frac{n}{2}}$, hence $w-y = o(|z|^{\frac{n-k}{2}-1-\frac{n}{2}})$ and

$$w = \epsilon_\nu z^{\frac{n}{2}} + \sum_{j=1}^{k+1} c_j^{[k]} z^{\frac{n-j}{2}} + o(|z|^{\frac{n-(k+1)}{2}}) = \psi_{k+1}(z) + o(|z|^{\frac{n-(k+1)}{2}}),$$

that is, (5.11) holds with k replaced with $k+1$.
To determine the total number of strings of poles we first assume that n is even, and remind the reader of the construction of the closed simple curve Γ_r in Exercise 4.17: Γ_r has winding number 1 and 0 about poles on $|p_k| \le r$ and $|p_k| > r$, respectively, $w = O(|z|^{\frac{n}{2}})$ holds on Γ_r, and the length of the sub-arc of Γ_r contained in any sector $|\arg z - \theta| < \delta$ is $< K\delta r$. Since all but finitely many poles have residue 1,

$$n(r,w) = \frac{1}{2\pi i}\int_{\Gamma_r} w(z)\,dz + O(1) \tag{5.12}$$

holds. The contribution of the sector Σ_ν to the integral in (5.12) is

$$\epsilon_\nu r^{\frac{n}{2}+1}\frac{1}{2\pi}\int_{\hat\theta_{\nu-1}}^{\hat\theta_\nu} e^{i(\frac{n}{2}+1)\theta}\,d\theta + O(\delta r^{\frac{n}{2}+1}) = (-1)^\nu \epsilon_\nu \frac{r^{\frac{n}{2}+1}}{(\frac{n}{2}+1)\pi} + O(\delta r^{\frac{n}{2}+1}),$$

this proving that w has

$$\ell(w) = \sum_{\nu=0}^{n+1}(-1)^\nu \epsilon_\nu \tag{5.13}$$

strings of poles. To determine the number of strings along a single Stokes ray we again apply the Residue Theorem. For ν fixed let γ_r denote the simple closed and positively oriented curve that consists of the line segment from $r_0 e^{i(\hat\theta_\nu-\delta)}$ to $re^{i(\hat\theta_\nu-\delta)}$, followed by the sub-arc of Γ_r in $|\arg z - \hat\theta_\nu| < \delta$ from $re^{i(\hat\theta_\nu-\delta)}$ to $re^{i(\hat\theta_\nu+\delta)}$, the line segment from $re^{i(\hat\theta_\nu+\delta)}$ to $r_0e^{i(\hat\theta_\nu+\delta)}$, and finally the (short!) circular arc from $r_0e^{i(\hat\theta_\nu+\delta)}$ to $r_0e^{i(\hat\theta_\nu-\delta)}$. Then

$$\frac{1}{2\pi i}\int_{\gamma_r} w(z)\,dz = \frac{(-1)^\nu}{2}(\epsilon_\nu - \epsilon_{\nu+1})\frac{r^{\frac{n}{2}+1}}{(\frac{n}{2}+1)\pi} + O(\delta r^{\frac{n}{2}+1})$$

holds and may be interpreted as follows: there is exactly one string of poles along σ_ν if $\epsilon_\nu = -\epsilon_{\nu+1} = (-1)^\nu$, no such string if $\epsilon_{\nu+1} = \epsilon_\nu$, while $\epsilon_\nu = -\epsilon_{\nu+1} = (-1)^{\nu+1}$ never occurs. If $n = 2m-1$ is odd, the proof runs along the same lines. Since, however, $\sqrt{z}$ is discontinuous across the Stokes ray $\arg z = \pi$, it becomes necessary to enumerate the Stokes rays σ_ν from $\nu = -m$ to $\nu = m$, so that $\sigma_m : \arg z = \frac{(2m+1)\pi}{n+2}$ denotes the negative real axis. Then $\epsilon_m = \epsilon_{-m}$ means that the asymptotic expansions on Σ_m and Σ_{-m} are *different*. The total number of strings is

$$\ell(w) = \sum_{\nu=-m}^{m} (-1)^\nu \epsilon_\nu \tag{5.13'}$$

rather than given by (5.13). The final assertion follows from $m(r, w) = O(\log r)$. ☕

5.2.2 Value Distribution

Since $\coth \mathfrak{z}$ has poles $\mathfrak{z} = k\pi i$ and zeros $(k + \frac{1}{2})\pi i$, the zeros of the solutions w to the canonical equation (5.10) 'separate' the poles at distance $\frac{\pi}{2}$ w.r.t. the metric $ds = |z|^{\frac{n}{2}}|dz|$: the zeros form $(\pm\pi i, \frac{\pi}{2})$-strings (q_k) satisfying $q_k = p_k + (\frac{\pi i}{2} + o(1))p_k^{-\frac{n}{2}}$. The value distribution of w takes place in an arbitrarily small neighbourhood of the set $\mathfrak{Q}$ of zeros: $\mathfrak{Q}_\delta = \bigcup_{q \neq 0} \Delta_\delta(q)$ with $\Delta_\delta(q) = \{z : |z - q| < \delta|q|^{-\frac{n}{2}}\}$. To determine the distribution of arbitrary κ-points of transcendental solutions to the general Riccati equation (5.9) it suffices to determine the distribution of the zeros of $w - \rho$ for solutions w to the canonical equation, where $\rho(z) = Cz^s + \cdots$ ($C \neq 0$ and $z \to \infty$) may be any rational function. There are five cases to be considered.

1. $s > \frac{n}{2}$. Since $w = O(|z|^{\frac{n}{2}})$ holds outside and $|z|^{\frac{n}{2}} = O(|w(z)|)$ inside the union of discs $\Delta_\delta(p) = \{z : |z - p| < \delta|p|^{-\frac{n}{2}}\}$ about the poles, all but finitely many zeros of $w - \rho$ are contained in these discs. To prove that for $|p|$ sufficiently large there is exactly one zero in $\Delta_\delta(p)$ consider $f(z) = \rho(z)/w(z)$, which is holomorphic on $\Delta_\delta(p)$ and has exactly one zero, namely p. Since $f(z)$ tends to infinity as $p \to \infty$, uniformly on $\partial\Delta_\delta(p)$, Rouché's Theorem applies to f and $f - 1$ on $\Delta_\delta(p)$ ($|p|$ sufficiently large), hence $w - \rho = \rho(1 - f)/f$ has exactly one zero on $\Delta_\delta(p)$. In other words, the zeros of $w - \rho$ form strings that 'follow' the strings of poles; at infinity they become indistinguishable from the poles in the metric $ds = |z|^{\frac{n}{2}}|dz|$.
2. $s < \frac{n}{2}$. In the same manner it can be proved that the zeros of $w - \rho$ form strings just like the zeros of w; up to finitely many, all zeros of $w - \rho$ are contained in the union of discs $\Delta_\delta(q)$ about the zeros of w, exactly one in each disc.
3. $s = \frac{n}{2}$, $C \neq \pm 1$. Re-scaling along any sequence of zeros ζ_k of $w - \rho$ leads to the initial value problem $\mathfrak{w}' = 1 - \mathfrak{w}^2$, $\mathfrak{w}(0) = C \neq \pm 1$, hence $\mathfrak{w}(\mathfrak{z}) = \coth(\mathfrak{z} + \mathfrak{z}_0)$ with $\mathfrak{z}_0 = \frac{1}{2}\log\frac{C+1}{C-1}$. Thus the zeros of $w - \rho$ form strings that are 'parallel' to the strings of poles: $\zeta_k = p_k + \mathfrak{z}_0 p_k^{-\frac{n}{2}} + o(|p_k|^{-\frac{n}{2}})$.
4. $\rho' = P(z) - \rho^2$. Then $w - \rho$ has only finitely many zeros.
5. $s = \frac{n}{2}$, $C = \pm 1$, but $\rho' \neq P(z) - \rho^2$. Again re-scaling along any sequence (ζ_k) of zeros of $w - \rho$ leads to the initial value problem $\mathfrak{w}' = 1 - \mathfrak{w}^2$, $\mathfrak{w}(0) = \pm 1$, hence $\mathfrak{w} \equiv \pm 1$. Thus the zeros of $w - \rho$ are 'invisible' from the poles since $|\zeta_k|^{\frac{n}{2}} \operatorname{dist}(\zeta_k, \mathfrak{P})$ tends to infinity as $k \to \infty$. Now $u = 1/(w - \rho)$ solves the Riccati equation $u' = 1 + 2\rho(z)u + \phi(z)u^2$ ($\phi = \rho' - P + \rho^2 \not\equiv 0$) with normal

form $v' = P^*(z) - v^2$ and coefficient $P^* = P - 2\rho' + \frac{3}{4}(\phi'/\phi)^2 + \rho(\phi'/\phi) - \frac{1}{2}(\phi''/\phi) = z^n + \cdots$. Thus the zeros of $w - \phi$ form strings of the same kind.

The transformation $w \mapsto c(z)w + \frac{1}{2}b(z) + \frac{1}{2}(c'(z)/c(z))$ provides the essential step leading from (5.9) to the canonical equation (5.10), while the final transformation of the independent variable, which just stretches and rotates the plane, is inessential. We assume for simplicity that the first transformation already leads to the canonical form. Up to finitely many, the zeros of $y - \kappa$ are zeros of $w - \rho_\kappa$ with $\rho_\kappa = \kappa c(z) + \frac{1}{2}b(z) + \frac{1}{2}(c'(z)/c(z)$, and *vice versa.*

Exercise 5.11 Prove that the zeros q and poles p of the solutions to (5.9) are $\frac{n}{2}$-separated ($\inf_q \operatorname{dist}(q, \mathfrak{P})|q|^{\frac{n}{2}} > 0$) if and only if $\deg b \le \frac{n}{2}$. Assuming this, prove that the transcendental solutions belong to the Yosida class $\mathfrak{Y}_{-\deg c + \frac{n}{2}, \frac{n}{2}}$.

Exercise 5.12 (Continued) Assuming $\deg b \le \frac{n}{2}$, prove that all but finitely many zeros of $w - \kappa$ are contained in the discs

- $\Delta_\delta(p)$ about the poles of w if $\deg c > \frac{n}{2}$ and $\kappa \ne 0$, and
- $\Delta_\delta(\zeta)$ about the zeros ζ of $w - \lambda z^{\frac{n}{2}}$ if $\kappa c(z) + \frac{1}{2}b(z) = \lambda z^{\frac{n}{2}} + \cdots$,

exactly one zero in each disc.

If, however, $\deg b > \frac{n}{2}$, prove that

- almost all zeros of $w - \kappa$ belong to the discs $\Delta_\delta(p)$ about the poles of w, with at most one exceptional κ, and
- w does not belong to any Yosida class.

Exercise 5.13 Employ the previous exercise to discuss in detail the value distribution of the solutions to

- $w' = z^2 + z^3 + 2(z + z^2)w + zw^2$,
- $w' = -\frac{3}{4}z^2 + (2z + z^2)w + z^2w^2$,
- $w' = z + (2z + z^2)w + (z - \frac{3}{4}z^3)w^2$, and
- $w' = \frac{1}{4}z + (2z + z^2)w + (z + z^3)w^2$.

5.2.3 *Truncated Solutions*

Every Stokes sector Σ_ν is 'pole-free' in that *every* solution has only finitely many poles on the closed sub-sectors of Σ_ν. Conversely, a *fixed* solution is said to be pole-free on some sector S if it has only finitely many zeros on every closed sub-sector.

Exercise 5.14 Suppose w is a pole-free solution on some sector S. Adapt the proof of Theorem 5.7 to show that w has an asymptotic expansion $w \sim \epsilon z^{\frac{n}{2}} + \cdots$ on S, with $\epsilon = \epsilon(w) \in \{-1, 1\}$.

Exercise 5.14 is relevant only if the sector in question contains some Stokes ray σ_ν. It turns out that in this case the asymptotics holds on $\Sigma_{\nu-1} \cup \sigma_\nu \cup \Sigma_\nu$. Any solution that has no string of poles along the Stokes ray σ_ν is called *truncated* (along σ_ν). For $n = -1$ truncated solutions are rational. From now on we assume $n \geq 0$.

Theorem 5.8 *Given any Stokes ray σ_ν there exists a solution that is truncated along σ_ν. It is uniquely determined by its asymptotics $w \sim \epsilon z^{\frac{n}{2}} + \cdots$, which holds on $\Sigma_{\nu-1} \cup \sigma_\nu \cup \Sigma_\nu$ with central angle $\frac{4\pi}{n+2}$; $\epsilon \in \{-1, 1\}$ may be prescribed.*

Proof We choose $\epsilon \in \{-1, 1\}$ and set $y(z) = z^{-m}w(z) - \epsilon$ if $n = 2m$ is even, and $y(z) = z^{-n}w(z^2) - \epsilon$ if n is odd to obtain

$$\begin{aligned} z^{-m}y' + mz^{-m-1}(y + \epsilon) &= P(z)z^{-2m} - (y + \epsilon)^2, \\ z^{-n-1}y' + nz^{-n-2}(y + \epsilon) &= 2P(z^2)z^{-2n} - 2(y + \epsilon)^2, \end{aligned}$$

respectively. Then Theorem 1.10 applies to both equations when written as

$$z^{-q}y' = f(z, y) \quad \text{with } q = m \text{ resp. } q = n + 1 \text{ and } \lim_{z\to\infty} f_y(z, 0) \neq 0.$$

Given any sector with central angle $\frac{2\pi}{q+1}$ there exists some solution having an asymptotic expansion $y \sim c_1 z^{-1} + \cdots$. Hence to every sector $|\arg z - \theta_0| < \frac{\pi}{n+2}$ (in both cases) there exists some solution to Eq. (5.10) with asymptotic expansion $w \sim \epsilon z^{\frac{n}{2}} + \cdots$; ϵ may be prescribed. In particular, this applies to $|\arg z - \hat{\theta}_\nu| < \frac{\pi}{n+2}$, and then even to $S_\nu : |\arg z - \hat{\theta}_\nu| < \frac{2\pi}{n+2}$. To prove uniqueness we consider two solutions of this kind, set $u = w_1 - w_2$ and observe that

$$u' = -(w_1(z) + w_2(z))u = -2\epsilon z^{\frac{n}{2}}(1 + O(|z|^{-\frac{1}{2}}))u$$

holds on S. Since $u(z) \to 0$ this requires $\epsilon \operatorname{Re} z^{\frac{n}{2}+1} > 0$ on S, which, however, is impossible on sectors with central angle greater than $\frac{2\pi}{n+2}$; this proves $w_1 = w_2$. ☕

Exercise 5.15 For $n = 0$ the Stokes rays are $\arg z = \pm\frac{\pi}{2}$. Suppose w is truncated along $\arg z = \frac{\pi}{2}$, hence $w(z) = \epsilon + O(|z|^{-1})$ holds as $z \to \infty$ on $|\arg z + \frac{\pi}{2}| < \pi$. Prove that w is also truncated along $\arg z = -\frac{\pi}{2}$ and thus is a rational function. (**Hint.** The Phragmén–Lindelöf Principle applies to $F(z) = e^{-\epsilon z + \int w(z)\,dz}$ on every sector $|\arg z + \frac{\pi}{2}| < \delta < \frac{\pi}{4}$, $|z| > r_0$; note that w may have finitely many poles with residue $\neq 1$.) ☕

The number of strings is as large as possible if $\epsilon_\nu = (-1)^\nu$ holds for *every* ν. On the other hand, $\epsilon_\nu = -(-1)^\nu$ for *some* ν turns out to be exceptional.

Theorem 5.9 *Suppose w is any solution with the 'false' asymptotics*

$$w \sim -(-1)^\nu z^{\frac{n}{2}} + \cdots \quad \text{on } \Sigma_\nu.$$

Then w is truncated along the Stokes rays $\sigma_{\nu-1}$ and σ_ν which bound Σ_ν. Conversely, if w is truncated along σ_ν and has the 'right' asymptotics

$$w \sim (-1)^\nu z^{\frac{n}{2}} + \cdots \quad \text{on } \Sigma_\nu,$$

then w has the 'false' asymptotics $w \sim -(-1)^{\nu+1} z^{\frac{n}{2}} + \cdots$ on $\Sigma_{\nu+1}$ and is also truncated along $\sigma_{\nu+1}$.

Proof It follows from the proof of Theorem 5.7 that $\epsilon_\nu = -(-1)^\nu$ implies $\epsilon_{\nu+1} = \epsilon_\nu$ and also $\epsilon_{\nu-1} = \epsilon_\nu$, hence w is truncated along $\sigma_{\nu-1}$ and σ_ν; this proves the first part. If, on the other hand, w is truncated along σ_ν and $\epsilon_\nu = (-1)^\nu$, then again from the proof of Theorem 5.7 it follows that $\epsilon_{\nu+1} = \epsilon_\nu = -(-1)^{\nu+1}$, and we are in the first case with ν replaced with $\nu + 1$. (Note that $z^{\frac{n}{2}} = z^m \sqrt{z}$ with $\operatorname{Re} \sqrt{z} > 0$ is assumed if $n = 2m + 1$ is odd.)

Summary

- The asymptotic expansions on Σ_ν of non-truncated (generic) solutions have leading terms $(-1)^\nu z^{\frac{n}{2}}$.
- Truncated solutions are truncated along adjacent pairs of Stokes rays.
- To each ν there exists some uniquely determined solution with asymptotics $w_\nu \sim -(-1)^\nu z^{\frac{n}{2}}$ on Σ_ν; w_ν is truncated along $\sigma_{\nu-1}$ and σ_ν.
- The solutions w_ν are not necessarily distinct. In any case truncated solutions are truncated along $2d(w)$ Stokes rays with $\sum_w d(w) = n + 2$.
- The exceptional solutions w_ν correspond to those solutions to the linear differential equation $y'' = P(z)y$ that are *subdominant* on some sector S; $y = \exp \int w(z)\,dz$ is called subdominant on S if y tends to zero exponentially as $z \to \infty$ on S.

Example 5.4 Gundersen and Steinbart [74] considered the linear differential equation $f'' - z^n f = 0$. They proved among other things that certain contour integrals (generalised Laplace transformations)

$$f_\nu(z) = \frac{1}{2\pi i} \int_{C_\nu} e^{P(z,w)}\,dw$$

represent solutions having no zeros along given Stokes rays $\sigma_{\nu-1}$ and σ_ν. These solutions give rise to the truncated solutions $w_\nu = f_\nu'/f_\nu$ to the special Riccati equation $w' = z^n - w^2$, which is invariant under the transformations $w(z) \mapsto \eta w(\eta z)$, $\eta^{n+2} = 1$. Obviously there are exactly two solutions which themselves are invariant, namely those which either have a pole or else a zero at the origin. These solutions cannot be truncated, hence there exist $n + 2$ mutually distinct exceptional solutions. They are obtained from a single one by rotating the plane: $w_\nu(z) = e^{-\frac{2\nu\pi i}{n+2}} w_0\big(e^{-\frac{2\nu\pi i}{n+2}} z\big)$ has a single string of poles along every Stokes ray σ_μ except those which bound the Stokes sector Σ_ν.

Example 5.5 The eigenvalue problem $f'' + (z^4 - \lambda)f = 0, f \in L^2(\mathbb{R})$, has infinitely many solutions (λ_k, f_k), see Titchmarsh [188]. The eigenfunctions f_k have only finitely many non-real zeros. For every eigenpair $(\lambda, f) = (\lambda_k, f_k)$, $u(z) = f(e^{-i\pi/6}z)$ satisfies $u'' - (z^4 + e^{-i\pi/3}\lambda)u = 0$, and $w = u'/u$ solves

$$w' = z^4 + e^{-i\pi/3}\lambda - w^2.$$

In general there are $4 + 2 = 6$ exceptional solutions w_ν, $0 \le \nu \le 5$. They are, however, not mutually distinct: $w_2 = w_5$ is truncated along $\sigma_1, \sigma_2, \sigma_4$, and σ_5, has strings of poles along σ_0 and σ_3, and the asymptotics $w \sim -z^2 + \cdots$ on $\frac{\pi}{6} < \arg z < \frac{7}{6}\pi$ and $w \sim z^2 + \cdots$ on $-\frac{5}{6}\pi < \arg z < \frac{\pi}{6}$. The exceptional solutions w_0, w_1, w_3, and w_4 have four strings of poles and a 'pole-free' half-plane (Fig. 5.1).

Example 5.6 Eremenko and Gabrielov [39] considered the linear differential equation $y'' - (z^3 - az + \lambda)y = 0$. For certain real parameters a and λ there exist solutions with infinitely many zeros, almost all are real and negative. Thus

$$w' = z^3 - az + \lambda - w^2$$

has a solution $w = w_{-1} = w_1$ which is truncated along $\sigma_{-2}, \sigma_{-1}, \sigma_0$, and σ_1 with asymptotics $w \sim z^{\frac{3}{2}}$ ($\mathrm{Re}\sqrt{z} > 0$) on $|\arg z| < \pi$, and mutually distinct solutions w_0, w_2, and w_{-2}, which are truncated along σ_{-1} and σ_0, σ_1 and σ_2, and σ_2 and σ_{-2}, respectively (Fig. 5.2).

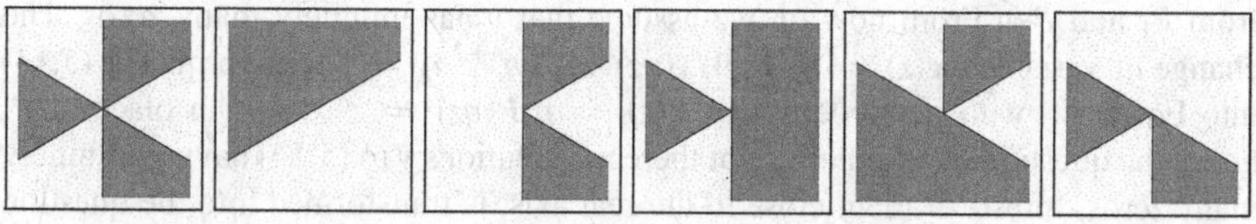

Fig. 5.1 Asymptotics and distribution of poles of generic and exceptional solutions in the eigenvalue case

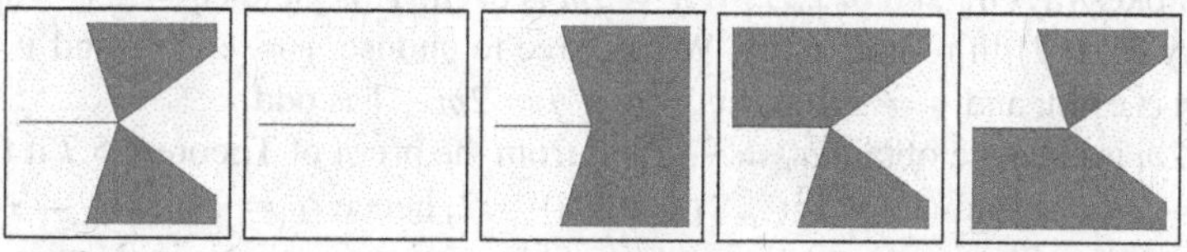

Fig. 5.2 Asymptotics and distribution of poles of generic and exceptional solutions in the eigenvalue case. The square-root $\sqrt{z}$ is discontinuous along the negative real axis; in the third and fourth case the ray $\arg z = \pi$ is truncated!

5.2.4 Poles Close to a Single Line

Several papers (Eremenko and Gabrielov [39], Eremenko and Merenkov [40], Gundersen [67, 70], Shin [156]) are devoted to the question under which circumstances linear differential equations

$$y'' - P(z)y = 0 \quad (P(z) = a_n z^n + \cdots \text{ some polynomial with } |a_n| = 1) \tag{5.14}$$

have solutions with all but finitely many zeros on the real axis. From Theorem 5.8 we obtain (see also [40, 67]):

Theorem 5.10 *Suppose that Eq.* (5.14) *has a solution whose zeros are asymptotic to the real axis. Then the following is true.*

- *If n is even, then either y has only finitely many zeros, or else $n \equiv 0 \bmod 4$, $a_n = -1$, y has exactly one string of zeros asymptotic to the negative and positive real axis, respectively, and $y'/y \sim \pm iz^{\frac{n}{2}} + \cdots$ holds on $\pm\operatorname{Im} z > 0$.*
- *If $n = 2m - 1$ is odd, then either $a_n = 1$ and y has exactly one string of zeros asymptotic to the negative real axis with asymptotics $y'/y \sim (-z)^m z^{-\frac{1}{2}} + \cdots$ on $|\arg z| < \pi$, or else $a_n = -1$ and y has exactly one string of zeros asymptotic to the positive real axis with asymptotics $y'/y \sim z^m(-z)^{-\frac{1}{2}} + \cdots$ on $|\arg(-z)| < \pi$.*

If P is a real polynomial, y is a (multiple of a) *real entire function with all but finitely many zeros real.*

Proof If $y(z) = P_1(z)e^{P_2(z)}$ has only finitely many zeros, then $n = 2\deg P_2 - 2$ is even, and hardly more can be said (of course, P can be computed explicitly from P_1 and P_2). From now on we assume that y has infinitely many zeros. The change of variables $w(z) = \eta y'(\eta z)/y(\eta z)$ with $\eta^{n+2}a_n = 1$ transforms Eq. (5.14) into Eq. (5.10) with new coefficient $\hat{P}(z) = \eta^2 P(\eta z) = z^n + \cdots$ in place of P, hence the question of whether or not there are solutions y to (5.14) having infinitely many zeros, 'most' of them close to the real axis, is transformed into the question of whether or not there are solutions w to Eq. (5.10) having just one string of poles asymptotic to some Stokes ray $\sigma_\nu : \arg z = \hat{\theta}_\nu = \frac{(2\nu+1)\pi}{n+2}$ if n is odd, and asymptotic to the Stokes rays σ_ν and $\sigma_{\nu+m+1}$ if $n = 2m$ is even. This yields $\bar{\eta} = \pm e^{i\hat{\theta}_\nu}$ up to an arbitrary $(n+2)$-th root of unity. We are free to choose $\eta = e^{-i\frac{\pi}{n+2}}$ and $\nu = 0$ if $n = 2m$ is even, and $\eta = \pm 1$ and $\nu = m$ if $n = 2m - 1$ is odd.
If $n = 2m$ is even we obtain $a_n = -1$, and from the proof of Theorem 5.7 it follows that $\epsilon_0 - \epsilon_1 = 2$ and $(-1)^{m+1}(\epsilon_{m+1} - \epsilon_{m+2}) = 2$, hence $\epsilon_0 = 1$ and $\epsilon_1 = -1$, this implying $\epsilon_1 = \epsilon_2 = \cdots = \epsilon_{m+1} = -1, \epsilon_{m+2} = \cdots = \epsilon_{2m+1} = \epsilon_0 = 1, m = 2k$ and $n = 4k$. This proves the first part of Theorem 5.10.
If $n = 2m - 1$ is odd we have $a_n = +1$ and $a_n = -1$ with zeros asymptotic to the negative and positive real axis, respectively, and asymptotic expansions

$$y'/y \sim (-1)^m(\pm z)^{m-\frac{1}{2}} + \cdots \text{ on } |\arg(\pm z)| < \pi.$$

If P is a real polynomial, the zeros of the solution(!) $y^*(z) = \overline{y(\bar{z})}$ are also asymptotic to the real axis, hence y and y^* are linearly dependent since y is uniquely determined up to a constant factor. Thus y is a multiple of a real entire function with all but finitely many zeros real.

5.2.5 *Locally Univalent Meromorphic Functions*

To construct meromorphic functions with sum of deficiencies $\sum_{a\in\widehat{\mathbb{C}}}\delta(a,f) = 2$, Nevanlinna [127] considered locally univalent meromorphic functions f of finite order. They are characterised by the fact that their Schwarzian derivative $S_f = (f''/f')' - \frac{1}{2}(f''/f')^2$ is a polynomial $2P$, say. Moreover, f equals the ratio $y(z;0)/y(z;\infty)$ of linearly independent solutions to the linear differential equation

$$y'' + P(z)y = 0 \quad (\deg P = n). \tag{5.15}$$

Using Hille's method of asymptotic integration ([85], §5.6.), Nevanlinna was able to show that f has finitely many deficient values a_ν with deficiencies $\delta(a_\nu,f) = \frac{2d_\nu}{n+2}$ ($d_\nu \in \mathbb{N}$) and $\sum_\nu d_\nu = n+2$. Since Eq. (5.15) is equivalent to the Riccati equation $w' = -P(z) - w^2$ via $w = y'/y$, an easier way is to 'count' poles. Generic solutions have counting function of poles and Nevanlinna characteristic $\sim Cr^{\frac{n}{2}+1}$. The exceptional (truncated) solutions w_ν, however, have counting function of poles and Nevanlinna characteristic $\sim C\frac{n+2-2d_\nu}{n+2}r^{\frac{n}{2}+1}$, where d_ν is some positive integer such that $\sum_\nu d_\nu = n+2$. Since $f-a$ and $y(z;a) = y(z;0) - ay(z;\infty)$ have the same zeros, which coincide with the poles of $w(z;a) = y'(z;a)/y(z;a)$, it follows that f has the desired properties with a_ν defined by $w_\nu(z) = w(z;a_\nu)$ and $\delta(a_\nu,f) = \frac{2d_\nu}{n+2}$; note that $T(r,f) \sim Cr^{\frac{n}{2}+1}$ holds by Cartan's Identity.

5.2.6 *A Problem of Hayman*

We return to a problem which we discussed in Sect. 3.2.5, namely to determine those meromorphic functions f such that ff'' has no zeros. This problem was settled by Langley [104] (solution $f(z) = e^{az+b}$ and $f(z) = (az+b)^{-n}$) in general, and by Mues [119] for functions of finite (lower) order. Mues' proof matches the topic just being discussed. He considered the meromorphic function $F(z) = z - f(z)/f'(z)$ with non-vanishing derivative

$$F'(z) = \frac{f(z)f''(z)}{f'(z)^2}.$$

Since F has only simple poles (at zeros of f'), F is locally univalent and its Schwarzian derivative $S_F = 2Q$ is an entire function, and even a polynomial: $T(r_k, Q) = m(r_k, Q) = O(\log r_k)$ holds on some sequence $r_k \to \infty$; F has the form w_1/w_2 with linearly independent solutions to $w'' + Q(z)w = 0$. We first consider the case $Q \not\equiv 0$ and may assume $Q(z) = -z^n + \cdots$. From $F'/F = w_1'/w_2 - w_2'/w_2 = y_1 - y_2$ with $y_j' = -Q(z) - y_j^2$ it follows that the plane is divided into finitely many sectors $\Sigma_\nu : \hat{\theta}_{\nu-1} < \arg z < \hat{\theta}_\nu$ such that either $F'/F \sim 0$ (that is, $F'/F = o(|z|^{-m})$ holds for every $m \in \mathbb{N}$) and $F \sim c_\nu$, or else $F'/F \sim 2\epsilon_\nu z^{\frac{n}{2}} + \cdots$ with $\epsilon_\nu = \pm 1$ and $F(z) \to 0$ or $F(z) \to \infty$, exponentially as $z \to \infty$ on Σ_ν. The entire function $g = 1/f$ has finite order of growth and logarithmic derivative $\frac{g'(z)}{g(z)} = \frac{1}{-z+F(z)} = O(|z|^{-1})$, hence g grows at most logarithmically along every ray $\arg z = \theta \neq \hat{\theta}_\nu$. From the Phragmén–Lindelöf Principle applied to the sectors $|\arg z - \hat{\theta}_\nu| < \delta$ it follows g is constant. This, however, contradicts $S_F \not\equiv 0$, and it remains to discuss the case when S_F vanishes identically.

Exercise 5.16 Suppose $F(z) = \frac{Az+B}{Cz+D}$ is a Möbius transformation, hence $\frac{f'(z)}{f(z)} = \frac{Cz+D}{Cz^2+(D-A)z-B}$ holds. From the fact that the residues of f'/f are negative integers deduce $f(z) = e^{az+b}$ or $f(z) = (az+b)^{-n}$.

5.3 First-Order Algebraic Differential Equations

This section is devoted to implicit first-order differential equations

$$P(z, w, w') = \sum_{\nu=0}^{q} P_\nu(z, w) w'^\nu = 0 \quad (q \geq 2), \tag{5.16}$$

where $P(z, x, y)$ is an irreducible polynomial over the field $\mathbb{C}(z)$. Equations of this type are completely understood. With the help of the so-called Second Theorem of Malmquist and the investigations of the previous section we will draw a comprehensive picture of the solutions.

5.3.1 *Malmquist's Second Theorem*

Malmquist's Second Theorem shows that the so-called *Fuchsian conditions* for the absence of movable singularities other than poles are necessary for the existence of some transcendental meromorphic solution. These conditions may be verified algebraically. Towards Malmquist's Theorem we will start with a simple example. By $D(z, x)$ we denote the discriminant of the polynomial $P(z, x, y)$ with respect to y.

Example 5.7 The differential equation $z^2w'^2 + P_1(z,w)w' + P_0(z,w) = 0$ with

$$\begin{aligned} P_1(z,w) &= -4z^2 - 4z^4 + (2z + 4z^3)w + (1+z^2)w^2,\\ P_0(z,w) &= 4z^2 + 8z^4 + 8z^6 - (4z + 8z^3 + 16z^5)w\\ &\quad -(6z^2 - 6z^4)w^2 + (4z + 2z^3)w^3 + w^4,\\ D(z,w) &= 16z^4(w^2 + 4zw - 4z^2)((1-z^2)w + 2z^3)^2 \end{aligned}$$

has three rational solutions: $\psi_0(z) = z$, which does not solve the discriminant equation $D(z,w) = 0$, and $\psi_{1,2}(z) = (-2 \pm 2\sqrt{2})z$, which do. On the other hand, the solution $\psi_3(z) = \frac{2z^3}{z^2-1}$ to $D(z,w) = 0$ does not solve the differential equation. The solutions $\psi_{1,2}$ are called *singular*. The general solution is transcendental and given by $w = \frac{\mathfrak{t}^2+z^2}{\mathfrak{t}+z}$ with $\mathfrak{t}' = z^2 + 1 - \mathfrak{t}^2$; ψ_0 is obtained from $\mathfrak{t}(z) = z$.[†]

Theorem 5.11 (Malmquist's Second Theorem) *Suppose Eq.* (5.16) *has a transcendental meromorphic solution. Then*

- $\deg_x P_\nu \le 2q - 2\nu$; *in particular,* P_q *is independent of* x;
- *if* $x = \psi(z)$ *solves* $D(z,x) = 0$, *and* $y = \sigma(z) + b(z)\xi^{k/m} + \cdots$ *with* $b(z) \not\equiv 0$, $(k,m) = 1$, *and* $m \ge 2$ *solves* $P(z, \psi(z) + \xi, y) = 0$ *in a neighbourhood of* $\xi = 0$, $k \ge m - 1$ *and* $\sigma = \psi'$ *hold.*
- *an analogous condition holds for* $\tilde{P}(z, v, v') = v^{2q}P(z, 1/v, -v'/v^2) = 0$ *with discriminant* $\tilde{D}(z,v)$; $\tilde{D}(z,0) = 0$ *corresponds to the 'solution'* $\psi = \infty$.

Conversely, these conditions are sufficient in order that the solutions have no movable singularities except poles, hence are meromorphic if no fixed singularities exist.

Remark 5.7 The necessity of the first condition was proved in Theorem 3.3 with the help of Eremenko's Lemma. For the second condition the reader is referred to Eremenko's 'algebraic' paper [34], and also to Golubev [58], II. §2, where the Fuchsian conditions for the absence of so-called 'movable singularities' other than poles are derived. The second condition says that certain solutions to $D(z,x) = 0$ also solve (5.16). Solutions of this kind are called *singular*. In Example 5.7 we have $k = 1$ and $m = 2$ for the singular solutions $\psi_{1,2}$, but $k = m = 1$ for ψ_3, which solves the discriminant equation but not the differential equation.

---

† MAPLE code — decoded

`t1:=1+z^2-t^2; r1:=t^2+z^2; r2:=t+z; r:=r1/r2;`	$\mathfrak{t}' = 1 + z^2 - \mathfrak{t}^2$; $w = \mathfrak{r}(z,\mathfrak{t}) = \frac{\mathfrak{r}_1(z,\mathfrak{t})}{\mathfrak{r}_2(z,\mathfrak{t})} = \frac{z^2+\mathfrak{t}^2}{z+\mathfrak{t}}$
`s:=diff(r,z)+diff(r,t)*t1; s1:=numer(s); s2:=denom(s);`	$w' = \mathfrak{s}(z,\mathfrak{t}) = \frac{\mathfrak{s}_1(z,\mathfrak{t})}{\mathfrak{s}_2(z,\mathfrak{t})}$
`P:=resultant(x*r2-r1,y*s2-s1,t);`	$P(z,w,w') = 0$
`Delta:=discrim(P,y); solve(Delta,x);`	compute the discriminant and its roots

5.3.2 *Binomial Differential Equations*

The so-called *binomial differential equations*

$$w'^q = R(z, w) \tag{5.17}$$

were considered in [6, 7, 160, 208, 211],[2] and the general degree-two equation

$$(w' + B(z, w))^2 = R(z, w)$$

in [161] (R resp. B and R rational). It turns out that there are five canonical equations as follows:

$$\begin{aligned}
&1.\ \ (w' + b(z)w)^2 = a(z)w(w - c(z))^2\\
&2.\ \ w'^2 = a(z)(w - \tau_1)(w - \tau_2)(w - \tau_3)(w - \tau_4)\\
&3.\ \ w'^3 = a(z)(w - \tau_1)^2(w - \tau_2)^2(w - \tau_3)^2\\
&4.\ \ w'^4 = a(z)(w - \tau_1)^2(w - \tau_2)^3(w - \tau_3)^3\\
&5.\ \ w'^6 = a(z)(w - \tau_1)^3(w - \tau_2)^4(w - \tau_3)^5
\end{aligned} \tag{5.18}$$

In any case a, b, and c are not identically vanishing rational functions, and $\tau_\nu \neq \tau_\mu$ ($\nu \neq \mu$) holds. The discriminant $D(z, w)$ is just given by the respective right-hand side. Note that in the first case the root $w = c(z)$ of $D(z, w)$ is not necessarily a singular solution. The first equation may be transformed into some Riccati equation for $\mathrm{t} = \frac{w'+b(z)w}{w-c(z)}$, hence $a(z)w = \mathrm{t}^2$, while for $a(z) =$ const. the solutions to the other equations are specific elliptic functions; they are closely related to the *Schwarz–Christoffel* conformal mappings of rectangles and certain regular triangles onto the upper half-plane. The Malmquist conditions are obviously fulfilled.

Example 5.8 Binomial differential equations occur in a quite different context. Let R be any non-linear rational function and suppose there exists some map $\nu : \widehat{\mathbb{C}} \longrightarrow \mathbb{N} \cup \{\infty\}$ such that $\nu(a) \deg_{(a)} R = \nu(R(a))$ and $\sum_{a\in\widehat{\mathbb{C}}} \big(1 - 1/\nu(a)\big) = 2$, where $\deg_{(a)} R$ denotes the local degree of R (the multiplicity of the solution to $R(z) = R(a)$ at $z = a$). Then $\mathfrak{O} = (\widehat{\mathbb{C}}, \nu)$ is called a *parabolic orbifold.* Any parabolic orbifold has a ramified covering map $F : \mathbb{C} \longrightarrow \mathfrak{O}$; it omits a if $\nu(a) = \infty$, is branched exactly over those a with $1 < \nu(a) < \infty$, and has local degree $\deg_{(z)} f = \nu(f(z))$. It turns out that every Zalcman function $f(z) = \lim_{k\to\infty} R^{n_k}(z_k + \rho_k z)$ (see Sect. 4.1.3) is a ramified covering map $f : \mathbb{C} \longrightarrow \mathfrak{O}$; it satisfies the binomial

[2]The authors seemed to be unaware of Malmquist's Second Theorem, or were in doubt about his reasoning. One reason might be that several elegant and transparent proofs of his First Theorem were known, but none for his Second. Eremenko's commented in [35]: "The paper [42] (Malmquist's paper [114] here) has had practically no influence on the work of other authors [...]".

differential equation

$$w'^m = c \prod_{1<\nu(a)<\infty} (w-a)^{m(a)}$$

with $m = \operatorname{lcm}\{\nu(a) : a \in \widehat{\mathbb{C}}\} \in \{2,3,4,6\}$ and $m(a) = m(1 - 1/\nu(a)) = m\vartheta(a,f)$; for $a = \infty$ the corresponding term has to be dropped.

Exercise 5.17 Determine some f with $c = 1$ in the following cases
1. $\nu(0) = \nu(\infty) = \infty$.
2. $\nu(1) = \nu(-1) = 2, \nu(\infty) = \infty$.
3. $\nu(0) = \nu(1) = \nu(-1) = \nu(\infty) = 2$.

5.3.3 Briot–Bouquet Equations

A second subclass of differential equations (5.16) is formed by the autonomous *Briot–Bouquet* equations

$$P(w, w') = 0; \tag{5.19}$$

the polynomial $P(x, y)$ is assumed to be irreducible over $\mathbb{C}$. Without reference to Malmquist's Second Theorem we will describe the nature of meromorphic solutions. By **W** (like Weierstraß) we denote the class of non-constant meromorphic functions that satisfy some algebraic addition theorem $Q(f(z), f(\zeta), f(z + \zeta)) = 0$; **W** consists of rational, trigonometric, and elliptic functions. Each $f \in$ **W** has a degree: the usual degree if f is rational, the degree of R if $f(z) = R(e^{az})$ is trigonometric, and the elliptic order if f is elliptic. In other words, $\deg f$ is the mapping degree of the proper mapping $f : D \longrightarrow f(D)$, where D is the Riemann sphere, a cylinder, and a torus, respectively. Each of these functions also satisfies some algebraic differential equation (5.19) (just differentiate with respect to ζ and set $\zeta = 0$ afterwards). Conversely, the following is true.

Theorem 5.12 *The meromorphic solutions to $P(w, w') = 0$ belong to* **W** *and have degree at most $q = \deg_y P$.*

Proof The classical proof is based on the Uniqueness Theorem for

$$w'' = f(w, w') \quad (f(x, y) = -P_x(x, y)y/P_y(x, y)), \tag{5.20}$$

with 'regular' initial values $w(0) = w_0$ and $w'(0) = w_0'$: w_0 is not a root of the discriminant of P with respect to y, and w_0' is any solution to $P(w_0, w_0') = 0$. Assume that w is not a rational function of degree at most q. Then there exist some regular value w_0 and $q + 1$ points z_j with $w(z_j) = w_0$, hence $P(w, w') = 0$ has $q + 1$

solutions $w(z+z_j)$. Since only q solutions (q values $w'(z_j)$) can be distinct, at least one difference $\vartheta = z_j - z_\ell \neq 0$ is a period of w. The same argument applies if w is periodic of degree greater than q: there exist $q+1$ points z_j with $w(z_j) = w_0$ in any period strip, hence w has some period $\vartheta' = z_j - z_\ell$ such that ϑ and ϑ' are $\mathbb{R}$-linearly independent. In the last step it follows that w is elliptic of order at most q, since otherwise there would exist a third $\mathbb{R}$-linearly independent period.

Exercise 5.18 Let Q be any polynomial of degree three or four with distinct zeros. Prove that neither the differential equation $w'^2 = Q(w)$ nor $a^2\zeta^2u'^2 = Q(u)$ ($u = u(\zeta)$, $a \neq 0$) has non-constant rational solutions. Deduce that the non-constant solutions to $w'^2 = Q(w)$ are elliptic functions of elliptic order two.

Remark 5.8 It is obvious that the above proof does not apply to higher-order Briot–Bouquet equations

$$P(w, w^{(k)}) = 0 \quad (k \geq 2). \tag{5.21}$$

The case $k = 2$ was settled by Picard (and rediscovered by several authors [8, 86, 87]). The example $w^{(k)} - w = 0$ with general solution $w = \sum_{\vartheta^k=1} a_\vartheta e^{\vartheta_\nu z}$ shows that Theorem 5.12 becomes false if $k \geq 3$. It holds, however, for non-entire meromorphic solutions: they are rational, trigonometric, or elliptic, see [34] for odd k, and [42] in the general case. The idea of the proof is to show in a first step that there are only finitely many formal solutions $\sum_{j=0}^{\infty} c_j z^{-n+j}$ with pole at $z = 0$, and secondly to prove that transcendental solutions w with at least one pole have infinitely many poles z_ν, hence some difference $\vartheta = z_\nu - z_\mu$ is a period. Once this is established, it turns out that w either is trigonometric or else has a second $\mathbb{R}$-linearly independent period ϑ'.

5.3.4 Back to Malmquist's Second Theorem

Theorem 5.11 has significant impact. It turns out that the freely interpreted essence of Theorem 5.12 also holds for non-autonomous equation (5.16) with transcendental meromorphic solutions. The algebraic curve $P(z, x, y) = 0$, where z is viewed as a parameter, either has genus zero for almost every z, or else genus one and the following is true. In case of

- *genus zero* there exist rational functions $\mathfrak{r}$ and $\mathfrak{s}$, such that every transcendental solution to (5.16) is given by

$$w = \mathfrak{r}(z, \mathfrak{t}) \quad \text{and} \quad w' = \mathfrak{s}(z, \mathfrak{t}),^3 \tag{5.22}$$

[3]This is a most non-trivial fact, known as *Tsen's Theorem* (C. Tsen, Divisionsalgebren über Funktionenkörpern, *Nachr. Ges. Wiss. Göttingen, Math.-Phys. Kl.* (1933), 335–339), communicated to

where $\mathfrak{t}$ solves some Riccati equation

$$\mathfrak{t}' = a(z) + b(z)\mathfrak{t} + c(z)\mathfrak{t}^2 \tag{5.23}$$

with rational coefficients a, b, and c. The latter follows from Malmquist's First Theorem applied to $\mathfrak{r}_z(z,\mathfrak{t}) + \mathfrak{r}_\mathfrak{t}(z,\mathfrak{t})\mathfrak{t}' = \mathfrak{s}(z,\mathfrak{t})$. Conversely, $\mathfrak{t}$ may be written as $\mathfrak{t}(z) = \rho(z, w(z), w'(z))$, where ρ is rational.[4]

- *genus one* the parametrisation is given by some bi-rational transformation

$$\left.\begin{aligned} w &= \mathfrak{r}(z,\mathfrak{t},\mathfrak{t}') \\ w' &= \mathfrak{s}(z,\mathfrak{t},\mathfrak{t}') \end{aligned}\right\} \quad \text{and} \quad \left\{\begin{aligned} \mathfrak{t} &= \rho(z,w,w') \\ \mathfrak{t}' &= \sigma(z,w,w') \end{aligned}\right. \tag{5.24}$$

where $\mathfrak{r}$, $\mathfrak{s}$, ρ, and σ are rational in the principal variables $\mathfrak{t}$, $\mathfrak{t}'$, and w, w', respectively, and $\mathfrak{t}$ satisfies some differential equation

$$\mathfrak{t}'^2 = a(z)(4\mathfrak{t} - g_2\mathfrak{t} - g_3) \quad (g_2^3 - 27g_3^2 \neq 0). \tag{5.25}$$

This time, however, a and the coefficients in $\mathfrak{r}$, $\mathfrak{s}$, ρ, and σ may be *algebraic* rather than rational functions.

5.3.5 Equations of Genus Zero

For almost every pair (z_0, w_0), the equation $\mathfrak{r}(z_0, \tau) = w_0$ has $\deg \mathfrak{r}$ distinct solutions $\tau = \tau_j$, hence from (5.22) we obtain $\deg \mathfrak{r}$ different solutions to Eq. (5.16), defined by the initial values $w(z_0) = w_0$, $w'(z_0) = \mathfrak{s}(z_0, \tau_j)$. Conversely, for $D(z_0, w_0) \neq 0$ the equation $P(z_0, w_0, \omega) = 0$ has q solutions $\omega = \omega_j$, and (5.16) has unique solutions satisfying $w(z_0) = w_0$ and $w'(z_0) = \omega_j$, since Picard's Existence and Uniqueness Theorem applies to the differential equation (5.20) with regular initial values $w(z_0) = w_0$, $w'(z_0) = w'_0$: $D(z_0, w_0) \neq 0$, $P(z_0, w_0, w'_0) = 0$, and all coefficients regular at z_0. This shows that $\deg \mathfrak{r} = q$. The parametrisation (5.22), (5.23) is not unique. We are free to choose (5.23) in normal form

$$\mathfrak{t}' = \hat{P}(z) - \mathfrak{t}^2, \quad \hat{P}(z) = z^n + \cdots. \tag{5.26}$$

The normalisation of $\hat{P}$ (leading term z^n instead of $c_n z^n$) is no real restriction. To avoid repeated separation into cases we assume $\deg_w P_0 = 2q$ to the effect that $m(r,w) = O(\log r)$ and almost all poles are simple. This can be obtained by

---

me by A. Eremenko. A local form generalising Tsen's Theorem was proved by Lang (On quasi algebraic closure, *Ann. of Math.* **55** (1952), 373–390).

[4]Equations (5.22) and (5.23) correspond to $w = R(e^{az})$ when (5.19) has genus zero.

replacing w with some appropriate Möbius transformation

$$\tilde{w} = \frac{a(z)w + b(z)}{c(z)w + d(z)} \quad (a, b, c, d \in \mathbb{C}(z), ad - bc \not\equiv 0). \tag{5.27}$$

The Möbius transformations (5.27) also allow us to transfer any information about the distribution of poles of the solutions into information about the distribution of the zeros of $w - \psi(z)$, where ψ is any rational function. The main result on transcendental meromorphic solutions to Eq. (5.16) of genus zero may now be stated as follows. We remind the reader that the non-singular solutions are given by (5.22) where $\mathfrak{t} = \mathfrak{t}_1/\mathfrak{t}_2$ is rational of degree $\deg_\tau \mathfrak{t} = q$, and $\mathfrak{t}$ is a solution to the Riccati equation (5.26). The integer $n \geq -1$, the rays $\sigma_\nu : \arg z = \hat{\theta}_\nu = \frac{(2\nu+1)\pi}{n+2}$, and the open sectors $\Sigma_\nu : \hat{\theta}_{\nu-1} < \arg z < \hat{\theta}_\nu$ are associated with (5.26).

Theorem 5.13 *For any transcendental meromorphic solution to Eq.* (5.16) *of genus zero the following is true.*

- *On each sector Σ_ν, w has an asymptotic expansion in $\sqrt{z}$ (in z if n is even).*
- *The poles of w form finitely many $(\pm\pi i, \frac{n}{2})$-strings (p_k); each string*
  - *has counting function $n(r, (p_k)) \sim \frac{r^{\frac{n}{2}+1}}{(\frac{n}{2}+1)\pi}$;*
  - *is asymptotic to some Stokes ray $\sigma_\nu : \arg z = \hat{\theta}_\nu$,*

  and w is either truncated or has q strings along σ_ν.
- *w has Nevanlinna characteristic $T(r, w) \sim q\,\dfrac{n + 2 - 2d_w}{(\frac{n}{2} + 1)^2\pi}\, r^{\frac{n}{2}+1}$, where $2d_w =$ $2d_{\mathfrak{t}}$ denotes the number of truncated Stokes rays ($\sum_w d_w = n + 2$).*
- *w has at most two deficient rational functions ψ; the deficiencies $\delta(\psi, w) = \delta(0, w - \psi)$ are integer multiples of $1/q$. Non-singular rational solutions are deficient, while singular rational solutions are not.*
- *Ramified rational functions also solve* (5.16), *and $\vartheta(\psi, w) = \vartheta(0, w - \psi)$ is an integer multiple of $1/q$. Singular rational solutions are ramified, and the number of ramified functions does not exceed $2q - 2$.*

Proof The first assertion follows from the subsequent exercises.

Exercise 5.19 Prove that any solution that is rational or algebraic at infinity is represented by one of the (then convergent) asymptotic series for w. Conversely, if this series converges on $|z| > r_0$, prove that it represents a solution to (5.26) that is rational or algebraic at ∞. (**Hint.** The asymptotic series are *formal* solutions, and actually are solutions (rational or algebraic at infinity) if and only if they converge.)

Exercise 5.20 Let $\mathfrak{p}(z, \tau)$ be any non-constant polynomial in τ with rational coefficients. Prove that $\mathfrak{p}(z, \mathfrak{t}(z))$ also has an asymptotic expansion in $\sqrt{z}$, which is trivial (all coefficients vanish) if and only if the equation $\mathfrak{p}(z, \tau) = 0$ has some solution $\tau = \phi(z)$ that solves the Riccati equation (5.26). In this case,

$$\mathfrak{t}(z) - \phi(z) = \exp\Big(-\frac{4\epsilon}{n+2}z^{\frac{n}{2}+1} + \sum_{k=0}^{n+1} a_k z^{\frac{k}{2}}\Big) z^{\kappa}(1+o(1)) \tag{5.28}$$

holds as $z \to \infty$, uniformly on each closed sub-sector of S; $\kappa = \kappa(\epsilon)$ is some complex constant and $\operatorname{Re} \epsilon z^{\frac{n}{2}+1} > 0$ holds on S (always $\epsilon \in \{-1, 1\}$).
(**Hint.** $y = \mathfrak{t} - \phi$ satisfies $y' = -(\mathfrak{t}(z) + \phi(z))y$ with $\mathfrak{t}(z) + \phi(z) \sim 2\epsilon z^{\frac{n}{2}} + \cdots$.) ☕

To proceed we note that the general assumption $\deg_w P_0 = 2q$ implies $m(r, w) = O(\log r)$, hence $\mathfrak{r}_2(z, \mathfrak{t}(z))$ has a non-trivial asymptotic expansion and Exercise 5.20 applies to $\mathfrak{p} = \mathfrak{r}_2$. Almost all poles of w are simple; they arise from poles of $\mathfrak{t}$ if $\deg_\tau \mathfrak{r}_1 = \deg_\tau \mathfrak{r}_2 + 1$ on one hand, and the solutions $\tau = \phi(z)$ to the equation $\mathfrak{r}_2(z, \tau) = 0$ which do not solve $\phi' = \hat{P}(z) - \phi^2$ on the other. The classification of zeros of $\mathfrak{t} - \phi$ and their distribution in the plane (see Sect. 5.2.2, where ϕ may be any *rational* function) remains valid for algebraic functions ϕ, and this proves the second assertion. The proof of the third assertion follows from Valiron's Lemma

$$T(r, w) \sim qT(r, \mathfrak{t}) \quad \text{and} \quad T(r, \mathfrak{t}) \sim \frac{n+2-2d_{\mathfrak{t}}}{(\frac{n}{2}+1)^2\pi} r^{\frac{n}{2}+1}.$$

To prove the fourth assertion we note that the Wittich–Mokhon'ko Lemma (see also Exercise 3.5) yields $m\big(r, 1/(w - \psi)\big) = O(\log r)$ and $\delta(\psi, w) = 0$, provided the rational function ψ does not solve Eq. (5.16). If, however, ψ solves (5.16) but is not singular, the algebraic equation $\mathfrak{r}(z, \tau) - \psi(z) = 0$, equivalently $\mathfrak{p}(z, \tau) = \mathfrak{r}_1(z, \tau) - \psi(z)\mathfrak{r}_2(z, \tau) = 0$, has solutions $\tau = \phi(z)$ that also solve Eq. (5.26). This implies that there are at most two deficient rational functions of this kind. Now $\mathfrak{p}$ factors into $\mathfrak{p}_1\mathfrak{p}_2$, such that all solutions to $\mathfrak{p}_1(z, \tau) = 0$, but none to $\mathfrak{p}_2(z, \tau) = 0$ also solve the Riccati equation (5.26). To proceed we refer to

Exercise 5.21 Let $\mathfrak{p}(z, \tau)$ be any polynomial over $\mathbb{C}(z)$ such that none of the solutions $\tau = \phi(z)$ to $\mathfrak{p}(z, \tau) = 0$ solves (5.26). Prove that $\mathfrak{p}(z, \mathfrak{t}(z))$ has equally many strings of poles and zeros (strings of k-fold zeros will be counted k times), hence $m\big(r, \frac{1}{\mathfrak{p}(z,\mathfrak{t}(z))}\big) = N(r, \mathfrak{p}(z, \mathfrak{t}(z))) - N\big(r, \frac{1}{\mathfrak{p}(z,\mathfrak{t}(z))}\big) + O(\log r) = o(r^{\frac{n}{2}+1})$ holds. ☕

Exercise 5.21 applies to $\mathfrak{p} = \mathfrak{p}_2$ and $\mathfrak{p} = \mathfrak{r}_2$, hence (note that $m(r, \mathfrak{t}) = O(\log r)$)

$$m\Big(r, \frac{\mathfrak{r}_2(z, \mathfrak{t}(z))}{\mathfrak{p}_2(z, \mathfrak{t}(z))}\Big) + m\Big(r, \frac{\mathfrak{p}_2(z, \mathfrak{t}(z))}{\mathfrak{r}_2(z, \mathfrak{t}(z))}\Big) = o(r^{\frac{n}{2}+1})$$

holds, and $w - \psi(z) = \frac{\mathfrak{p}_2(z,\mathfrak{t}(z))}{\mathfrak{r}_2(z,\mathfrak{t}(z))}\mathfrak{p}_1(z, \mathfrak{t}(z))$ implies

$$m\Big(r, \frac{1}{w - \psi}\Big) = m\Big(r, \frac{1}{\mathfrak{p}_1(z, \mathfrak{t}(z))}\Big) + o(r^{\frac{n}{2}+1}).$$

Since $\mathfrak{p}_1(z, \mathfrak{t}(z))$ has only finitely many zeros, the First Main Theorem yields

$$m\Big(r, \frac{1}{w-\psi}\Big) = T(r, \mathfrak{p}_1(z, \mathfrak{t}(z))) + o(r^{\frac{n}{2}+1}) = \tfrac{1}{q}\deg_\tau \mathfrak{p}_1 T(r, w) + o(r^{\frac{n}{2}+1}),$$

hence $\delta(\psi, w) = \frac{1}{q}\deg_\tau \mathfrak{p}_1$. On the other hand, singular solutions ψ have $\mathfrak{p}_1 \equiv 1$ and $\delta(\psi, w) = 0$. This proves the fourth assertion. To prove the fifth assertion we will now clarify the relation between ramified functions and rational solutions. It is obvious that $w - \psi(z)$ has infinitely many multiple zeros if and only if

$$\mathfrak{r}(z, \tau) = \psi(z) \tag{5.29}$$

has $(m_j + 1)$-fold solutions $\tau = \phi_j$ $(1 \le j \le s)$ which do not solve (5.26). The zeros of $\mathfrak{t} - \phi_j(z)$ are arranged in $n + 2 - 2d_\mathfrak{t}$ strings, and $w - \psi(z)$ has $n + 2 - 2d_\mathfrak{t}$ strings of $(m_j + 1)$-fold zeros. This shows $\vartheta(\psi, w) = \frac{1}{q}(m_1 + \cdots + m_s)$. Also for z fixed, the rational function $r(\tau) = \mathfrak{r}(z, \tau)$ has at most $2q - 2$ critical values, hence there are at most $2q - 2$ rational functions such that (5.29) has multiple roots.

Finally, let $\psi(z) \equiv 0$, say, be a singular solution, and recall that $P(z, x, y)$ is the resultant of the polynomials $x\mathfrak{r}_2(z, \tau) - \mathfrak{r}_1(z, \tau)$ and $y\mathfrak{s}_2(z, \tau) - \mathfrak{s}_1(z, \tau)$ w.r.t. τ; in particular, the resultant of $-\mathfrak{r}_1(z, \tau)$ and $-\mathfrak{s}_1(z, \tau)$, namely $P(z, 0, 0)$, vanishes identically, hence $\mathfrak{r}$ and $\mathfrak{s}$ have common roots. From $\mathfrak{s}(z, \tau) \equiv r_z(z, \tau) + \mathfrak{r}_\tau(z, \tau)(\hat{P}(z) - \tau^2)$ and $\mathfrak{r}_z(z, \phi(z)) + \mathfrak{r}_\tau(z, \phi(z))\phi'(z) \equiv 0$ it follows that

$$\mathfrak{s}(z, \phi(z)) \equiv \mathfrak{r}_\tau(z, \phi(z))(\hat{P}(z) - \phi(z)^2 - \phi'(z))$$

holds for any solution to $\mathfrak{r}(z, \tau) \equiv 0$. In other words, the common roots of $\mathfrak{r}$ and $\mathfrak{s}$ coincide with the multiple roots of $\mathfrak{r}$, and we are in the first case. This proves the fifth assertion and also finishes the proof of Theorem 5.13.

5.3.6 *Some Examples of Genus Zero*

We will now discuss several examples to illustrate various features of Theorem 5.13.

Example 5.9 Singular solutions need not be single-valued since the differential equation (5.16) may have *fixed singularities*. This is the case with

$$\begin{aligned} &w'^3 + 7z^2w'^2 + [54zw^2 - 21zw + z + 8z^4]w' \\ &\qquad + [27w^4 - 27w^3 + (9 - 104z^3)w^2 - (1 + 4z^3)w - (z^3 + 16z^6)] = 0, \\ &D(z, w) = (4z^3 + 27w^2)D_3(z, w)^2 \quad (\deg_w D_3 = 3). \end{aligned}$$

The non-singular solutions are given by $w = \mathfrak{t}(\mathfrak{t}^2 + z)$, where $\mathfrak{t}$ is any solution to $\mathfrak{t}' = z - \mathfrak{t}^2$. The roots of $D_3(z, w)$ do not solve the differential equation, and the singular solutions $\psi(z) = \sqrt{-\frac{4}{27}z^3}$ are not single-valued. The origin is a fixed singularity of the differential equation, which, however, is invisible at first glance. Regarding ψ we refer to the second condition in Malmquist's Second Theorem. ☕

Example 5.10 For $q = 2$, regular (non-singular) rational solutions are deficient, while ramified rational functions and singular rational solutions coincide and are not deficient. This is not necessarily true if $q > 2$, where 'regular-singular' rational solutions may occur. With the exception of $\phi(z) = z+1$, the solutions to the Riccati equation $\mathfrak{t}' = z^2 + 2z + 2 - \mathfrak{t}^2$ are transcendental meromorphic functions, and $w = \mathfrak{r}(z, \mathfrak{t}) = (\mathfrak{t} - \phi(z))\mathfrak{t}^2$ satisfies some irreducible differential equation

$$w'^3 - 6(z+1)ww'^2 + \hat{P}_1(z,w)ww' + \hat{P}_0(z,w)w^2 = 0$$

with $\hat{P}_1(z,0) \not\equiv 0$, $\deg_w \hat{P}_1 = 1$, and $\hat{P}_0(z,0) \not\equiv 0$, $\deg_w \hat{P}_0 = 2$. The discriminant is $D(z,w) = w^3(27w + 4(z+1)^3)D_1(z,w)^2$. Rational solutions are 0 and $\psi(z) = -\frac{4}{27}(z+1)^3$; ψ is singular since $\mathfrak{r}(z,\tau) - \psi(z) = (\tau + \frac{1}{3}(z+1))(\tau - \frac{2}{3}(z+1))^2$, while 0 is simultaneously deficient and singular. It is easily seen that $\delta(0,w) = \vartheta(0,w) = \frac{1}{3}$, $\Theta(0,w) = \frac{2}{3}$, $\delta(\psi,w) = 0$, $\vartheta(\psi,w) = \Theta(\psi,w) = \frac{1}{3}$, and $\delta(\infty,w) = 0$, $\vartheta(\infty,w) = \Theta(\infty,w) = \frac{2}{3}$; ∞ is completely ramified, and may be viewed as a 'singular solution', while 0 is deficient and completely ramified. The procedure to be described in Sect. 5.4 will show that $w \in \mathfrak{Y}_{3,1}$ can be immediately read off the differential equation. ☕

Exercise 5.22 For $\mathfrak{t}' = 1 + z^2 - \mathfrak{t}^2$ and $w = \mathfrak{t}^2\frac{\mathfrak{t}-z}{\mathfrak{t}+z}$ compute the corresponding differential equation $z^3w'^3 + \hat{P}_2(z,w)ww' + \hat{P}_1(z,w)ww' + \hat{P}_0(z,w)w^2 = 0$; prove that $\psi_{1,2}(z) = \frac{1}{2}(11 \pm 5\sqrt{5})z^2$ and ∞ are singular solutions and 0 is a regular-singular solution. Verify $\vartheta(0,w) = \delta(0,w) = \vartheta(\infty,w) = \Theta(\infty,w) = \vartheta(\psi_{1,2},w) = \Theta(\psi_{1,2},w) = \frac{1}{3}$ and $\Theta(0,w) = \frac{2}{3}$. ☕

Exercise 5.23 Singular solutions need not be single-valued (cf. Example 5.9), so it makes no sense to ask whether a many-valued singular solution may be deficient or ramified for w. Many-valued singular solutions, however, form cycles $\psi_1, \ldots, \psi_s$ under analytic continuation, and $\mathfrak{p}(z,\tau) = \prod_{j=1}^{s}(\tau - \psi_j(z))$ is a polynomial in τ over $\mathbb{C}(z)$. Prove that $v = \mathfrak{p}(z, \mathfrak{r}(z,\mathfrak{t}))$ satisfies some first-order equation $Q(z,v,v') = 0$ of degree qs, and 0 is a singular solution that is ramified for v. ☕

Example 5.11 The strings of poles need not be $\frac{\eta}{2}$-separated from each other. Let $\mathfrak{t}$ be any solution to $\mathfrak{t}' = z^2 - \mathfrak{t}^2$. Then $w = \frac{\mathfrak{t}^4}{(\mathfrak{t}-z)(\mathfrak{t}-2z)(\mathfrak{t}-z^2)}$ solves some Eq. (5.16) of degree four. According to the classification of the zeros of $\mathfrak{t} - \phi(z)$, ϕ rational, w has four different types of poles distributed in $(\pm\pi i, 1)$-strings along (some of) the Stokes rays $\arg z = (2\nu+1)\frac{\pi}{4}$. They correspond to

1. the poles p of $\mathfrak{t}$; they form the set $\mathfrak{P}(\mathfrak{t})$ and are arranged in (two or four) strings;
2. the zeros ζ of $\mathfrak{t}(z) - z^2$; they are contained in $\mathfrak{P}_\delta(\mathfrak{t})$, exactly one belongs to $\Delta_\delta(p) = \{z : |z-p| < \delta|p|^{-1}\}$ for $|p|$ sufficiently large.

3. the zeros $\tilde{\zeta}$ of $\mathfrak{t}(z) - 2z$; they form strings that are 'parallel' to the strings of the first kind, $\tilde{\zeta}_k = p_k - (\frac{1}{2}\log 3 + o(1))p_k^{-1}$;
4. the zeros $\hat{\zeta}$ of $\mathfrak{t}(z) - z$; they are 'invisible' from the poles of the first kind in the metric $ds = |z||dz|$, that is, $|\hat{\zeta}_k|\operatorname{dist}(\hat{\zeta}_k, \mathfrak{P}(\mathfrak{t})) \to \infty$ as $k \to \infty$.

5.3.7 *Equations of Genus One*

The solutions to first-order algebraic differential equations of genus one are locally given by

$$w(z) = \mathfrak{r}(z, \mathfrak{t}(z), \mathfrak{t}'(z)),\ \mathfrak{t}(z) = \wp(\Phi(z)), \text{ and } \Phi'(z) = \sqrt{a(z)}.$$

The main problem arising here is that the coefficients are algebraic rather than rational functions, and Φ may (and will) have algebraic and logarithmic singularities. On the other hand, all coefficients are functions of some fixed root $\sqrt[p]{z}$ in some neighbourhood of infinity, and this also holds for $\mathfrak{t}(z) = \rho(z, w(z), w'(z)) = f(\sqrt[p]{z})$. Then $W = f(\zeta)$ satisfies $W'^2 = A(\zeta)(4W^3 - g_2W - g_3)$ with $A(\zeta) = p^2\zeta^{2p-2}a(\zeta^p) = A_0\zeta^{2\beta} + \cdots$ on $|\zeta| > r_0$, and from $\wp^\sharp \asymp 1$, which holds outside arbitrarily small euclidean discs about the critical points of $\wp$, it follows that

$$\sup_{|\xi-\zeta|<\delta|\zeta|^{-\beta}} f^\sharp(\xi) \asymp |\zeta|^\beta.$$

Thus f belongs to the Yosida class $\mathfrak{Y}^0_{0,\beta}$ on $|\zeta| > r_0$ and $T(r, w) \asymp \log^2 r$ ($\beta = -1$) resp. $T(r, w) \asymp r^{\frac{2\beta+2}{p}}$ ($\beta > -1$) holds. Exercise 4.7 shows that the first case actually may occur. In the second case, however, the order of growth $\frac{2\beta+2}{p}$ is restricted to integer multiples of $\frac{1}{2}$ or $\frac{1}{3}$, see Eremenko [35] for a proof. This result is related to the question of whether elliptic functions may be written as $f \circ Q$, where f is meromorphic and Q is a non-linear polynomial. The answer given by Mues [118] is that $Q(z) = a + b(z - z_0)^d$ and $d \in \{2, 3, 4, 6\}$ is necessary. Each case may occur:

Exercise 5.24 Let $\wp(z; g_2, g_3)$ denote the Weierstraß P-function with differential equation $\wp'^2 = 4\wp^3 - g_2\wp - g_3$. Prove that

1. $\wp(z; g_2, g_3) = f_2(z^2)$ and $w = f_2$ satisfies $w'^2 = \frac{1}{4}z^{-1}(4w^3 - g_2w - g_3)$.
2. $\wp'(z; 0, g_3) = f_3(z^3)$ and $w = f_3$ satisfies $w'^3 = \frac{1}{2}z^{-2}(w^2 - g_3)^2$.
3. $\wp^2(z; g_2, 0) = f_4(z^4)$ and $w = f_4$ satisfies $w'^4 = \frac{1}{16}z^{-3}w^3(4w - g_2)^3$.
4. $\wp^3(z; 0, g_3) = f_6(z^6)$ and $w = f_6$ satisfies $w'^6 = \frac{1}{64}z^{-5}w^4(4w - g_3)^3$.

Remark 5.9 First-order equations are discussed in [185] in more detail. Eremenko considered the same class of equations (5.16). Among other things he gave a proof of Malmquist's Second Theorem in [34]. He also proved in [35] that the solutions have order of growth $\varrho = k/2$ mean type ($k \in \mathbb{N}$) in case of genus zero, and $\varrho = k/2$ or $\varrho = k/3$ mean type ($k \in \mathbb{N}_0$) otherwise, where $k = 0$ means $T(r, w) \asymp \log^2 r$.

5.4 Differential Equations and the Yosida Classes

Let w be any transcendental solution to the algebraic differential equation

$$Q(z, w, w', \ldots, w^{(n)}) = 0.$$

We set $z = h + h^{-\beta}\mathfrak{z}$, $w_h(\mathfrak{z}) = h^{-\alpha}w(h + h^{-\beta}\mathfrak{z})$, $w'_h(\mathfrak{z}) = h^{-\alpha-\beta}w'(h + h^{-\beta}\mathfrak{z})$, etc., this leading to $Q(h + h^{-\beta}\mathfrak{z}, h^\alpha w_h, h^{\alpha+\beta}w'_h, \ldots, h^{\alpha+n\beta}w_h^{(n)}) = 0$. Taking the limit $\lim_{h\to\infty} h^{-\gamma}Q(h, h^\alpha \mathfrak{w}_0, h^{\alpha+\beta}\mathfrak{w}_1, \ldots, h^{\alpha+n\beta}\mathfrak{w}_n)$ for suitably chosen γ then yields some autonomous differential equation

$$\mathfrak{Q}(\mathfrak{w}, \mathfrak{w}', \ldots, \mathfrak{w}^{(n)}) = 0,$$

which is solved by every limit function $\mathfrak{w} = \lim_{h_k\to\infty} w_{h_k}$. Apart from the fact that the real parameters α and β are arbitrary, the method is by no means justified. Nevertheless it can be justified if w belongs to the Yosida class $\mathfrak{Y}_{\alpha,\beta}$.

5.4.1 *Application to First-Order Differential Equations*

For Eq. (5.16) the formal re-scaling process leads to Briot–Bouquet equations

$$\mathfrak{P}(\mathfrak{w}, \mathfrak{w}') = 0. \tag{5.30}$$

In the case of genus zero, neither the parametrisation (5.22) nor the Riccati equation (5.26) are at hand. Thus the problem arises how to determine the parameter n as well as the asymptotic expansions exclusively from (5.16), and also possible values of α, if any, such that $w \in \mathfrak{Y}_{\alpha,\frac{n}{2}}$.

1. *Asymptotic expansions.* To determine the possible leading terms of the asymptotic series $w \sim az^{\frac{m}{2}} + \cdots$, hence $w' \sim \frac{m}{2}az^{\frac{m}{2}-1} + \cdots \sim \frac{m}{2}z^{-1}w$, exclusively from the differential equation in question (it follows from (5.22) that m must be some integer), consider $\tilde{P}(x, y) = P(x, y, \frac{m}{2}x^{-1}y) = \sum_{\nu=0}^{2q} A_\nu(1 + O(|x|^{-1}))x^{k_\nu}y^\nu$; $\tilde{P}(z, w)$ may be viewed as an approximation to $P(z, w, w')$. The Newton–Puiseux method applies to the simplified equation

$$\sum_{\nu=0}^{2q} A_\nu x^{k_\nu} y^\nu = 0. \tag{5.31}$$

As $x \to \infty$ the solutions have leading terms $a_j x^{\rho_j}$ ($a_j \neq 0$), and the possible leading terms of the asymptotic expansions for w are among the terms $a_j z^{\rho_j}$ with $2\rho_j \in \mathbb{Z}$. Some terms may also point to singular solutions $\psi(z) = a_j z^{\rho_j} + \cdots$.

2. *The parameters.* To select the 'true' parameters α and β out of a variety of candidates (the procedure is not unique) it is reasonable to postulate that Eq. (5.30) has maximal degree $\deg_{\mathfrak{w}'} \mathfrak{P} = q$ and $\beta = \frac{n}{2}$ is as small as possible (the 'local unit discs' $\Delta_1(p) : |z-p| < |p|^{-\beta}$ should be as large as possible). We assume $P_q(z,w) \equiv 1$. For $\mathfrak{w}_0$ and $\mathfrak{w}_1$ fixed, consider

$$\Phi(h, \mathfrak{w}_0, \mathfrak{w}_1) = h^{-q(\alpha+\beta)} P(h, h^\alpha \mathfrak{w}_0, h^{\alpha+\beta}\mathfrak{w}_1) = \mathfrak{P}(\mathfrak{w}_0, \mathfrak{w}_1) + \phi(h, \mathfrak{w}_0, \mathfrak{w}_1)$$

with $\mathfrak{P}(\mathfrak{w}_0, \mathfrak{w}_1) = \mathfrak{w}_1^q + \cdots$ and $\deg_{\mathfrak{w}_1} \phi < q$. Then α and β (as small as possible) have to be adjusted in such a way that $\phi(h, \mathfrak{w}_0, \mathfrak{w}_1)$ tends to zero as $h \to \infty$.

3. *Is $w \in \mathfrak{Y}_{\alpha,\beta}$?* Having determined the possible parameters one has to prove

$$w^{\sharp_\alpha}(z) = \frac{|z|^\alpha |w'(z)|}{|z|^{2\alpha} + |w(z)|^2} = O(|z|^\beta).$$

This may be done by using well-known estimates for the roots of ordinary polynomials applied to (5.16), where P is regarded a polynomial in w'.

Example 5.12 Consider $z^2 w'^2 + P_1(z,w)w' + P_0(z,w) = 0$ with

$$\begin{aligned} P_1(z,w) &= (2z - 2z^3)w - \tfrac{1}{4}(2 - z^2)w^2, \\ P_0(z,w) &= 2z^5 w + (1 + \tfrac{31}{4}z^4)w^2 + (4z + 5z^3)w^3 - \tfrac{1}{4}(2 + 3z^2)w^4, \\ D(z,w) &= 16z^4 w(w - 8z)((2 + 7z^2)w + 4z^3)^2. \end{aligned} \tag{5.32}$$

1. The reduced equation $yx^2(32x^3 + 124x^2y + 80xy^2 - 12y^3) = 0$ has solutions $y = 0, 8x, -\frac{1}{3}x, -x$. The first pair corresponds to the singular solutions $w = 0$ and $w = 8z$, while the second pair determines the principal terms of the asymptotic expansions $w \sim -\frac{1}{3}z + \cdots$ and $w \sim -z + \cdots$

2. For any choice of α and β the principal part of $\Phi(h, \mathfrak{w}_0, \mathfrak{w}_1)$ has the form

$$\mathfrak{w}_1^2 - (2h^{1-\beta}\mathfrak{w}_0 - \frac{1}{4}h^{\alpha-\beta}\mathfrak{w}_0^2)\mathfrak{w}_1 + 2h^{3-\alpha-2\beta}\mathfrak{w}_0$$
$$+ \frac{31}{4}h^{2-2\beta}\mathfrak{w}_0^2 + 5h^{1+\alpha-2\beta}\mathfrak{w}_0^3 - \frac{3}{4}h^{2\alpha-2\beta}\mathfrak{w}_0^4.$$

Obviously, $\beta \geq 1$ is necessary. Choosing $\beta = 1$, the terms $-\frac{1}{4}h^{\alpha-1}\mathfrak{w}_0^2\mathfrak{w}_1$ and $2h^{1-\alpha}\mathfrak{w}_0$ enforce $\alpha = 1$. The corresponding Briot–Bouquet equation is

$$\mathfrak{P}(\mathfrak{w}, \mathfrak{w}') = \mathfrak{w}'^2 - (2\mathfrak{w} - \tfrac{1}{4}\mathfrak{w}^2)\mathfrak{w}' + 2\mathfrak{w} + \tfrac{31}{4}\mathfrak{w}^2 + 5\mathfrak{w}^3 - \tfrac{3}{4}\mathfrak{w}^4 = 0$$

with solutions $\mathfrak{w} = 0, 8, -\frac{1}{3}, -1$ (again!), and $\mathfrak{w} = \frac{\coth^2 \mathfrak{z}}{\coth \mathfrak{z} - 2}$ (normalised).

3. To prove $|w'| = O(|z|^2 + |w|^2)$ ($\alpha = \beta = 1$) we start with

$$\begin{aligned}|w'| &= O(\max\{|P_1(z,w)/z^2|, |P_0(z,w)/z^2|^{\frac{1}{2}}\})\\ &= O(\max\{|z||w|, |w|^2|, |z|^{\frac{3}{2}}|w|^{\frac{1}{2}}, |z|^{\frac{1}{2}}|w|^{\frac{3}{2}}\}).\end{aligned}$$

Applying various Hölder inequalities then shows $|w'| = O(|z|^2 + |w|^2)$. Re-scaling along any sequence of poles ($w_p(\mathfrak{z}) = p^{-1}w(p + p^{-1}\mathfrak{z})$) yields non-constant limit functions $\mathfrak{w} = \rho/\mathfrak{z} + \cdots$ (the residues satisfy $\rho^2 - \frac{1}{4}\rho^3 - \frac{3}{4}\rho^4 = 0$). To exclude $w_{h_k} \to \infty$ as $h_k \to \infty$ we consider the differential equation

$$\mathfrak{v}'^2 + (2\mathfrak{v} - \tfrac{1}{4})\mathfrak{v}' + 2\mathfrak{v}^3 + \tfrac{31}{4}\mathfrak{v}^2 + 5\mathfrak{v} - \tfrac{3}{4} = 0$$

for $\mathfrak{v} = 1/\mathfrak{w}$, which has no trivial solution. This proves $w \in \mathfrak{Y}_{1,1}$.[5]

4. *The distribution of poles.* Almost all poles of w are simple ($\deg_w P_0 = 4$) and are distributed in $(\pm\pi i, 1)$-strings asymptotic to the rays $\arg z = (2\nu + 1)\frac{\pi}{2}$. The poles of w occur in pairs p and $\tilde{p} = p + (\frac{1}{2}\log 3 + o(1))p^{-1}$, this following from the distribution of the poles of $\coth \mathfrak{z}$, the zeros of $\coth \mathfrak{z} - 2$, and Hurwitz' Theorem. ☕

Example 5.13 The same procedure applied to

$$w'^2 - z^2(4w - w^2)w' + 4z^4w + (8z^2 - z^2)w^2 + (4 - 6z^2)w^3 - (1 - z^2)w^4 = 0$$

yields $\alpha = 0$ and $\beta = 2$, which, however, does not reflect the properties of the transcendental solutions $w = \frac{\mathfrak{t}^2}{\mathfrak{t}-1}$ with $\mathfrak{t}' = z^2 - \mathfrak{t}^2$, hence $\mathfrak{t} \in \mathfrak{Y}_{1,1}$. The reason for the failure of the method is that the zeros q and poles p of w that correspond to the zeros and one-points of $\mathfrak{t}$ 'collide' ($\inf_p |q||q - p| \to 0$ as $q \to \infty$). ☕

Exercise 5.25 Discuss the differential equation

$$\begin{aligned}&z^2w'^2 + [(2z - 4z^2)w - (1 - z)w^2]w'\\ &\quad + [4z^3w + (1 - z^2 + 8z^3)w^2 + (4z - 6z^2 + 4z^3)w^3 - (1 - z + z^2)w^4] = 0\end{aligned}$$

in the spirit of Example 5.12. In particular, prove that the transcendental solutions belong to $\mathfrak{Y}_{0,\frac{1}{2}}$ by showing that $\alpha = 0$, $\beta = \frac{1}{2}$ (as small as possible) is necessary, and $w^\sharp(z) = O(|z|^{\frac{1}{2}})$ follows exclusively from the differential equation. Derive $\mathfrak{w}'^2 = -4\mathfrak{w}(\mathfrak{w} + 1)^2$ for every limit function (solutions $\mathfrak{w} = 0, -1$ and

[5] One cannot expect to re-construct the parametrisation $w = \mathfrak{r}(z, \mathfrak{t})$ and $\mathfrak{t}' = \hat{P}(z) - \mathfrak{t}^2$. The fact that $\mathfrak{t}, w \in \mathfrak{Y}_{1,1}$ and $\mathfrak{w} = \frac{\coth^2 \mathfrak{z}}{\coth \mathfrak{z} - 2}$ suggests $w = \frac{\mathfrak{t}^2}{\mathfrak{t} - 2z - c_0}$ and $\hat{P}(z) = z^2 + a_1 z + a_0$, actually $a_1 = a_0 = c_0 = 0$.

$\mathfrak{w} = -\coth^2 \mathfrak{z}$), and deduce the essential properties exclusively from the differential equation, namely:

- the poles of w occur in pairs p and $\tilde{p} = p + o(|p|^{-\frac{1}{2}})$ and are distributed in $(\pm\pi i, \frac{1}{2})$-strings along $\arg z = (2\nu + 1)\pi/3$;
- $\vartheta(0, w) = \vartheta(4z, w) = \frac{1}{2}$;
- the possible asymptotic expansions are $w \sim \mp z^{-\frac{1}{2}} - z^{-1} \pm \cdots$.

Solution $D(z, w) = z^4 w(w - 4z)((1 - z + 2z^2)w + 2z^2)^2$, $w = \frac{\mathfrak{t}^2}{\mathfrak{t}-z}$, $\mathfrak{t}' = z - \mathfrak{t}^2$.

Resume There is a clear distinction between the solutions to equations of genus zero and genus one. In the first case the poles are arranged in finitely many strings along regularly distributed rays, while in the second case the poles locally form a lattice and globally 'fill' the plane. The trigonometric and elliptic functions, solutions to certain Briot–Bouquet equations $P(w, w') = 0$, are prototypal.

Chapter 6
Higher-Order Algebraic Differential Equations

In this chapter we will extend the investigations of the previous chapter to second-order algebraic differential equations and two-dimensional Hamiltonian systems whose solutions are meromorphic functions. Having established the Painlevé property for distinguished equations and systems, we will draw a comprehensive picture of the solutions. This includes detecting and describing the distribution of zeros and poles, zero- and pole-free regions, and asymptotic expansions on pole-free regions, and characterising the so-called sub-normal solutions. As in the preceding chapter, a crucial role is played by the method of Yosida Re-scaling. It establishes the central discovery that the first, second, and fourth Painlevé transcendents belong to the Yosida classes $\mathfrak{Y}_{\frac{1}{2},\frac{1}{4}}$, $\mathfrak{Y}_{\frac{1}{2},\frac{1}{2}}$, $\mathfrak{Y}_{1,1}$, respectively.

6.1 Introduction

Higher-order algebraic differential equations

$$\Omega(z, w, w', \ldots, w^{(n)}) = 0 \tag{6.1}$$

and systems

$$x'_\nu = P_\nu(z, x_1, \ldots, x_n) \quad (1 \le \nu \le n) \tag{6.2}$$

all of whose solutions admit unrestricted analytic continuation up to 'fixed singularities' are said to have the *Painlevé property.* In the absence of fixed singularities the solutions are meromorphic functions. In most cases the evidence of the Painlevé property follows from Painlevé's Theorem 1.3.9, applied to the equation under consideration itself or to some often ingeniously transformed equation or system. This will exemplarily be discussed in Sect. 6.2. For distinguished equations and

N. Steinmetz, *Nevanlinna Theory, Normal Families, and Algebraic Differential Equations*, Universitext, DOI 10.1007/978-3-319-59800-0_6

systems having the Yosida property we will determine the value distribution of generic and exceptional solutions.

6.1.1 *Painlevé Tests*

As a rule, within any proper class of nonlinear algebraic differential equations and systems, the equations having the Painlevé property are still the exception. To sort the wheat from the chaff there are several Painlevé tests in common, which provide *necessary* conditions for the validity of the Painlevé property.

A simple version is to test whether or not for all (or countably many) p there exists some (formal) solution $\sum_{j=-k}^{\infty} c_j(z-p)^j$ with pole at p ($c_{-k} \neq 0$, $c_j = c_j(p)$). Equations and systems with this property are said to pass the Painlevé test for poles. Usually it is not hard to determine the principal terms $c_{-k}(z-p)^{-k}$ from the differential equation. For example, the meromorphic solutions (if any) to $w'' = a(z) + b(z)w + 2w^3$ with polynomial coefficients have simple poles with residues 1 or -1. The differential equation passes the Painlevé test if and only if $a'(z) = b''(z) \equiv 0$ (cf. Example 3.1).

Exercise 6.1 Realise the Painlevé test for

- the Hamiltonian system $x' = -H_y(z,x,y)$, $y' = H_x(z,x,y)$ with Hamiltonian $H(z,x,y) = \frac{1}{3}(x^3+y^3) + c(z)xy + b(z)x + a(z)y$, a, b, and c polynomials.
  Solution. $a' = b' = c'' = 0$ [99, 100] is necessary and sufficient.
- $w'' + 3ww' + w^3 + z(w' + w^2) = 0$ (special case of Eq. (VI). in [94]).
- $w'' + 3ww' + w^2 + zw - z = 0.$

Example 6.1 In Painlevé's fourth equation

$$w'' = \frac{w'^2}{2w} + \tfrac{3}{2}w^3 + 4zw^2 + 2(z^2-\alpha)w + \frac{\beta}{w} \tag{IV}$$

$w = 0$ may be singular. One way to check this is to apply the Painlevé pole test to the differential equation for $v = 1/w$. The direct and natural way is to prove or disprove that there are always (formal) solutions $w = c_1(z-q) + c_2(z-q)^2 + \cdots$. This could be called the Painlevé test for zeros. Actually, Eq. (IV) passes the Painlevé test for zeros; the coefficient $c_2 = \frac{1}{2}w''(q)$ remains undetermined and is completely *free*, while $c_1 = w'(q)$ is restricted to $c_1^2 + 2\beta = 0$. Prescribing q, $w(q) = 0$, $w'(q) = \sqrt{-2\beta}$, and $w''(q)$ determines a unique solution.

6.1.2 The Painlevé Story

Painlevé [134, 135] classified the second-order differential equations

$$w'' = R(z, w, w') \quad (R \text{ rational in } w, w') \tag{6.3}$$

without *movable singularities* other than poles. We will not make use of the terms 'fixed' and 'movable' singularity, and thus will not give a precise definition. To give an idea, suppose the solution defined by initial values $w_0 = w(z_0)$ and $w_0' = w'(z_0)$ has a singularity at z_*. Then solutions with initial values close to w_0 and w_0' likewise have a singularity of the same type in some neighbourhood of z_*: the singularity 'moves' with the initial values. For example, $w'' = \frac{2w-1}{w^2+1}w'^2$ is solved by $w = \tan\log(z-c)$, with poles $z = c + e^{(k+\frac{1}{2})\pi}$ and a logarithmic singularity at $z = c$; all singularities 'move' with c.

Like many other stories, ours does not tell the whole truth, since Painlevé did not detect the corresponding class of differential equations in its entirety. At any rate, he and subsequently Fuchs, Gambier, and others established a list of 50 canonical differential equations which should have this property. The list in [94] starts with I. $w'' = 0$ and ends up with the master equation L., which is now known as Painlevé's equation (VI). Actually 44 equations in this list may be reduced to known equations which have this property, such as linear and Riccati equations, equations for elliptic functions, etc., or to one of the remaining six Eqs. (I)–(VI). For example,

$$2ww'' = w'^2 + 4aw^3 - 2zw^2 - 1 \qquad \text{XXXIV.}$$

(the numbering follows Ince [94]) may be reduced to $v'' = 2v^3 + zv - a - \frac{1}{2}$ via $aw = v' + v^2 + \frac{1}{2}z$ if $a \neq 0$, and to $w''' + 2zw' + w = 0$ if $a = 0$.

6.1.3 The Painlevé Transcendents

Painlevé's equations (I)–(VI) form two subgroups: (III), (V), and (VI), which have fixed singularities at $z = 0$, and $z = 1$ additionally for (VI), and

$$w'' = z + 6w^2 \tag{I}$$

$$w'' = \alpha + zw + 2w^2 \tag{II}$$

$$2ww'' = w'^2 + 3w^4 + 8zw^3 + 4(z^2 - \alpha)w^2 + 2\beta \tag{IV}$$

without fixed singularities. The Painlevé property will be verified in the next section. Any transcendental solution to one of these equations is called Painlevé *transcendent*, more precisely, *first, second,* and *fourth* transcendent.[1] Our treatment follows the classical ideas until the method of Yosida Re-scaling enters the stage. It will turn out that the respective *first integrals W* :

$$w'^2 = 4w^3 + 2zw - 2W, \quad W' = w \tag{I'}$$

$$w'^2 = w^4 + zw^2 + 2\alpha w - W, \quad W' = w^2 \tag{II'}$$

$$w'^2 = w^4 + 4zw^3 + 4(z^2 - \alpha)w^2 - 2\beta - 4wW, \quad W' = w^2 + 2zw \tag{IV'}$$

are even more interesting than the transcendents itself.

We note that Painlevé's equations may be treated in a quite different way, based on the *Riemann–Hilbert* method and the method of *isomonodromic deformations.* This approach is strongly linked to the fields of *Special Functions* and *Mathematical Physics* and differs from ours concerning methods and issues. The reader is referred to the monograph by Fokas, Its, Kapaev, and Novokshënov [44].

Exercise 6.2 For first, second, and fourth transcendents compute in detail the first terms of the formal series expansions about poles:

$$\begin{aligned} w &= \;\; (z-p)^{-2} - \tfrac{1}{10}p(z-p)^2 - \tfrac{1}{6}(z-p)^3 + \mathbf{h}(z-p)^4 + \cdots \\ W &= -(z-p)^{-1} - 14\mathbf{h} - \tfrac{1}{30}p(z-p)^3 + \cdots \end{aligned} \tag{I\&I'}$$

$$\begin{aligned} w &= \;\; \epsilon(z-p)^{-1} - \tfrac{1}{6}\epsilon p(z-p) - \tfrac{1}{4}(\alpha+\epsilon)(z-p)^2 + \mathbf{h}(z-p)^3 + \cdots \\ W &= -(z-p)^{-1} + 10\epsilon\mathbf{h} - \tfrac{7}{36}p^2 - \tfrac{1}{3}p(z-p) + \cdots \end{aligned} \tag{II\&II'}$$

$$\begin{aligned} w &= \;\; \epsilon(z-p)^{-1} - p + \tfrac{1}{3}\epsilon(p^2 + 2\alpha - 4\epsilon)(z-p) + \mathbf{h}(z-p)^2 + \cdots \\ W &= -(z-p)^{-1} + 2\mathbf{h} + 2(\alpha-\epsilon)p + \tfrac{1}{3}(p^2 + 4\alpha - 2\epsilon)(z-p)\cdots \end{aligned} \tag{IV\&IV'}$$

($\epsilon = \pm 1$). All equations have one degree of freedom: the coefficient $\mathbf{h} = \mathbf{h}(p)$ remains undetermined and may be prescribed.

Exercise 6.3 Prove that Painlevé's equations (I) and (II) pass the Painlevé test for poles. Write $z = t + p$, $w = \sum_{n=0}^{\infty} a_n t^{n-2}$ ($a_0 = 1$) and $w = \sum_{n=0}^{\infty} b_n t^{n-1}$ ($b_0^2 = 1$) to obtain

$$\begin{aligned} (n^2 - 5n - 6)a_n &= \sum_{k=2}^{n-2} 6a_k a_{n-k} && (n \geq 7), \\ (n^2 - 3n - 4)b_n &= \sum_{\substack{j+k+\ell=n \\ j,k,\ell<n}} 2b_j b_k b_\ell + p b_{n-2} + b_{n-3} && (n \geq 5), \end{aligned}$$

[1] The original meaning of *transcendent* was different, namely: *The solutions are transcendental functions of the 'two constants of integration'*.

respectively. Assuming $|p| > 1$ and $|a_6| \le K|p|^{\frac{3}{2}}$ resp. $|b_4| \le K|p|^2$, prove that

$$|a_n| \le M^{n-\frac{1}{2}}|p|^{\frac{n}{4}} \quad \text{and} \quad |b_n| \le M^{n-\frac{1}{2}}|p|^{\frac{n}{2}} \quad (n \ge 1)$$

hold for some $M = M(K)$.

6.1.4 Hamiltonian Systems

Painlevé's equations are strongly connected with certain Hamiltonian systems

$$x' = -H_y(z, x, y),\; y' = H_x(z, x, y).$$

This will be discussed in several exercises.

Exercise 6.4 Consider the Hamiltonian system with solution (x, y) for $H(z, x, y) = -yx^2 - \frac{1}{2}zy - \frac{1}{2}y^2 + (\alpha - \frac{1}{2})x$ and let p be a pole of x. Prove that p is simple, with residue either -1 and p is a zero of y, or else with residue 1 and p is a double pole of y. Prove that x and y separately solve

$$\begin{aligned} x'' &= \alpha + zx + 2x^3 && \text{(Painlevé (II))} \\ 2yy'' &= y'^2 - 4y^3 - 2zy - (\alpha - \tfrac{1}{2})^2 && \text{(related to XXXIV. in [94]).} \end{aligned}$$

For $\eta(z) = H(z, x(z), y(z))$ prove $\eta''^2 - 4\eta'^3 + 2z\eta'^2 + 2\eta\eta' = \frac{1}{4}(\alpha - \frac{1}{2})^2$.

Exercise 6.5 (Continued, see [60], p. 100) For

$$u = -x + \frac{\alpha - \frac{1}{2}}{x' - x^2 - \frac{1}{2}z} \quad (\alpha \ne \tfrac{1}{2}) \tag{6.4}$$

deduce $u' + u^2 + \frac{1}{2}z = -y$ and $u'' = \alpha - 1 + zu + 2u^3$. In other words, the *Bäcklund transformation* (6.4) transforms Painlevé's equation $(\mathrm{II})_\alpha$ into $(\mathrm{II})_{\alpha-1}$.

Exercise 6.6 For $H(z, x, y) = xy^2 + yx^2 + 2zxy + 2\lambda x + 2\kappa y$, hence

$$x' = -2xy - x^2 - 2zx - 2\kappa,\; y' = 2xy + y^2 + 2zy + 2\lambda,$$

prove that x and y separately solve

$$\begin{aligned} 2xx'' &= x'^2 + 3x^4 + 8zx^3 + 4(z^2 + \kappa - 2\lambda)x^2 - 4\kappa^2 \\ 2yy'' &= y'^2 + 3y^4 + 8zy^3 + 4(z^2 + \lambda - 2\kappa)y^2 - 4\lambda^2 \end{aligned}$$

(Painlevé (IV)). In other words, the *Bäcklund transformations*

$$y = -\frac{x' + 2\kappa + 2zx + x^2}{2x} \quad \text{and} \quad x = \frac{y' - 2\lambda - 2zy - y^2}{2y}$$

transform one Painlevé (IV) equation into the other.

The system

$$x' = -y^2 - zx - \alpha, \; y' = x^2 + zy + \beta \tag{IV}$$

with time-dependent Hamiltonian

$$H(z,x,y) = \tfrac{1}{3}(x^3 + y^3) + zxy + \beta x + \alpha y, \tag{6.5}$$

was recently discovered by Kecker [99, 100]. It admits trivial and non-trivial *Bäcklund transformations* $\mathsf{M}_\omega : (x, y, \alpha, \beta) \mapsto (\omega x, \bar{\omega} y, \omega\alpha, \bar{\omega}\beta)$ $(\omega^3 = 1)$ and

$$\mathsf{B}_\omega : (x, y, \alpha, \beta) \mapsto \begin{cases} x(z) - \bar{\omega}\dfrac{\omega\alpha - \bar{\omega}\beta + 1}{\omega x(z) + \bar{\omega} y(z) - z} \\ y(z) + \omega\dfrac{\omega\alpha - \bar{\omega}\beta + 1}{\omega x(z) + \bar{\omega} y(z) - z} \\ (\omega\beta - \bar{\omega}, \bar{\omega}\alpha + \omega) \end{cases} \tag{6.6}$$

The poles of solutions (x, y), if any, have residues $(\omega, -\bar{\omega})$.

Remark 6.1 In [100] systems with more general Hamiltonians

$$H(z,x,y) = a_{n0}(z)x^n + a_{0m}(z)y^m + \sum_{\mu n + \nu m < mn} a_{\nu\mu}(z)x^\nu y^\mu$$

$(a_{n0}(z)a_{0m}(z) \not\equiv 0$, slightly simplified) are considered, with a focus on conditions implying that analytic continuation of the solutions along rectifiable curves leads to nothing else but algebraic singularities and poles. The generalised Painlevé test using Puiseux series $\sum_{j=-k}^{\infty} c_j(z-p)^{\frac{j}{h}}$ yields certain 'resonance conditions' for the coefficients and predicts algebraic poles with $h = \frac{(m-1)(n-1)-1}{\gcd(m,n)}$. For small m and n the resonance conditions lead to certain normalised Hamiltonian systems. Besides Hamiltonians like $H(z,x,y) = \frac{1}{2}y^2 + \frac{1}{2}x^n + P(z,x)$ with $\deg_x P < n \le 3$, just

$$H(z,x,y) = \tfrac{1}{3}(x^3 + y^3) + c(z)xy + b(z)x + a(z)y$$

(after normalisation) has to be considered ($h = 1$). Here the resonance conditions $a'(z) = b'(z) = c''(z) = 0$ (see Exercise 6.1) lead to the autonomous Hamiltonian $H(z,x,y) = \frac{1}{3}(x^3 + y^3) + \gamma xy + \beta x + \alpha y$ or to (6.5) normalised by $c(z) = z$.

System (IV) is strongly connected with Painlevé's fourth equation via the transformation $v = x + y - z$, which leads to

$$2vv'' = v'^2 - v^4 - 4zv^3 - (3z^2 + 2\alpha + 2\beta)v^2 - (1 + \alpha - \beta)^2. \qquad (\text{I}\Lambda)$$

Exercise 6.7 Compute in detail formal series expansions about poles of the solutions to (IV):

$$\begin{aligned} x(z) = {} & (z-p)^{-1} + \tfrac{1}{2}p + \left[1 + \tfrac{1}{3}(\alpha - 2\beta) - \tfrac{1}{4}p^2\right](z-p) \\ & + \left[\mathbf{h} - \left(\tfrac{5}{8} + \tfrac{1}{4}(\alpha - \beta)\right)p\right](z-p)^2 + \cdots, \\ y(z) = {} & -(z-p)^{-1} + \tfrac{1}{2}p + \left[1 + \tfrac{1}{3}(2\alpha - \beta) + \tfrac{1}{4}p^2\right](z-p) \\ & + \left[\mathbf{h} + \left(\tfrac{5}{8} + \tfrac{1}{4}(\alpha - \beta)\right)p\right](z-p)^2 + \cdots, \\ H(z) = {} & (z-p)^{-1} + \left[2\mathbf{h} + \tfrac{1}{3}p^3 + \tfrac{1}{2}(\alpha + \beta)p\right] + \left[\tfrac{1}{3}(\alpha + \beta) + \tfrac{3}{4}p^2\right](z-p) + \cdots \end{aligned}$$

for $H(z) = H(z, x(z), y(z))$, and

$$x(z) + y(z) - z = [1 + \alpha - \beta](z-p) + 2\mathbf{h}(z-p)^2 + \cdots$$

Remark 6.2 Noting that $u = \bar{\omega}x$, $v = \omega y$ ($\omega = e^{\pm 2\pi i/3}$), solve the system with parameters $\bar{\omega}\alpha$ and $\omega\beta$ in place of α and β, derive similar expansions at poles with residues $(\omega, -\bar{\omega})$. The parameter $\mathbf{h}$ cannot be determined and is *free*; in combination with p and ω it determines a unique solution. The expansion for H also holds with $\omega\alpha$ and $\bar{\omega}\beta$ in place of α and β at poles with residues $(\omega, -\bar{\omega})$. In particular, H is the logarithmic derivative of an entire function with simple zeros at the poles of (x, y), and no others.

Exercise 6.8 Suppose p ($|p| > 1$) is a pole with residues $(\omega, -\bar{\omega})$, hence $x(z) = \sum_{n=0}^{\infty} a_n t^{n-1}$ and $y(z) = \sum_{n=0}^{\infty} b_n t^{n-1}$ hold with $z = p + t$, $a_0 = \omega$, and $b_0 = -\bar{\omega}$. Derive the nonlinear recursion

$$\left.\begin{aligned} (n-1)a_n - 2\bar{\omega}b_n &= -\sum_{k=1}^{n-1} b_k b_{n-k} - pa_{n-1} - a_{n-2} \\ -2\omega a_n + (n-1)b_n &= \sum_{k=1}^{n-1} a_k a_{n-k} + pb_{n-1} + b_{n-2} \end{aligned}\right\} \quad (n \geq 3).$$

Then a_1, b_1, a_2, and b_2 may be computed explicitly by solving $2\omega a_1 = p\bar{\omega}$, $2\bar{\omega}b_1 = p\omega$, $a_2 - 2\bar{\omega}b_2 = -b_1^2 - pa_1 - a_0 - \alpha$, and $-2\omega a_2 + b_2 = a_1^2 + pb_1 + b_0 + \beta$, while a_3 and b_3 remain undetermined (cf. Exercise 6.7), but satisfy $|a_3| + |b_3| = O(|p|^3)$ (this will be proved in Lemma 6.5). Assuming

$$|a_n| < M^{n-\frac{1}{2}}|p|^n \quad \text{and} \quad |b_n| < M^{n-\frac{1}{2}}|p|^n \tag{6.7}$$

for $1 \leq n \leq 3$, where M is independent of p, prove that (6.7) also holds for $n \geq 4$ if $M > 1$ is chosen sufficiently large.
(**Hint.** Estimate the right-hand side and use Cramer's rule.)

6.1.5 *Theorems of Malmquist Type*

The Painlevé property for Eq. (6.3) means that *every* solution admits analytic continuation in $\mathbb{C} \setminus S$, where S denotes the set of 'fixed singularities'. If S is empty, *every* solution is meromorphic on the plane. On the other hand, Theorems of Malmquist type require *just one* admissible resp. transcendental solution (in case of rational coefficients). One obstacle for theorems of Malmquist type is that meromorphic solutions that also solve some first-order algebraic differential equation $P(z, w, w') = 0$ have to be excluded. Accounting for the complexity of (6.3), the fact that the output should contain at least 50 normalised equations, and the above mentioned side condition, the problem seems inaccessible. However, the first step in Painlevé's analysis leads to equations

$$w'' = L(z, w)w'^2 + M(z, w)w' + N(z, w), \tag{6.8}$$

where L, M, N are rational functions with respect to w. It is thus quite natural to start with equations of this type, assuming that L, M, and N are rational in both variables and (6.8) has a transcendental meromorphic solution that does not satisfy any first-order algebraic differential equation. A first general result concerning the degrees of L, M and N w.r.t. w is as follows: If (6.8) is written in lowest terms as

$$D(z, w)w'' = C(z, w)w'^2 + B(z, w)w' + A(z, w)$$

with polynomials A, B, C, D w.r.t. w, then

$$\deg_w D \le 4, \ \deg_w(wC - 2D) \le 3, \ \deg_w B \le 4, \text{ and } \deg_w A \le 6$$

holds; the bounds are sharp, see [170] and also the next example.

Example 6.2 Let $v \not\equiv$ const satisfy $v'^2 = \prod_{j=1}^4 (v - c_j)$ ($c_j \ne c_k$ for $j \ne k$). Then $w(z) = 1/(v(e^z) + z)$ satisfies some Eq. (6.8) with coefficients

$$L(z, w) = \frac{P_w(z, w)}{2P(z, w)}, \ M(z, w) = 1 + \frac{P_z(z, w)}{P(z, w)}, \ N(z, w) = w^2 + \frac{w^2 P_z(z, w)}{2P(z, w)},$$

and $P(z, w) = \prod_{j=1}^4 ((z + c_j)w - 1)$. Since w has infinite order of growth it cannot satisfy some first-order algebraic differential equation.

If additional conditions are imposed on the coefficients L, M, and N, the state of the art is as follows.

1. $w'' = L(z, w)w'^2$ is impossible [167].
2. $w'' = N(z, w)$ implies $\deg_w N \le 3$. If $N(z, w) = \sum_{j=0}^3 a_j(z)w^j$ is a polynomial, the Painlevé test for poles leads to resonance conditions for the coefficients ([201]). If a_3 is a non-zero constant resp. $a_3 \equiv 0$ and a_2 is a non-zero constant,

these conditions lead to equations with constant coefficients or Painlevé's second resp. first equation as canonical representatives.

3. $w'' = M(z,w)w' + N(z,w)$, where $M \not\equiv 0$ and N are polynomials w.r.t. w. Necessary conditions are $\deg_w M \le 1$ and $\deg_w N \le 3$. In case of $\deg_w M = 1$ there are two canonical equations:

$$\text{(i) } w'' + 2ww' = A(z) + (B(z)^2 - B'(z))w + B(z)w^2,$$
$$\text{(ii) } w'' + 3ww' = A(z) + B(z)w + C(z)w^2 + D(z)w^3 \quad (D \not\equiv 0).$$

The first equation is equivalent to the weakly coupled system $u' = A(z) + B(z)u$, $w' = u - B(z)w - w^2$; if A and B are polynomials, then u is entire and w is meromorphic [102, 163]. The second equation (also [163]) with $D(z) \equiv -1$ is closely related to equation VI. in the list of 50 equations in [94].

Remark 6.3 There are three canonical equations

$$W'' = M(z,W)W' + N(z,W) \quad (M \not\equiv 0)$$

in the list of 50 equations in [94]. Adapted to our preferred form (transformation $w = W + \mu(z)$ to obtain $M(z,w) = -kw$, $k = 1, 2, 3$) these are

$$w'' + 2ww' = \tfrac{1}{2}(qq' - q'') \quad (\text{hence } w' + w^2 = \tfrac{1}{4}q^2 - \tfrac{1}{2}q' + \text{const}). \qquad \text{V.}$$
$$w'' + 3ww' = (2q^3 - q'') + 3(q^2 - q')w - w^3. \qquad \text{VI.}$$
$$w'' + ww' = 12q' - 12qw + w^3. \qquad \text{X.}$$

In V. and VI., q is an arbitrary meromorphic function, while in X., q is a solution to some equation $q'' = 6q^2 + \eta$ with $\eta = 0$, $\eta = 1$, or $\eta = z$ (Painlevé (I)). It is apparent that none of the Eqs. V., VI., and X. covers Eq. 3.(i) with $C \not\equiv 0$. Also the equation $w_1 w_1'' = 2w_1'^2 - 2w_1' - C(z)w_1 + (C'(z) - C(z)^2)w_1^2 - A(z)w_1^3$, which is obtained from 3.(i) by the transformation $w_1 = 1/w$, cannot be reduced to one of the remaining 40 equations with $L \not\equiv 0$.[2]

4. $w'' = M(z,w)w' + N(z,w)$, where at least one of $M \not\equiv 0$ and N is not a polynomial w.r.t. w. There is only one canonical equation,

$$ww'' = [1 + Aw + 3Cw^2]w' - 1 - Aw + Bw^2 + (C' - AC)w^3 - C^2w^4 \quad \text{(LI.)}$$

(A, B, C rational, $C \not\equiv 0$), equivalent to the weakly coupled Riccati system

$$u' = B + 2C + Au - u^2, \; w' = 1 + uw + Cw^2.$$

[2]This could be a hint that the list of 50 is incomplete. We note, however, that in [94] and other texts like [58], 'many-valued' transformations and solutions are on the agenda, while here only meromorphic solutions and rational transformations are admitted.

Remark 6.4 Meromorphic solutions to LI., if any, have finite order of growth [165]. For the proof just set $v = 1/w$ to obtain

$$vv'' = 2v'^2 + (3C + Av + v^2)v' + C^2 + (AC - C')v - Bv^2 + Av^3 + v^4,$$

which can be written as $v^4 = \Omega(z, v, v', v'')$ with degree $d_\Omega < 4$ and weight $\mathsf{d}_\Omega = 4$. Since v has simple poles with residue 1 ($w = 0$ implies $w' = 1$), Theorem 3.4 applies. Now assume that A, B, and C are polynomials. Then u is transcendental meromorphic with infinitely many simple poles with residue 1, and these poles may cause difficulties: w is meromorphic if and only if the poles of u are zeros of v with $v' = -C/2$. **Problem.** Determine all polynomials A, B, C such that Eq. LI. has meromorphic solutions or even the Painlevé property.

The proofs of the Malmquist-type theorems are too elaborate and involved to be presented here; the interested reader is referred to the literature cited above. We will just discuss the case

$$w'' = N(z, w) = P(z, w)/Q(z, w).$$

From Valiron's Lemma and $T(r, w'') \le 3T(r, w) + S(r, w)$ it follows that $\deg_w N = \max\{\deg_w P, \deg_w Q\} \le 3$. We want to show that Q is independent of w. To this end we assume $\deg_w Q = q \ge 1$, and note that $Q(z, w(z))$ can have only finitely many zeros (they are also zeros of the resultant of P and Q w.r.t. w). By the Tumura–Clunie Theorem for $Q(z, w(z))$, the algebraic equation $Q(z, \omega) = 0$ can have at most two mutually distinct solutions. If this number is two, w solves some Riccati equation $w' = a(z) + b(z)w + c(z)w^2$ with 'small' coefficients, again by the Tumura–Clunie Theorem, and equating coefficients shows that a, b, and c are rational functions. Otherwise we may assume that $Q(z, w) = w^q$ with $1 \le q \le 3$, hence we have to consider the differential equation $w^q w'' = P(z, w)$ ($\deg_w P \le 3$, $P(z, 0) \not\equiv 0$) having a transcendental solution with finitely many zeros. This, however, contradicts $m(r, 1/w) = S(r, w)$, which easily follows from $P(z, 0) \not\equiv 0$.

6.2 The Painlevé Property

The step from Painlevé tests to the Painlevé property is enormous and realised only for a few differential equations and systems. In this section we will show, representative for other equations, that System (IV), hence also Painlevé (IV), and Painlevé's equations (I) and (II) have this property. Whenever Painlevé's Theorem fails, the argument is more or less the same; we refer to [58, 88–90, 99, 100, 132, 151, 152, 175]. For example, the pattern to prove that an arbitrary polynomial system

$$x' = P(z, x, y) = \sum a_{jk}(z)x^j y^k, \quad y' = Q(z, x, y) = \sum b_{jk}(z)x^j y^k \tag{6.9}$$

has the Painlevé property is as follows. Suppose (x, y) is a meromorphic solution on $|z| < R$ with singularity at $z_0 = Re^{i\theta}$. Then

$$|x(z)| + |y(z)| \to \infty \quad \text{holds as } z \to z_0 \text{ on } [0, z_0), \tag{6.10}$$

and the goal is to prove that z_0 is a pole. To reach the goal one tries to determine a bi-rational transformation $u = R(z, x, y)$, $v = S(z, x, y)$, $x = R_1(z, u, v)$, $y = S_1(z, u, v)$, such that Painlevé's Theorem applies to the transformed polynomial system

$$u' = \hat{P}(z, u, v), \; v' = \hat{Q}(z, u, v).$$

To this end one tries to construct a so-called *Lyapunov function*

$$\Phi(z) = V(z, x(z), y(z))$$

that is bounded on $[0, z_0)$; usually this is implied by some differential inequality

$$|\Phi'| \le A + B|\Phi|. \tag{6.11}$$

The procedure, if any, is not unique. As a rule, the rational functions R, S, and V have to be constructed in such a way that they remain analytic at the poles of $x = x(z)$ and $y = y(z)$. The proof that Φ is bounded requires some side conditions like $|x(z)/y(z)| \le K$ on $[0, z_0)$. If this is not the case, $[0, z_0)$ has to be deformed into some path Γ from 0 to z_0 on which these side conditions hold. In order that a bound for Φ can be derived from (6.11) it is required that Γ is rectifiable.

6.2.1 The Painlevé Property for System (IV) and Painlevé (IV)

For System (IV) we will now implement the details in the above pattern.

Theorem 6.1 *The solutions to System* (IV) *are meromorphic in the plane.*

Proof Suppose (x, y) is a solution to System (IV) on the disc $D : |z| < R$, but has a singularity at $z_0 = Re^{i\theta}$ on the boundary. Then (6.10) holds. Following [99] we consider the Lyapunov function

$$\Phi(z) = H(z, x(z), y(z)) + x(z)^2/y(z) \;^{\ddagger} \tag{6.12}$$

‡$H(z, x, y) + x^2/y$ and also $H(z, x, y) - y^2/x$ are regular at every (common) pole.

to obtain $\Phi' + 3(q(z)/y(z))\Phi = 3q(z)^3 + 2\beta q(z)^2 + \alpha q(z)$ with $q(z) = x(z)/y(z)$. Suppose for the moment that

$$\frac{|x(z)|}{|y(z)|} \quad \text{is bounded on } [0, z_0). \tag{6.13}$$

Then also $1/|y(z)| = O\big((|x(z)| + |y(z)|)/|y(z)|\big) = O(|q(z)| + 1)$ is bounded, hence (6.11) holds and Φ is bounded. This implies $|x(z)^3 + y(z)^3| = O(|x(z)|^2) + O(|y(z)|^2)$ and $y(z) \sim -\omega x(z)$ as $z \to z_0$ for some third root of unity; we may assume $\omega = 1$. Now $H(z, x, y) + x^2/y$ with $y \sim -x$ as $x \to \infty$ can only be bounded if

$$y = -x + z + (\alpha + 1 - \beta)/x + O(1/|x|^2) \quad (x \to \infty).$$

Then $u = 1/y$ tends to 0 and $v = y(\alpha + 1 - \beta + y(x + y - z))$ remains bounded as $z \to z_0$ on $[0, z_0)$,[§] and Painlevé's Theorem applies to the polynomial system[¶]

$$u' = P(z, u, v), \ v' = Q(z, u, v);$$

$u, v, y = 1/u$ and $x = vu^2 - (\alpha + 1 - \beta)u + z - 1/u$ admit analytic continuation along $[0, z_0]$, u has a zero and x a pole at z_0.

It is, however, quite unlikely that (6.13) holds on $[0, z_0)$. On the other hand, the above argument works if (6.13) holds on some *rectifiable* path Γ joining 0 to z_0 in $D \cup \{z_0\}$. For the existence of Γ, which finishes the proof, see Lemma 6.1.

Lemma 6.1 *There exists some rectifiable path Γ joining 0 to z_0 in $D \cup \{z_0\}$ on which x/y is bounded.*

Proof The deformation of $[0, z_0)$ into Γ usually requires quite technical procedures. We will give a proof using a simple re-scaling technique, which also works for general polynomial systems. We assume that (x, y) solves System (IV) on $|z| < R$ and (6.10) holds, and set $E = \{\tau \in [0, z_0) : |y(\tau)| \le \frac{1}{2}|x(\tau)|\}$. We may assume that E accumulates at z_0 since otherwise we are done. Then $x(\tau) \to \infty$ as $\tau \to z_0$ on E, and E contains no pole. For $\tau \in E$ set $\mathfrak{x}(\mathfrak{z}) = \mathfrak{x}(\mathfrak{z}; \tau, \lambda) = \lambda x(\tau + \lambda\mathfrak{z})$ and

---

[§]Substituting $y = -x + z + (\alpha + 1 - \beta)/x + C/x^2 + \cdots$, where v is viewed as a function of the complex variables x, y, and z gives $v = -(\alpha + 1 - \beta)z + C + O(|x|^{-1})$ as $x \to \infty$.

[¶]$P(z, u, v) = -1 + zu - (\alpha + 1 + \gamma + z^2)u^2 + 2z\gamma u^3 - \gamma^2 u^4 + 2vu^3 - 2zvu^4 + 2\gamma u^5 - v^2 u^6$ and $Q(z, u, v) = -\gamma(\alpha + 1 + z^2) + 2\gamma^2 zu - zv + 2(\alpha + 1 + \gamma + z^2)vu - \gamma^3 u^2 - 6\gamma zvu^2 - 3v^2u^2 + 4\gamma^2 vu^3 + 4zv^2u^3 - 5\gamma v^2 u^4 + 2v^3u^5$ with $\gamma = \alpha + 1 - \beta$ are obtained with the help of MAPLE as follows

(x1, y1, v1 symbolise x', y', v'; 10 is just a large number to estimate the total degrees of P and Q):

```
beta:=alpha+1-gamma; x1:=-y^2-z*x-alpha; y1:=x^2+z*y+beta;
y:=1/u; x:=v*u^2-(alpha+1-beta)*u+z-1/u; P:=mtaylor(-y1*u^2,[u,v],10);
Q:=mtaylor(solve(x1-diff(x,z)-diff(x,v)*v1-diff(x,u)*P,v1),[u,v],10).
```

$\mathfrak{y}(\mathfrak{z}) = \mathfrak{y}(\mathfrak{z}; \tau, \lambda) = \lambda y(\tau + \lambda\mathfrak{z})$ to obtain the system

$$\mathfrak{x}' = -\mathfrak{y}^2 - \lambda(\tau + \lambda\mathfrak{z})\mathfrak{x} + \alpha\lambda^2, \ \mathfrak{y}' = \mathfrak{x}^2 + \lambda(\tau + \lambda\mathfrak{z})\mathfrak{y} + \beta\lambda^2$$

with initial values $\mathfrak{x}(0) = \lambda x(\tau)$, $\mathfrak{y}(0) = \lambda y(\tau)$, which depends on the parameters $\tau \in E$ and λ. For $\lambda = 1/x(\tau)$ we obtain by taking the limit $\tau \to z_0$

$$\mathfrak{x}' = -\mathfrak{y}^2, \ \mathfrak{y}' = \mathfrak{x}^2 \quad (\mathfrak{x}(0) = 1, \ |\mathfrak{y}(0)| \le \tfrac{1}{2}).$$[‖]

The solutions are elliptic functions which parametrise the algebraic curve $\xi^3 + \eta^3 = 1 + \mathfrak{y}(0)^3$ and depend analytically on $\mathfrak{y}(0)$. Thus given $r > 0$ sufficiently small there exists a $K > 1$ such that $|\mathfrak{x}(\mathfrak{z})/\mathfrak{y}(\mathfrak{z})| \le K$ holds on $|\mathfrak{z}| = r$, uniformly w.r.t. $\mathfrak{y}(0)$, $|\mathfrak{y}(0)| \le \frac{1}{2}$. In other words, the family $(\mathfrak{x}_\tau, \mathfrak{y}_\tau)_{\tau \in E}$ with $\mathfrak{x}_\tau(\mathfrak{z}) = \mathfrak{x}(\mathfrak{z}; \tau, 1/x(\tau))$ and $\mathfrak{y}_\tau(\mathfrak{z}) = \mathfrak{y}(\mathfrak{z}; \tau, 1/x(\tau))$ is normal on the disc $|\mathfrak{z}| < 2r$, say, and $|\mathfrak{x}_\tau(\mathfrak{z})/\mathfrak{y}_\tau(\mathfrak{z})| \le 2K$ holds on $|\mathfrak{z}| = r$ for $\tau \in E_0 = E \cap \{\tau : \tau_0 \le |\tau| < R\}$. The discs $|z - \tau| \le r|\lambda(\tau)|$ are contained in $|z| < R$, since otherwise z_0 would not be a singularity. For $\tau \in E_0$ this implies $|x(z)/y(z)| \le 2K$ on $|z - \tau| = r|\lambda(\tau)|$. To obtain a curve of finite length on which x/y is bounded we set $C = [\tau_0 e^{i\theta}, z_0) \cup \bigcup_{\tau \in E_0} \{z : |z - \tau| \le r|\lambda(\tau)|\}$. The boundary of C consists of two curves $\Gamma_\pm$ (envelopes), symmetric to $[0, z_0)$ and of length at most πR, and $|x(z)/y(z)|$ is bounded on $\Gamma_\pm$; also $\Gamma_\pm$ terminates at z_0 since $x(\tau) \to \infty$ as $\tau \to z_0$ on E; zeros of y on $[0, \tau_0 e^{i\theta})$ may be avoided by small semi-circles, and poles are irrelevant since $|x/y| = 1$ holds at every pole.

Remark 6.5 Lemma 6.1 can be adapted to polynomial systems (6.9) as follows. Suppose one has to prove that x/y is bounded on some rectifiable path Γ, where $|x(z)| + |y(z)| \to \infty$ as $z \to z_0$ on $[0, z_0)$ is assumed. Set $\alpha_{jk} = a_{jk}(z_0)$, $\beta_{jk} = b_{jk}(z_0)$, $I = \{(j,k) : (a_{jk}(z), b_{jk}(z)) \not\equiv (0,0)\}$, $E = \{\tau \in [0, z_0) : |y(\tau)| \le \frac{1}{2}|x(\tau)|\}$, $p = \max_I(j+k)$, $\lambda = x(\tau)^{-1}$, $\mu = \lambda^{p-1}$, $\mathfrak{x}(\mathfrak{z}) = \lambda x(\tau + \mu\mathfrak{z})$, and $\mathfrak{y}(\mathfrak{z}) = \lambda y(\tau + \mu\mathfrak{z})$ to obtain $\mathfrak{x}' = \tilde{P}(\mathfrak{x}, \mathfrak{y}) = \sum_{j+k=p} \alpha_{jk}\mathfrak{x}^j\mathfrak{y}^k$, $\mathfrak{y}' = \tilde{Q}(\mathfrak{x}, \mathfrak{y}) = \sum_{j+k=p} \beta_{jk}\mathfrak{x}^j\mathfrak{y}^k$ as $\tau \to z_0$ on E, $\lambda \to 0$, with $\mathfrak{x}(0) = 1$ and $|\mathfrak{y}(0)| \le \frac{1}{2}$. We just need $\tilde{Q}(1,0) = \beta_{p,0} \ne 0$ to obtain $|x(z)/y(z)| \le K$ on each circle $|z - \tau| = r|\lambda(\tau)|^{p-1}$ ($\tau \in E_0$, r fixed).

From Theorem 6.1 we obtain

Theorem 6.2 *The solutions to Painlevé (IV) are meromorphic.*

Proof By $v = x + y - z$, System (IV) is transformed into Eq. (IΛ) on p. 179, and every solution is obtained this way. Hence the solutions are meromorphic in the plane, and so are the solutions to Painlevé (IV) when written as

$$2ww'' = w'^2 + 3w^4 + 8zw^3 + 4(z^2 - \tfrac{i}{\sqrt{3}}(\alpha + \beta))w^2 + \tfrac{4}{9}(1 + \alpha - \beta)^2 \qquad \text{(IV}'')$$

(in (IΛ) set $w(z) = av(bz)$ with $b = \sqrt[4]{-4/3}$ and $a = -\frac{1}{2}b^3$).

[‖]The point z_0 is 'blown up' to a finite circle $|\mathfrak{z}| < r$. This is a common technique in differential equations, which also forms the basis of the Zalcman- and Yosida Re-scaling and Painlevé's so-called 'α-method'. *Nihil novi sub sole* [there is nothing new under the sun].

6.2.2 The Painlevé Property for Painlevé (I) and (II)

Theorem 6.3 *The solutions to Painlevé (I) and Painlevé (II) are meromorphic.*

Proof Let w be any solution to Painlevé (I) and (II), respectively, defined in some neighbourhood of the origin, say. If w is assumed not to be meromorphic in $\mathbb{C}$ then there exists a circle $|z| = R$ such that w is meromorphic inside, but has a singularity at z_0 on that circle. Then

$$|w(z)| + |w'(z)| \to \infty \quad \text{as } z \to z_0 \text{ on } [0, z_0). \tag{6.14}$$

The functions $\Phi = 2W - w'/w$ and $\Phi = W - w'/w$, where W denotes the respective first integral, are regular at poles of w and will serve as Lyapunov functions; they satisfy

$$\Phi' + \frac{\Phi}{w^2} = \frac{z - q(z)}{w} \quad \text{and} \quad \Phi' + \frac{\Phi}{w^2} = \frac{\alpha - q(z)}{w} \qquad (q = w'/w^2),$$

respectively. If q is unbounded on $[0, z_0)$ we have to show in both cases that $[0, z_0)$ can be deformed in $\{z : |z| < R\} \cup \{z_0\}$ into some rectifiable path Γ joining 0 to z_0, on which q is bounded; then also $1/|w|^2 = O((|w'|+|w|^2)/|w|^2) = O(|q(z)|+1)$, *a fortiori* $1/|w|$ and also Φ is bounded. In the first case we set $\lambda(\tau) = 1/\sqrt[3]{w'(\tau)}$ and $\mathfrak{w}(\mathfrak{z}) = \lambda^2 w(\tau + \lambda\mathfrak{z})$, and $\lambda(\tau) = 1/\sqrt{w'(\tau)}$ and $\mathfrak{w}(\mathfrak{z}) = \lambda w(\tau + \lambda\mathfrak{z})$ in the second, and assume that $\tau \in E = \{\tau : |w(\tau)|^2 \le \frac{1}{4}|w'(\tau)|\}$ accumulates at z_0. As $\tau \to z_0$ and $\lambda(\tau) \to 0$ we obtain $\mathfrak{w}'' = 6\mathfrak{w}^2$ with $\mathfrak{w}(0) = 0$, $\mathfrak{w}'(0) = 1$), and $\mathfrak{w}'' = 2\mathfrak{w}^3$ with $|\mathfrak{w}(0)| \le \frac{1}{2}$, $\mathfrak{w}'(0) = 1$), respectively. In both cases we obtain $|\mathfrak{w}'(\mathfrak{z})|/|\mathfrak{w}(\mathfrak{z})|^2 \le K$ on some circle $|\mathfrak{z}| = r$ independent of $\mathfrak{w}(0)$, and $|w'(z)|/|w(z)|^2 \le 2K$, say, holds on $|z - \tau| = r|\lambda(\tau)|$. Just as in the proof of Lemma 6.1 this permits us to deform $[0, z_0)$ into some rectifiable path Γ on which Φ is bounded. Moreover, not only does (6.14) hold, but $w(z)$ even tends to infinity as $z \to z_0$ on Γ. In the second case we set $u = 1/w$ to obtain

$$(u' - \tfrac{1}{2}u^3)^2 = 1 + zu^2 + 2\alpha u^3 - \Phi(z)u^4 + \tfrac{1}{4}u^6$$

from $w'^2 = w^4 + zw + 2\alpha w - (\Phi + w'/w)$. Since $u \to 0$ as $z \to z_0$ and Φ is bounded on Γ, u' tends to 1, say, and

$$u' = 1 + \tfrac{1}{2}zu^2 + (\alpha + \tfrac{1}{2})u^3 + vu^4 = P(z, u, v)$$

holds, where $v = -\frac{1}{2}\Phi + O(|u|)$ is bounded and u tends to zero as $z \to z_0$. In combination with $w'' = \alpha + zw + 2w^3$ we obtain

$$v' = -\tfrac{1}{2}(\alpha + \tfrac{1}{2})z - (\alpha + \tfrac{1}{2})^2 u - zuv - 3(\alpha + \tfrac{1}{2})vu^2 - 2v^2u^3 = Q(z, u, v).$$

Painlevés Theorem applies to the polynomial system $u' = P(z,u,v)$, $v' = Q(z,u,v)$, and z_0 is a pole of w with residue 1 (assuming $u' \to -1$ leads to a pole of w with residue -1). In the first case the final proof runs along the same lines. This time, $u = 1/\sqrt{w}$ is meromorphic on some neighbourhood of $\Gamma \setminus \{z_0\}$ (note that Γ contains no zeros of w, and poles may be avoided by small circular arcs), tends to zero as $z \to z_0$ on Γ, and satisfies $u'^2 - \frac{1}{2}u^5u' = 1 + \frac{1}{2}zu^4 - \frac{1}{4}\Phi(z)u^6$. Again assuming $u' \to 1$ as $z \to z_0$ along Γ leads to

$$u' = 1 + \tfrac{1}{4}zu^4 + \tfrac{1}{4}u^5 + vu^6 = P(z,u,v);$$

the function $v = -\frac{1}{8}\Phi(z) + O(|u|^2)$ is bounded as $z \to z_0$ on Γ and satisfies

$$v' = Q(z,u,v) = -\tfrac{1}{16}z^2u - \tfrac{3}{16}zu^2 - zvu^3 - \tfrac{1}{8}u^3 - \tfrac{5}{4}vu^4 - 3v^2u^5.$$

By Painlevé's Theorem, u has a simple zero and w has a double pole at z_0. ☕

Remark 6.6 The Painlevé property for equation (I) and (II) conceivably may also be deduced from the Painlevé property of the master equation (IV) resp. (VI) by a method called 'coalescence'. Actually there is a whole cascade of coalescence

$$\begin{array}{ccccc} \text{(VI)} & \longrightarrow & \text{(V)} & \longrightarrow & \text{(III)} \\ & & \downarrow & & \downarrow \\ & & \text{(IV)} & \longrightarrow & \text{(II)} & \longrightarrow & \text{(I)} \end{array}$$

([94], p. 346). We will describe the step (II) $\longrightarrow$ (I). Let w be the local solution to

$$w'' = z + 6w^2,\ w(z_0) = w_0,\ w'(z_0) = w_0', \tag{I}$$

λ any non-zero complex number, and y the solution to

$$y'' = 4\lambda^{-15} + xy + 2y^3,\ y(x_0) = \lambda w_0 + \lambda^{-5},\ y'(x_0) = \lambda^{-1}w_0', \tag{II$_\lambda$}$$

where $x_0 = \lambda^2 z_0 - 6\lambda^{-10}$. Clearly $y(x) = y(x;\lambda)$ depends on λ and

$$\omega(z) = w(z;\lambda) = \lambda^{-1}y(x;\lambda) - \lambda^{-6},\ z = \lambda^{-2}x + 6\lambda^{-12},$$

solves $\omega'' = z + 6\omega^2 + \lambda^6(z\omega + 2\omega^3)$ and has the same initial values as w at $z = z_0$. Applying the theorem on analytic dependence on parameters, $w(z) = w(z;0) = \lim_{\lambda\to 0} w(z;\lambda)$ follows. One has, however, to take care about the accumulation points of the poles of $y(x;\lambda)$ as $\lambda \to 0$, which can hardly be controlled. ☕

6.3 Algebraic Differential Equations and the Yosida Classes

Just as in the first-order case, the distribution of values of the solutions to higher-order differential equations and systems is strongly coupled with the distribution of poles. To determine the Nevanlinna functions, the distribution of poles, and the asymptotic expansions of transcendental solutions, our strategy will be to prove that the solutions belong to certain Yosida classes $\mathfrak{Y}_{\alpha,\beta}$. There are two possible ways to get an idea of which parameters (α, β) are worth considering: to study special solutions, and to apply the so-called Yosida test. Our goal will be to prove the following theorem, a homage to Painlevé and Yosida, and certainly the main result of the whole chapter.

Painlevé–Yosida Theorem *The solutions to System (IV) and Painlevé (IV) belong to the Yosida class $\mathfrak{Y}_{1,1}$, while first and second Painlevé transcendents belong to $\mathfrak{Y}_{\frac{1}{2},\frac{1}{4}}$ and $\mathfrak{Y}_{\frac{1}{2},\frac{1}{2}}$, respectively.*

The proof will be given in Sect. 6.3.4. The step from the Yosida and Painlevé test to the Yosida and Painlevé property is likewise technical and intricate.

6.3.1 Special Solutions

Airy Solutions The solutions to the so-called *Airy equations*

$$w' = \pm(\tfrac{1}{2}z + w^2) \tag{6.15}$$

also solve Painlevé (II) with parameters $\alpha = \pm\frac{1}{2}$. In particular, these solutions belong to $\mathfrak{Y}_{\frac{1}{2},\frac{1}{2}}$. By $w = \mp u'/u$, Eq. (6.15) is transformed into the linear differential equation $u'' + \frac{1}{2}zu = 0$, which is also known as an Airy equation when written as $v'' + zv = 0$.

Exercise 6.9 Prove that $w'' = \alpha + zw + 2w^3$ and $w' = a(z) + b(z)w + c(z)w^2$ (rational coefficients) simultaneously hold if and only if $\alpha = \pm\frac{1}{2}$ and w solves some Airy equation (6.15). (**Hint.** $c(z) = \pm 1$ is necessary and simplifies matter.)

For $\alpha \neq \pm\frac{1}{2}$ the Bäcklund transformations (see also Exercise 6.5)

$$w \mapsto -w + \frac{\alpha - \frac{1}{2}}{w' - w^2 - \frac{1}{2}z} \quad \text{and} \quad w \mapsto -w - \frac{\alpha + \frac{1}{2}}{w' + w^2 + \frac{1}{2}z} \tag{6.16}$$

transform Painlevé's equation (II) into the same equation with parameter $\alpha \mp 1$ in place of α. Any solution to Painlevé (II) that is obtained from some solution to (6.15) by repeated application of (6.16) and the trivial transformations $w \mapsto \overline{w(\bar{z})}$, $w \mapsto -w$, and $w \mapsto \omega w(\omega z)$ with $\omega^3 = 1$ is called an *Airy solution*.

Exercise 6.10 Prove that Airy solutions have the form $w = R(z, w_0)$, where R is some rational function and w_0 solves $w_0' = \frac{1}{2}z + w_0^2$, say. Thus w satisfies some first-order algebraic differential equation $P(z, w, w') = 0$ of genus zero and has Nevanlinna characteristic $T(r, w) \asymp r^{\frac{3}{2}}$. ☕

It is obvious (and algebraically trivial) that second transcendents satisfy some first-order algebraic differential equation if and only if their first integrals do.

Weber–Hermite Solutions The poles of any solution (x, y) to System (IV) have residues $(\omega, -\bar{\omega})$ with $\omega^3 = 1$. We are interested in the distribution of residues, and, in particular, what will happen if one or even two kinds of residues are missing.

Example 6.3 To determine the transcendental solutions to (IV) having just poles with residues $(1, -1)$ we note that $x + y - z$ is holomorphic and vanishes at every pole of (x, y) (see Exercise 6.7), hence vanishes identically. Thus x and y separately solve Riccati equations

$$x' = -\alpha - z^2 + zx - x^2 \quad \text{and} \quad y' = \beta + z^2 - zy + y^2 \tag{6.17}$$

with $\alpha - \beta + 1 = 0$. In particular, x and y belong to the Yosida class $\mathfrak{Y}_{1,1}$. ☕

The differential equations (6.17) are closely related to the *Weber–Hermite equations*

$$w' = \gamma \pm (2zw + w^2) \tag{6.18}$$

Any solution to System (IV) that is obtained from some solution to (6.17) by repeated application of the Bäcklund transformations (6.6) combined with the trivial transformations $(x, y) \mapsto (\bar{\omega}x, \omega y)$ with $\omega^3 = 1$ is called a *Weber–Hermite solution*. Weber–Hermite solutions have Nevanlinna characteristic $T(r, x) \asymp r^2$, $T(r, y) \asymp r^2$, and x and y separately satisfy first-order algebraic differential equations. Equivalently, x and y are algebraically dependent over $\mathbb{C}(z)$.

6.3.2 The Yosida Test

To test whether the solutions to the algebraic differential equation

$$\Omega(z, w, w', \ldots, w^{(n)}) = 0 \tag{6.19}$$

belong to some Yosida class, real parameters α and $\beta > -1$ have to be determined such that $h^{-\gamma}\Omega(h, h^\alpha \mathfrak{w}_0, h^{\alpha+\beta}\mathfrak{w}_1, \ldots, h^{\alpha+n\beta}\mathfrak{w}_n) \sim Q(\mathfrak{w}_0, \mathfrak{w}_1, \ldots, \mathfrak{w}_n)$ as $h \to \infty$ holds for some γ. This formally leads to the autonomous differential equation

$$Q(\mathfrak{w}, \mathfrak{w}', \ldots, \mathfrak{w}^{(n)}) = 0. \tag{6.20}$$

The procedure is far from being unique. A reasonable postulate is not to destroy the 'structure' of (6.19). In particular, Eq. (6.20) should also have order n. Moreover, and even more important, β should be as small as possible. Taking this into account the procedure will be called *Yosida test*. It was already discussed in Sect. 5.4.1 in the context of implicit first-order differential equations.

Example 6.4 For Painlevé's first equation $w'' = z + 6w^2$ the procedure yields

$$\mathfrak{w}_2 - h^{1-\alpha-2\beta} - 6\mathfrak{w}_0^2 h^{\alpha-2\beta} = Q(\mathfrak{w}_0, \mathfrak{w}_2) + o(1) \quad (h \to \infty),$$

hence $2\beta \ge \max\{1-\alpha, \alpha\}$. The smallest value of β is obtained if $\alpha = 1 - \alpha$, thus $\alpha = \frac{1}{2}$, $\beta = \frac{1}{4}$, and $\mathfrak{w}'' = 1 + 6\mathfrak{w}^2$. **Question.** Does w belong to $\mathfrak{Y}_{\frac{1}{2},\frac{1}{4}}$?

In the context of algebraic differential systems (6.2) it is appropriate to admit different parameters α_ν for different components, but one and the same β.

Example 6.5 Consider the Hamiltonian system (see Exercise 6.4)

$$x' = -y + x^2 + \tfrac{1}{2}z, \; y' = -2xy - (a - \tfrac{1}{2})$$

with Hamiltonian $H(z, x, y) = \frac{1}{2}y^2 - (x^2 + \frac{1}{2}z)y - (a - \frac{1}{2})x$. It is assumed that x and y are meromorphic (which is, indeed, the case). Replace z, x, x', y, and y' with h, $\mathfrak{x}h^{\alpha_1}$, $\mathfrak{x}_1 h^{\alpha_1+\beta}$, $\mathfrak{y}h^{\alpha_2}$, and $\mathfrak{y}_1 h^{\alpha_2+\beta}$ to obtain

$$\mathfrak{x}_1 h^{\alpha_1+\beta} + \mathfrak{y}h^{\alpha_2} - \mathfrak{x}^2 h^{2\alpha_1} - \tfrac{1}{2}h = o(h^{\alpha_1+\beta}),$$
$$\mathfrak{y}_1 h^{\alpha_2+\beta} + 2\mathfrak{x}\mathfrak{y}h^{\alpha_1+\alpha_2} + (a - \tfrac{1}{2}) = o(h^{\alpha_2+\beta}).$$

To preserve the structure we need $\beta \ge \max\{\alpha_2 - \alpha_1, \alpha_1, 1 - \alpha_1, -\alpha_2\}$; the smallest value $\beta = \frac{1}{2}$ is obtained if $\alpha_1 = \frac{1}{2}$ and $\alpha_2 = 1$. With this choice we obtain the autonomous Hamiltonian system $\mathfrak{x}' = \mathfrak{x}^2 - \mathfrak{y} + \frac{1}{2}$, $\mathfrak{y}' = -2\mathfrak{x}\mathfrak{y}$ with Hamiltonian $\mathfrak{H}(\mathfrak{x}, \mathfrak{y}) = \frac{1}{2}\mathfrak{y}^2 - (\mathfrak{x}^2 + \frac{1}{2})\mathfrak{y}$. **Question.** Is it true that $x \in \mathfrak{Y}_{\frac{1}{2},\frac{1}{2}}$ and $y \in \mathfrak{Y}_{1,\frac{1}{2}}$?

Exercise 6.11 Realise the Yosida test for

- $w'' = -3ww' - w^3 - z(w' + w^2)$ (special case of VI. in [94]).
- $w'' + 3ww' + w^2 + zw - z = 0$.

Solution $\alpha = \beta = 1$; $\alpha = \beta = \frac{1}{2}$. **Question.** Is $w \in \mathfrak{Y}_{\alpha,\beta}$?

6.3.3 *Yosida–Zalcman–Pang Re-scaling*

The Yosida test as well as the properties of the special solutions indicate that the solutions to (IV) and Painlevé (IV), first and second transcendents belong to $\mathfrak{Y}_{1,1}$, $\mathfrak{Y}_{\frac{1}{4},\frac{1}{2}}$ and $\mathfrak{Y}_{\frac{1}{2},\frac{1}{2}}$, respectively. To prove this we need a preliminary stage of the method of Yosida Re-scaling. Throughout this section, (x, y) denotes some fixed solution to System (IV). For $z \neq 0$ not a pole (of x and y) we set

$$r(z) = \min\{|z|^{-1}, |x(z)|^{-1}, |y(z)|^{-1}\}$$

(convention $0^{-1} = \infty$). By $\mathfrak{Q}$ we denote the set of non-zero zeros of $y - z$, and set

$$\Delta^\delta(z_0) = \{z : |z - z_0| < \delta r(z_0)\} \quad \text{and} \quad \mathfrak{Q}^\delta = \bigcup_{q\in\mathfrak{Q}} \Delta^\delta(q).$$

Given any sequence $z_n \to \infty$, the re-scaled functions

$$x_n(\mathfrak{z}) = r_n x(z_n + r_n\mathfrak{z}),\ y_n(\mathfrak{z}) = r_n y(z_n + r_n\mathfrak{z}) \quad (r_n = r(z_n))$$

satisfy

$$\begin{aligned} x_n' = -y_n^2 - (r_n z_n + r_n^2\mathfrak{z})x_n - \alpha r_n^2,\ y_n' = x_n^2 + (r_n z_n + r_n^2\mathfrak{z})y_n + \beta r_n^2 \\ x_n(0) = r_n x(z_n),\ y_n(0) = r_n y(z_n). \end{aligned} \tag{6.21}$$

Passing to a subsequence, if necessary, we may assume that $r_n z_n \to c$, $x_n(0) \to x_0$, and $y_n(0) \to y_0$, with

$$\max\{|c|, |x_0|, |y_0|\} = 1. \tag{6.22}$$

Then (6.21) may be considered as a system depending on parameters $r_n z_n$, r_n^2, and initial values $x_n(0) = r_n x(z_n)$, $y_n(0) = r_n y(z_n)$ on some neighbourhood of $(c, 0, x_0, y_0)$. The role of the Zalcman–Pang Lemma (which does not necessarily apply) is taken by the theorem on analytic dependence on parameters and initial values (cf. Sect. 1.3.7). From this theorem it follows that $x_n \to \mathfrak{x}$ and $y_n \to \mathfrak{y}$, where $\mathfrak{x}$ and $\mathfrak{y}$ satisfy the Hamiltonian system

$$\mathfrak{x}' = -\mathfrak{y}^2 - c\mathfrak{x},\ \mathfrak{y}' = \mathfrak{x}^2 + c\mathfrak{y} \quad \text{with} \quad \mathfrak{x}(0) = x_0,\ \mathfrak{y}(0) = y_0 \tag{6.23}$$

and Hamiltonian $H(\mathfrak{x}, \mathfrak{y}) = \frac{1}{3}(\mathfrak{x}^3 + \mathfrak{y}^3) + c\mathfrak{x}\mathfrak{y}$. We claim that $\mathfrak{y} \not\equiv c$. Otherwise the second equation yields $\mathfrak{x}^2 + c^2 \equiv 0$, while $c^2 + c\mathfrak{x} \equiv 0$ follows from the first equation. This is possible if and only if $c = 0$, $x_0 = 0$, and $y_0 = 0$, in contrast to (6.22). The non-constant solutions are either elliptic or trigonometric functions, see also the detailed discussion in Sect. 6.4.1. To proceed we need some technical results.

Lemma 6.2 $|z| + |x(z)| = O(|y(z) - z|)$ *holds as* $z \to \infty$ *outside* $\mathfrak{Q}^\delta$.

Proof We first assume $y(z_n) - z_n = o(|z_n|)$ on some sequence $z_n \to \infty$ and obtain the system (6.23) with $y_0 = c$. Since $\mathfrak{y}(\mathfrak{z}) \not\equiv c$, Hurwitz' Theorem implies that there exists $q_n \in \mathfrak{Q}$ such that $z_n - q_n = o(r_n)$. To prove $r_n = O(r(q_n))$, hence $z_n - q_n = o(r(q_n))$ and $z_n \in \mathfrak{Q}^\delta$ $(n \geq n_\delta)$, we set $\tilde{r}_n = r(q_n)$ and assume to the contrary that $\tilde{r}_n = o(r_n)$ holds (at least on some subsequence). We re-scale along

(q_n), that is, set $\tilde{x}_n(\mathfrak{z}) = \tilde{r}_n x(q_n + \tilde{r}_n\mathfrak{z})$ and $\tilde{y}_n(\mathfrak{z}) = \tilde{r}_n y(q_n + \tilde{r}_n\mathfrak{z})$ to obtain limit functions satisfying $\tilde{\mathfrak{x}}' = -\tilde{\mathfrak{y}}^2 - \tilde{c}\tilde{\mathfrak{x}}$ and $\tilde{\mathfrak{y}}' = \tilde{\mathfrak{x}}^2 + \tilde{c}\tilde{\mathfrak{y}}$, now with $\tilde{c} = \lim_{q_n\to\infty} \tilde{r}_n q_n = \lim_{z_n\to\infty} \frac{\tilde{r}_n}{r_n} r_n z_n = 0$ and $\tilde{y}_0 = 0$. Again $\tilde{\mathfrak{y}}$ is non-constant with pole τ, that is, y has some pole $p_n = q_n + o(r_n)$. This, however, contradicts $y_n \to \mathfrak{y} \not\equiv$ const on the disc $|\mathfrak{z}| < 1$, say, which contains a pole $\mathfrak{p}_n \to 0$ corresponding to p_n, and a zero $\mathfrak{q}_n \to 0$ of y_n corresponding to q_n. To prove the second assertion we assume $|y(z_n) - z_n| = o(|x(z_n)|)$ along $z_n \to \infty$; the same argument yields (6.23) with $\mathfrak{y}(0) = y_0 = c$ and $z_n \in \mathfrak{P}^\delta$ $(n \ge n_\delta)$.

Lemma 6.3 *For $\delta > 0$ sufficiently small, each disc $\Delta^\delta(q)$ about some $q \in \mathfrak{P}$ contains at most one more $q' \in \mathfrak{P}$.*

Proof Let $q_n \in \mathfrak{P}$ tend to infinity and assume that q_n is accompanied by $q_n', q_n'' \in \mathfrak{P}$, that is, q_n' and q_n'' are contained in $\Delta^{\delta_n}(q_n)$ with $\delta_n \to 0$. Then the system (6.23) has a solution such that $\mathfrak{y} - c$ vanishes at the origin at least with multiplicity three. From the second equation (6.23) and $\mathfrak{y}'(0) = \mathfrak{y}''(0) = 0$ we obtain $x_0^2 + c^2 = 2x_0\mathfrak{x}'(0) = 0$, while the first equation gives $\mathfrak{x}'(0) = -c^2 - cx_0$, hence $x_0^2 + c^2 = -2x_0(c^2 + cx_0) = y_0 - c = 0$ and $x_0 = y_0 = c = 0$. This again contradicts (6.22).

We will use this property of $\mathfrak{P}$ to redefine the discs $\Delta^\delta(q)$ as follows: for $\delta > 0$ sufficiently small, $\Delta^{3\delta}(q)$ contains at most one $q' \neq q$. If $q' \notin \Delta^{2\delta}(q)$ (or if no such q' exists) nothing has to be done, since then $\Delta^\delta(q) \cap \Delta^\delta(q') = \emptyset$. If $q' \in \Delta^{2\delta}(q) \setminus \Delta^\delta(q)$ we replace $\Delta^\delta(q)$ and $\Delta^\delta(q')$ with $\Delta^{\delta/2}(q)$ and $\Delta^{\delta/2}(q')$, respectively, and finally replace $\Delta^\delta(q) \cup \Delta^\delta(q')$ with $\Delta^{2\delta}(q)$ if $q' \in \Delta^\delta(q)$. Then $\mathfrak{P}$ is covered by mutually disjoint discs $\Delta^{\theta(q)\delta}(q)$ with $\theta(q)$ varying between $\frac{1}{2}$ and 2, and Lemma 6.2 holds for the 'new' $\mathfrak{P}^\delta$; each disc $\Delta^{\theta(q)\delta}(q)$ contains at most two zeros of $y - z$.

The main step towards the proof of $x, y \in \mathfrak{Y}_{1,1}$ now consists in estimating the auxiliary function

$$V(z) = H(z, x(z), y(z)) + \frac{x(z)^2}{y(z) - z}$$

outside $\mathfrak{P}^\delta$ (note the close relationship to the Lyapunov function (6.12)); V has poles at the zeros of $y - z$ and is regular elsewhere; estimating V also means estimating the unknown coefficients $\mathbf{h}(p)$.

Exercise 6.12 Prove that $V(p) = 2\mathbf{h}(p) + \frac{1}{3}p^3 + \frac{1}{2}(1 + \alpha + \beta)p$ holds at every pole with $\operatorname{res}_p x = 1$ (and a similar result if $\operatorname{res}_p x = e^{\pm 2\pi i/3}$).

Differentiating V and arranging the terms appropriately yields

$$V' = A(z) + B(z)V \qquad (6.24)$$

with $A(z) = \dfrac{x(z^2+\alpha)}{y - z} + \dfrac{x(z^3 + 3\alpha z + (1 + 2\beta + 2z^2)x)}{(y - z)^2} + \dfrac{3x^3}{(y - z)^3}$, $B(z) = \dfrac{-3x}{(y - z)^2}$, $x = x(z)$ and $y = y(z)$. Equation (6.24) will be regarded as a linear differential equation for V with 'known' coefficients A and B.

Exercise 6.13 Employ Lemma 6.2 to prove

$$|A(z)| = O(|z|^2) \quad \text{and} \quad |B(z)| = O(|z|^{-1}) \quad (z \notin \mathfrak{Q}^\delta).$$

We need a better result.

Lemma 6.4 *Given $\sigma > 0$ and $\delta > 0$ there exists a $K > 0$ such that*

$$|V'(z)| \le K|z|^2 + \frac{\sigma}{|z|}|V(z)| \quad (z \notin \mathfrak{Q}^\delta).$$

Proof Consider any sequence $z_n \to \infty$, $z_n \notin \mathfrak{Q}^\delta$, such that $|z_n|^2 = o(|V'(z_n)|$ holds. If no such sequence exists, $V'(z) = O(|z|^2)$ holds outside $\mathfrak{Q}^\delta$, and we are done. From $|B(z_n)| = O(|z_n|^{-1})$, hence $|V'(z_n)| \le O(|z_n|^2) + O(|z_n^{-1}V(z_n)|$, it follows that $|z_n|^3 = o(|z_nV'(z_n)|) = o(|z_n|^3 + |V(z_n)|)$, hence

$$|z_n|^3 = o(|V(z_n)|) = o(|y(z_n) - z_n|^3);$$

the latter follows from the definition of V and Lemma 6.2. This eventually implies

$$|B(z_n)| = o(|z_n|^{-1}) \quad \text{and} \quad |V'(z_n)| = o(|z_n|^{-1})|V(z_n)|.$$

Lemma 6.5 *For $\delta > 0$ sufficiently small, $V(z) = O(|z|^3)$ holds outside $\mathfrak{Q}^\delta$, and, in particular, $\mathbf{h}(p) = O(|p|^3)$.*

Proof The estimate of V via some differential inequality is strongly reminiscent of Gronwall's Lemma. Let $z_0 \notin \mathfrak{Q}^\delta$, $z_0 \neq 0$ be fixed and set

$$M(r) = \max\{|z^{-3}V(z)| : |z_0| \le |z| \le r,\ z \notin \mathfrak{Q}^\delta\}.$$

The maximum is attained at some point z_r, and $M(r)$ and $|z_r|$ increase with r. We may assume that $|z_r| \to \infty$ as $r \to \infty$, since there is nothing to do otherwise. We join z_0 and z_r by some path γ_r in $\{z : |z_0| \le |z| \le |z_r|,\ z \notin \mathfrak{Q}^\delta\}$ of length at most $L|z_r|$, where L is a universal constant (for example, $L = 2\pi$); this is possible since $\mathfrak{Q}^\delta$ consists of mutually disjoint discs. Then

$$\begin{aligned} M(r)|z_r|^3 = |V(z_r)| &\le |V(z_0)| + \int_{\gamma_r} \left(K|t|^2 + \sigma\frac{|V(t)|}{|t|}\right)|dt| \\ &\le |V(z_0)| + KL|z_r|^3 + \sigma M(r)\int_{\gamma_r} |t|^2\,|dt| \\ &\le |V(z_0)| + KL|z_r|^3 + \sigma M(r)L|z_r|^3 \end{aligned}$$

holds, hence $M(r)$ is bounded (choose $\sigma < 1/L$).

Exercise 6.14 It follows from Exercise 6.3 and $\mathbf{h}(p) = O(|p|^3)$ that for $\delta > 0$ sufficiently small the discs $|z - p| < \delta|p|^{-1}$ are mutually disjoint. By a simple geometric argument prove $\int_1^r t^{-2}\,dn(t,\mathfrak{P}) = O(r^2)$ and $T(r,x) = O(r^4)$.

We will now sketch the essential steps in the above argument for Painlevé's first and second equation, leaving the details to the interested reader.

1. Given any sequence $z_n \to \infty$, set $r_n = r(z_n) = \min\{|z_n|^{-\frac{1}{4}}, |w(z_n)|^{-\frac{1}{2}}, |w'(z_n)|^{-\frac{3}{4}}\}$ in the first case, and $r_n = r(z_n) = \min\{|z_n|^{-\frac{1}{2}}, |w(z_n)|^{-\frac{1}{2}}, |w'(z_n)|^{-1}\}$ in the second, and define sequences $w_n(\mathfrak{z}) = r_n^2 w(z_n + r_n\mathfrak{z})$ resp. $w_n(\mathfrak{z}) = r_n w(z_n + r_n\mathfrak{z})$ to obtain limit functions $\mathfrak{w} = \lim_{n\to\infty} w_n$ (take subsequences if necessary) satisfying

$$\begin{array}{c} \mathfrak{w}'' = c + 6\mathfrak{w}^2 \\ c = \lim_{n\to\infty} r_n^4 z_n \end{array} \quad \text{resp.} \quad \begin{array}{c} \mathfrak{w}'' = c\mathfrak{w} + 2\mathfrak{w}^3 \\ c = \lim_{n\to\infty} r_n^2 z_n. \end{array}$$

We note that $\mathfrak{w} \not\equiv 0$ resp. $\mathfrak{w} \not\equiv \pm\sqrt{c}$ follows from $\max\{|c|, |\mathfrak{w}(0)|, |\mathfrak{w}'(0)|\} = 1$.
2. Let $\mathfrak{Q}$ denote the set of non-zero zeros of w and $w^2 - z$, respectively. Then for $\delta > 0$ sufficiently small, $\mathfrak{Q}$ may be covered by countably many mutually disjoint discs $\Delta^\delta(q) = \{z : |z-q| < \delta\theta(q)r(q)\}$ ($\frac{1}{2} \le \theta(q) \le 2$), and $|z| = O(|w|^2)$, $|w'| = O(|w|^{\frac{3}{2}})$ resp. $|z| = O(|w^2 - z|)$, $|w'| = O(|w^2 - z|)$ hold outside $\mathfrak{Q}^\delta = \bigcup_q \Delta^\delta(q)$.
3. The functions $V = 2W - \dfrac{w'}{w}$ and $V = 2W - \dfrac{2ww' - 1}{w^2 - z}$, where W denotes the respective first integral, are regular at the poles of w, and satisfy

$$V' = \frac{zw(z)^2 - w'(z)}{w(z)^3} - \frac{1}{w(z)^2}V \quad \text{and} \quad V' = A(z) - \frac{z + w(z)^2}{(w(z)^2 - z)^2}V,$$

respectively, with $A(z) = \frac{(2zw' - 6\alpha z^2)w + (2 - 6z^3)w^2 + (4\alpha z - 6w')w^3 + 4z^2w^4 + 2\alpha w^5 + 2zw^6}{(w^2 - z)^3}$.
4. Given $\sigma > 0$ there exists a $K > 0$ such that

$$|V'| \le K|z|^{\frac{1}{2}} + \sigma\frac{|V|}{|z|} \quad \text{resp.} \quad |V'| \le K|z| + \sigma\frac{|V|}{|z|}$$

holds outside $\mathfrak{Q}^\delta$, this implying $V(z) = O(|z|^{\frac{3}{2}})$ resp. $V(z) = O(|z|^2)$ outside $\mathfrak{Q}^\delta$. In particular, $\mathbf{h}(p) = O(|p|^{\frac{3}{2}})$ resp. $\mathbf{h}(p) = O(|p|^2)$ holds by

Exercise 6.15 Let w be any first resp. second Painlevé transcendent. Prove that $V(p) = -28\mathbf{h}(p)$ resp. $V(p) = 10\epsilon\mathbf{h}(p) - \frac{7}{36}p^2$ holds at every pole p. In combination with Exercise 6.3 deduce

$$\int_1^r t^{-\frac{1}{2}}\, dn(t, \mathfrak{P}) = O(r^2) \quad \text{resp.} \quad \int_1^r t^{-1}\, dn(t, \mathfrak{P}) = O(r^2).$$

6.3.4 Yosida Re-scaling

We are now prepared to prove the Painlevé–Yosida Theorem (p. 188) for System (IV), say, and note first that for $\delta > 0$ sufficiently small the discs

$$\Delta_\delta(p) = \{z : |z - p| < \delta |p|^{-1}\}\ {}^{**}$$

about the poles $p \neq 0$ are mutually disjoint by Exercise 6.8; they form the δ-neighbourhood

$$\mathfrak{P}_\delta = \bigcup_{p \in \mathfrak{P}} \Delta_\delta(p).$$

We will now prove

$$|x(z)| + |y(z)| = O(|z|) \quad (z \notin \mathfrak{P}_\delta). \tag{6.25}$$

Assume that $|z_n| = o(|x(z_n)| + |y(z_n)|)$ holds for some sequence $z_n \to \infty$. Then $r(z_n) = o(|z_n|^{-1})$, and Yosida–Zalcman–Pang Re-scaling yields

$$\mathfrak{x}' = -\mathfrak{y}^2,\ \mathfrak{y}' = \mathfrak{x}^2, \text{ and } c = \lim_{n\to\infty} r(z_n) z_n = 0;$$

$\mathfrak{x}$ and $\mathfrak{y}$ are not constant, this following from

$$\max\{|\mathfrak{x}'(0)|, |\mathfrak{y}'(0)|\} = \max\{|\mathfrak{x}(0)|^2, |\mathfrak{y}(0)|^2\} = 1.$$

Moreover, from $\mathfrak{x}^2\mathfrak{x}' + \mathfrak{y}^2\mathfrak{y}' = 0$ it follows that $\mathfrak{x}^3 + \mathfrak{y}^3 \equiv \mathfrak{x}(0)^3 + \mathfrak{y}(0)^3$, hence $\mathfrak{x}$ and $\mathfrak{y}$ have some pole τ.[3] Then x and y have some pole $p_n = z_n + r(z_n)\tau + o(r(z_n)) = z_n + o(|z_n|^{-1})$; this implies $z_n = p_n + o(|p_n|^{-1}) \in \mathfrak{P}_\delta$ and proves (6.25). To finish the proof we note that the discs $\Delta_\delta(p)$ are mutually disjoint and the residues of x and y have modulus one. By Theorem 4.18, x and y are candidates for the Yosida class $\mathfrak{Y}_{1,1}$. The Yosida limit functions

$$\mathfrak{x}(\mathfrak{z}) = \lim_{h_n\to\infty} x_{h_n}(\mathfrak{z}) \quad \text{and} \quad \mathfrak{y}(\mathfrak{z}) = \lim_{h_n\to\infty} y_{h_n}(\mathfrak{z}) \tag{6.26}$$

are finite ($\not\equiv \infty$) since $\mathfrak{x}(0)$ and $\mathfrak{y}(0)$ are finite if $\inf |h_n| \operatorname{dist}(h_n, \mathfrak{P}) > 0$, and $\mathfrak{x}$ and $\mathfrak{y}$ have a simple pole at $\mathfrak{z} = 0$ if $|h_n| \operatorname{dist}(h_n, \mathfrak{P}) \to 0$. For fourth transcendents the proof follows from

---

**The discs $\Delta_\delta(z_0) = \{z : |z - z_0| < \delta |z_0|^{-1}\}$ are defined for arbitrary centres z_0, and should not be mixed up with the discs $\Delta^\delta(z_0) = \{z : |z - z_0| < \delta r(z_0)\}$.

[3]For $\mathfrak{x}(0)^3 + \mathfrak{y}(0)^3 = 0$, $\mathfrak{x} = \omega/(\mathfrak{z} - \tau)$ holds for some $\tau \neq 0$ and ω with $\omega^3 = 1$. In any other case $\mathfrak{x}$ and $\mathfrak{y}$ are elliptic functions.

Exercise 6.16 Prove that $x, y \in \mathfrak{Y}_{1,1}$ implies $v = x + y - z \in \mathfrak{Y}_{1,1}$. (**Hint.** Note that v is regular at poles of (x, y) with residues $(1, -1)$, and use Theorem 4.18.)

The proof for first and second transcendents is practically the same. For $\delta > 0$ sufficiently small, the discs $\Delta_\delta(p) = \{z : |z - p| < \delta|p|^{-\beta}\}$ ($\beta = \frac{1}{4}$ and $\beta = \frac{1}{2}$, respectively; again note the footnote on p. 195) about the non-zero poles are mutually disjoint (see also Exercise 6.3). For second transcendents, $w^{\sharp\frac{1}{2}}(z) = O(|z|^{\frac{1}{2}})$ follows from $|w| = O(|z|^{\frac{1}{2}})$ outside $\mathfrak{P}_\delta$, $|\operatorname{res}_p w| = 1$ and Theorem 4.18. First transcendents satisfy $|w| = O(|z|^{\frac{1}{2}})$ and $|w'| = O(|z|^{\frac{3}{4}})$ outside $\mathfrak{P}_\delta$, and the inequality $|w| \geq |z - p|^{-2}(1 + O(\delta^2)) > 0$ on $\Delta_\delta(p)$, which follows from Exercise 6.3, implies that both branches of $w^{\frac{3}{2}}$ are meromorphic on $\Delta_\delta(p)$. The Maximum Principle applied to $f(z) = w' \pm 2w^{\frac{3}{2}}$ on $\Delta_\delta(p)$ with boundary values $f(z) = O(|z|^{\frac{3}{4}})$ yields $|w'(z)| = O(|z|^{\frac{3}{4}} + |w(z)|^{\frac{3}{2}})$ and

$$w^{\sharp\frac{1}{2}}(z) = O(|z|^{\frac{1}{2}})\frac{|z|^{\frac{3}{4}} + |w(z)|^{\frac{3}{2}}}{|z| + |w(z)|^2} = O(|z|^{\frac{1}{4}}) \quad \text{on} \, \Delta_\delta(p).$$

Remark 6.7 The main step in the proof consisted in an appropriate estimate of the respective auxiliary functions V, which were obtained by modifying the corresponding Lyapunov functions in the proof of the Painlevé property. It is quite reasonable that the intimate connection between the Painlevé and the Yosida property holds for a wide class of algebraic differential equations and systems. This could be the topic of future research.

6.4 Value Distribution

The Nevanlinna characteristic and counting function of poles of functions $f \in \mathfrak{Y}_{\alpha,\beta}$ have magnitude $r^{2+2\beta}$, and the limit functions $\mathfrak{f}(\mathfrak{z}) = \lim_{h_n \to \infty} h_n^{-\alpha} f(h_n + h_n^{-\beta}\mathfrak{z})$ belong to the universal class $\mathfrak{Y}_{0,0}$; hardly more can be said. Meromorphic solutions to algebraic differential equations that belong to some Yosida class $\mathfrak{Y}_{\alpha,\beta}$ have limit functions that solve autonomous algebraic differential equations and belong to $\mathfrak{Y}_{0,0}$. In many cases they are trigonometric or elliptic functions whose value distribution is completely understood. All one has to do is to retrace their properties to the solutions of the original equation. This will be done exemplarily for the solutions to System (IV) and the Painlevé transcendents.

6.4.1 *The Cluster Set of the Solutions to (IV)*

In the case of System (IV) the limit functions (6.26) solve the Hamiltonian system

$$\mathfrak{x}' = -\mathfrak{y}^2 - \mathfrak{x},\ \mathfrak{y}' = \mathfrak{x}^2 + \mathfrak{y} \tag{6.27}$$

with *constant* Hamiltonian $\mathfrak{H}(\mathfrak{x},\mathfrak{y}) = \frac{1}{3}(\mathfrak{x}^3 + \mathfrak{y}^3) + \mathfrak{x}\mathfrak{y} = \frac{1}{3}\mathfrak{c}$. The constant solutions are $(\mathfrak{x},\mathfrak{y}) = (0,0)$ with $\mathfrak{c} = 0$, and $(\mathfrak{x},\mathfrak{y}) = (-\omega, -\bar{\omega})$ with $\mathfrak{c} = 1$, $\omega^3 = 1$.

The algebraic curve

$$\xi^3 + \eta^3 + 3\xi\eta = \mathfrak{c} \tag{6.28}$$

- is *reducible* if $\mathfrak{c} = 1$: $\xi^3 + \eta^3 + 3\xi\eta - 1 = \prod_{\omega^3=1}(\eta + \omega\xi - \bar{\omega})$ holds, and the autonomous Hamiltonian system (6.27) may be reduced to $\mathfrak{x}' = -1 + \mathfrak{x} - \mathfrak{x}^2$ via $\mathfrak{y} = 1 - \mathfrak{x}$, with non-constant trigonometric solution $\mathfrak{x} = \frac{1}{2} - \frac{\sqrt{3}}{2}\tan\big(\frac{\sqrt{3}}{2}\mathfrak{z}\big)$, and similar solutions if $\mathfrak{y} = \bar{\omega} - \omega\mathfrak{x}$ ($\omega = e^{\pm 2\pi i/3}$). It is just important to know that the poles form a $\frac{2\pi}{\sqrt{3}}$-periodic sequence with fixed residue.
- has *genus zero* if $\mathfrak{c} = 0$, with rational parametrisation $\xi = r(t) = \frac{-3t^2}{t^3+1}$ and $\eta = s(t) = \frac{-3t}{t^3+1}$. The Hamiltonian system (6.27) is solved by the trigonometric functions $\mathfrak{x} = r(e^{\mathfrak{z}})$ and $\mathfrak{y} = s(e^{\mathfrak{z}})$ with poles forming a $\frac{2\pi i}{3}$-periodic sequence, this time with alternating residues.
- has *genus one* if $\mathfrak{c} \neq 0, 1$, and is parametrised by elliptic functions $\xi = \mathfrak{x}(\mathfrak{z})$ and $\eta = \mathfrak{y}(\mathfrak{z})$ of elliptic order three; they solve $\mathfrak{x}'^3 - 3\mathfrak{x}\mathfrak{x}'^2 + \mathfrak{x}^6 + (4-2\mathfrak{c})\mathfrak{x}^3 + \mathfrak{c}^2 = 0$ (just compute the resultant of the polynomials $\xi^3 + \eta^3 + 3\xi\eta - \mathfrak{c}$ and $\xi_1 + \eta^2 + \xi$ with respect to η) and $\mathfrak{y}'^3 + 3\mathfrak{y}\mathfrak{y}'^2 - \mathfrak{y}^6 - (4-2\mathfrak{c})\mathfrak{y}^3 - \mathfrak{c}^2 = 0$.

To determine the possible constants $\mathfrak{c}$ in (6.28) consider any sequence $h_n \to \infty$ that stays away from $\mathfrak{P}$, that is, assume $\inf |h_n| \operatorname{dist}(h_n, \mathfrak{P}) > 0$ and also that the limits $\mathfrak{x} = \lim_{h_n\to\infty} x_{h_n}$ and $\mathfrak{y} = \lim_{h_n\to\infty} y_{h_n}$ exist. Then $\mathfrak{x}^3 + \mathfrak{y}^3 + 3\mathfrak{x}\mathfrak{y} \equiv \mathfrak{c}$ holds with

$$\mathfrak{c} = \lim_{h_n\to\infty} 3h_n^{-3}H(h_n) \quad (H(z) = H(z, x(z), y(z))). \tag{6.29}$$

The limits (6.29) form the *cluster set* $\mathfrak{C}(x,y)$. The cluster set determines the growth of the Nevanlinna characteristic. Far beyond the estimate $T(r,x) + T(r,y) = O(r^4)$ the following holds.

Theorem 6.4 *Suppose that $\mathfrak{C}(x,y)$ contains some $\tilde{\mathfrak{c}} \neq 0, 1$. Then*

$$T(r,x) \asymp r^4 \quad \text{and} \quad T(r,y) \asymp r^4 \tag{6.30}$$

hold, at least on some sequence $r = r_n \to \infty$.

Proof Let $\tilde{\mathfrak{c}} = \lim_{\tilde{h}_n\to\infty} 3\tilde{h}_n^{-3}H(\tilde{h}_n) \neq 0, 1$ and set $\epsilon = \frac{1}{2}\min\{|\tilde{\mathfrak{c}}|, |\tilde{\mathfrak{c}} - 1|\}$. From

$$\frac{d}{dz}(z^{-3}H(z)) = -3z^{-4}H(z) + z^{-3}x(z)y(z) = O(|z|^{-1})$$

outside $\mathfrak{P}_\delta$ it follows that there exists some $\eta > 0$ such that $|z^{-3}H(z) - \tilde{\mathfrak{c}}| < \epsilon$ holds on $D_n = \{z : |z - \tilde{h}_n| < \eta|\tilde{h}_n|\} \setminus \mathfrak{P}_\delta$. Re-scaling along any sequence (h_n) with $h_n \in D_n$ yields non-constant limit functions $\mathfrak{x}$ and $\mathfrak{y}$ satisfying $\mathfrak{x}^3 + \mathfrak{y}^3 + 3\mathfrak{x}\mathfrak{y} = \mathfrak{c}$ with $|\mathfrak{c} - \tilde{\mathfrak{c}}| \le \epsilon$, $|\mathfrak{c}| \ge \epsilon/2$ and $|\mathfrak{c} - 1| \ge \epsilon/2$. In other words, $\mathfrak{x}$ and $\mathfrak{y}$ are elliptic functions with common pair of primitive periods $(\vartheta, \vartheta') = (\vartheta_{\mathfrak{c}}, \vartheta'_{\mathfrak{c}})$ such that $|\vartheta| \le |\vartheta'| \le |\vartheta \pm \vartheta'| < L$. Hence there exists an $R > 0$ such that every disc $|z - z_0| < R|\tilde{h}_n|^{-1}$ centred at some z_0 with $|z_0 - \tilde{h}_n| < \eta|\tilde{h}_n|$ $(n \ge n_0)$ contains at least one pole of x and y. This implies that the disc $|z - \tilde{h}_n| < \eta|\tilde{h}_n|$ contains at least $\lambda|\tilde{h}_n|^4$ poles, where $\lambda > 0$ is independent of n (it is almost the same to say that the disc $|z| < r$ contains $\sim \pi r^2$ gaussian integers $k + i\ell$). This proves the assertion with $r_n = 2|\tilde{h}_n|$, say.

To obtain more and better information about the distribution of poles, the cluster set $\mathfrak{C}(x, y)$ has to be analysed in detail.

Exercise 6.17 For $\delta > 0$ let $\mathfrak{C}_\delta(x, y)$ denote the cluster set of (x, y) as $z \to \infty$ restricted to $\mathbb{C} \setminus \mathfrak{P}_\delta$, that is, in (6.29) only sequences (h_n) with $|h_n|\mathrm{dist}\,(h_n, \mathfrak{P}) \ge \delta$ are admitted. Prove that $\mathfrak{C}_\delta(x, y)$ is closed, bounded, and connected. The last assertion is non-trivial and relies on the fact that $\mathbb{C} \setminus \mathfrak{P}_\delta$ is *locally connected* at infinity.

Exercise 6.18 (Continued) Prove that $\mathfrak{c} = \lim_{h_n\to\infty} 3h_n^{-3}H(h_n)$ remains unchanged if $h_n \in \mathbb{C}\setminus\mathfrak{P}_\delta$ is replaced with $\tilde{h}_n \in \mathbb{C}\setminus\mathfrak{P}_\delta$ such that $|\tilde{h}_n - h_n| < K|h_n|^{-1}$. (**Hint.** Prove that $\tilde{\mathfrak{x}}(\mathfrak{z}) = \lim_{j\to\infty} x_{\tilde{h}_{k_j}}(\mathfrak{z}) = \mathfrak{x}(\mathfrak{z} + \mathfrak{z}_0)$ and $\tilde{\mathfrak{y}}(\mathfrak{z}) = \lim_{j\to\infty} y_{\tilde{h}_{k_j}}(\mathfrak{z}) = \mathfrak{y}(\mathfrak{z} + \mathfrak{z}_0)$ holds for every convergent subsequence and appropriate $\mathfrak{z}_0$.)

Exercise 6.19 (Continued) For $\delta > \eta$, the inclusion $\mathfrak{C}_\delta(x, y) \subset \mathfrak{C}_\eta(x, y)$ is trivial. Prove that $\mathfrak{C}_\delta(x, y) \supset \mathfrak{C}_\eta(x, y)$ if $\delta > \eta$ is sufficiently small, and conclude that $\mathfrak{C}(x, y) = \bigcup_{\delta>0} \mathfrak{C}_\delta(x, y)$ is closed, bounded, and connected. Prove also that $\mathfrak{C}_\delta(x, y)$ contains the limits $1 + \lim_{p_n\to\infty} 6p_n^{-3}\mathbf{h}(p_n)$ for appropriately chosen sequences $p_n \in \mathfrak{P}$. (**Hint.** If $\mathrm{dist}\,(h_n, \mathfrak{P}) = \delta|h_n|^{-1}$ replace h_n with $p_n \in \mathfrak{P}$ such that $\mathrm{dist}\,(h_n, \mathfrak{P}) = |h_n - p_n|$. Conversely, replace $p_n \in \mathfrak{P}$ with $h_n \in \partial\Delta_\delta(p_n)$.)

We mention without proof that the cluster set coincides with the set of accumulation points of the sequence $(1 + 6p_n^{-3}\mathbf{h}(p_n))$.

6.4.2 *Strings and Lattices*

The value distribution of any trigonometric function (a rational function of $e^{2\pi iz/\vartheta}$) is parallel to the ray $z = t\vartheta$, while for elliptic functions the distribution of values is regularly spread over the whole plane. Suppose $f \in \mathfrak{Y}_{\alpha,\beta}$ has only ϑ-periodic trigonometric limit functions $\mathfrak{f}(\mathfrak{z}) = \lim_{h_n \to \infty} h_n^{-\alpha} f(h_n + h_n^{-\beta}\mathfrak{z})$. Then the poles of $\mathfrak{f}$ are arranged in arithmetic sequences $(k\vartheta)_{k\in\mathbb{Z}}$, so by Hurwitz' Theorem the poles of f are arranged in sequences (p_k) satisfying an asymptotic recursion

$$p_{k+1} = p_k + (\vartheta + o(1))p_k^{-\beta}. \tag{6.31}$$

Any such sequence is called a *string*, more precisely, a (ϑ, β)-string. For example, the poles of the solutions to the Riccati equation $w' = z^n - w^2$ form finitely many $(\pm\pi i, \frac{n}{2})$-strings.

Exercise 6.20 Prove that for any $(\vartheta, \frac{s}{t})$-string $\sigma = (p_k)$ the following holds:

- $p_k = \big(k(1+\frac{s}{t})\vartheta\big)^{\frac{t}{s+t}}(1+o(1))$; (**Hint.** Consider $q_k = p_k^{1+\frac{s}{t}}$.)
- $(s+t)\arg p_k = t\arg\vartheta + o(1) \bmod 2\pi$ hold as $k \to \infty$;
- $n(r,\sigma) \sim \dfrac{r^{1+\frac{s}{t}}}{(1+\frac{s}{t})|\vartheta|}$.

Remark 6.8 For β some positive integer the analogy with the dynamics of the rational map $R(p) = p + \vartheta p^{-\beta}$ is evident (see [174]); R has a *parabolic fixed point* at infinity with $\beta + 1$ *invariant petals*, and for any p_0 in some petal, the iterates $p_k = R^k(p_0)$ converge to infinity along some ray $(\beta+1)\arg z = \arg\vartheta \bmod 2\pi$.

Exercise 6.21 Returning to (IV), set $\mathfrak{c} = 1 + \lim_{p_n\to\infty} 6p_n^{-3}\mathbf{h}(p_n)$ for some sequence (p_n) of poles and $\mathfrak{P}_n = \{z : |z - p_n| < R|p_n|^{-1}\} \cap \mathfrak{P}$. Prove that up to an error term $o(|p_n|^{-1})$, $\mathfrak{P}_n$ is a finite

- lattice $p_n + (k\vartheta + \ell\vartheta')p_n^{-1}$ if $\mathfrak{c} \neq 0, 1$, where (ϑ, ϑ') span the period lattice of the elliptic functions $\mathfrak{x} = \lim_{p_n\to\infty} x_{p_n}$ and $\mathfrak{y} = \lim_{p_n\to\infty} y_{p_n}$;
- arithmetic sequence $p_n + k\vartheta p_n^{-1}$ ($\vartheta = \frac{2\pi i}{3}$) with alternating residues $(1,-1)$, $(\omega, -\bar\omega)$, $(\bar\omega, -\omega)$ with $\omega = e^{2\pi i/3}$, if $\mathfrak{c} = 0$;
- arithmetic sequence $p_n + k\vartheta p_n^{-1}$ ($\vartheta = \frac{2\pi}{\sqrt{3}}$) with fixed residues $(\omega, -\bar\omega)$ for some ω with $\omega^3 = 1$, if $\mathfrak{c} = 1$.

We say that $\mathfrak{P}$ locally has a 'lattice' and a 'string structure', respectively. Two cases are of particular interest.

$\mathfrak{C}(x,y) = \{0\}$. The poles are arranged in $(\pm\frac{2\pi i}{3}, 1)$-strings with counting function $n(r,(p_k)) \sim \frac{3}{4\pi}r^2$, alternating residues, and $\arg p_k \sim (2\nu+1)\frac{\pi}{4}$ for some ν.

$\mathfrak{C}(x,y) = \{1\}$. The poles are arranged in $(\pm\frac{2\pi}{\sqrt{3}}, 1)$-strings with fixed residues, counting function $n(r,(p_k)) \sim \frac{\sqrt{3}}{4\pi}r^2$, and $\arg p_k \sim \nu\frac{\pi}{2}$ for some ν.

Remark 6.9 In both cases the question of whether there are only *finitely many* strings is still open. It is conjectured that there are no solutions with Nevanlinna characteristic strictly between $T(r,x) \asymp r^2$ and $T(r,x) \asymp r^4$.

Exercise 6.22 Prove that $\mathfrak{C}(x,y) = \{\mathfrak{c}\}$ implies $T(r,x) + T(r,y) = o(r^4)$ and $\mathfrak{c} \in \{0,1\}$. Thus $\mathfrak{C}(x,y)$ is either a continuum or else reduces to $\{0\}$ or $\{1\}$.

(**Hint.** Compute $\frac{1}{2\pi i}\int_{\Gamma_r} H(z, x(z), y(z))\, dz$ asymptotically; for the definition of Γ_r we refer to Exercise 4.17.)

Exercise 6.23 For $r > 1$ and $z = re^{i\theta} \notin \mathfrak{P}_\delta$ set $\eta_r(\theta) = z^{-3}H(z)$, and interpolate linearly in the θ-intervals corresponding to the intersection of $|z| = r$ and $\mathfrak{P}_\delta$. Prove that the family $(\eta_r)_{r>1}$ is bounded and equi-continuous, and deduce that $\mathfrak{C}(x, y)$ is the union of closed curves $C_{(r_n)} : \theta \mapsto \eta(\theta) = \lim_{r_n\to\infty} \eta_{r_n}(\theta)$. The question of whether $\mathfrak{C}(x, y)$ consists of a single curve (that is, $\lim_{r\to\infty} \eta_r(\theta)$ exists) is open.

Now we can reap the benefits of our work and obtain the following completions of the Painlevé–Yosida Theorem.

Theorem 6.5 *The transcendental solutions to System (IV) have the following properties:*

- $T(r, x) + T(r, y) = O(r^4)$;
- $r^2 = O(T(r, x) + T(r, y))$;
- $m(r, x) + m(r, y) = O(\log r)$;
- $m(r, 1/x) = O(\log r)$ *if* $\alpha \neq 0$, *and* $m(r, 1/y) = O(\log r)$ *if* $\beta \neq 0$;
- $x(z) - z \asymp |z|$ *and* $y(z) - z \asymp |z|$ *hold outside* $\mathfrak{P}_\delta \cup \mathfrak{Q}_\delta$
  ($\mathfrak{Q}$ *denotes the set of non-zero zeros of* $(x - z)(y - z)$).

Proof Only the second and fourth assertion requires a proof. For the latter, see Exercise 6.24. To prove the second, re-scale along any sequence of poles to obtain elliptic or trigonometric limit functions with periods (ϑ, ϑ') and ϑ, respectively; in the latter case ϑ is either real or purely imaginary. In any case, a simple geometric argument shows that to any pole p of sufficiently large modulus there exists some pole p' such that $|\text{Re}\, p| + |\text{Im}\, p| < |\text{Re}\, p'| + |\text{Im}\, p'|$ and $|p'| < |p| + O(|p|^{-1})$. The assertion $n(r, x) \geq \text{const} \cdot r^2$ then follows from the next exercise.

Exercise 6.24 Let (ξ_k) be a sequence of positive numbers that tends to infinity and satisfies $\xi_{k+1} \leq \xi_k + L\xi_k^{-1}$. Prove that $\xi_k \leq (2L + o(1))^{\frac{1}{2}} k^{\frac{1}{2}}$.

Exercise 6.25 To prove $m(r, 1/x) = O(\log r)$ for $\alpha \neq 0$ derive some second-order differential equation $\Omega(z, x, x', x'') = 0$ with $\Omega(z, 0, 0, 0) = 4\alpha^2 z^2 + 4\alpha\beta^2$. By symmetry, $m(r, 1/y) = O(\log r)$ holds if $\beta \neq 0$. In this case and if $\alpha = 0$ we obtain $m(r, 1/x) \leq m(r, y^2/x) + 2m(r, 1/y) \leq m(r, x'/x) + m(r, z) + O(\log r) = O(\log r)$, hence $m(r, 1/x) + m(r, 1/y) = O(\log r)$ holds whenever $|\alpha| + |\beta| > 0$.

The analogue to Theorem 6.5 is

Theorem 6.6 *Fourth Painlevé transcendents have the following properties.*

- $w \in \mathfrak{Y}_{1,1}$ *and* $w^{(k)} \in \mathfrak{Y}_{1+k,1}$;
- $T(r, w) = O(r^4)$;
- $r^2 = O(T(r, w))$;
- $m(r, w) = O(\log r)$;
- $m(r, 1/w) = O(\log r)$ *if* $\beta \neq 0$;
- $w(z) + z \asymp |z|$ *holds outside* $\mathfrak{P}_\delta \cup \mathfrak{Q}_\delta$
  ($\mathfrak{Q}$ *denotes the set of non-zero zeros of* $w + z$).

6.4.3 Weber–Hermite Solutions

Weber–Hermite solutions to System (IV) or Painlevé (IV) arise from solutions to certain Riccati equations (which may be reduced to the Weber–Hermite equation (6.18)) by repeated application of Bäcklund transformations for System (IV) resp. Painlevé (IV). Our focus will be on System (IV). The reader will not have any difficulty in transferring the results to the Weber–Hermite solutions to Painlevé (IV). The smallest number of non-trivial Bäcklund transformations (6.6) that are needed to obtain (x, y) from solutions $(\tilde{x}, \tilde{y})$ with $\tilde{x} + \tilde{y} - z = 0$ and $\tilde{\alpha} - \tilde{\beta} + 1 = 0$ is called the order, written $\mathrm{ord}(x, y)$. We will briefly write $H(z) = H(z, x(z), y(z))$.

Exercise 6.26 Prove that the coordinates of Weber–Hermite solutions separately solve first-order algebraic differential equations of genus zero and are algebraically dependent over $\mathbb{C}(z)$. (**Hint.** Prove that $x = R(z, \tilde{x})$, where R is rational and $\tilde{x}$ is an appropriate solution to some equation $\tilde{x}' = -\tilde{\alpha} - z^2 + z\tilde{x} - \tilde{x}^2$.)

Theorem 6.7 *Every solution (x, y) to System (IV) such that*

- *x and y are algebraically dependent over $\mathbb{C}(z)$, or*
- *x and y separately satisfy first-order algebraic differential equations, or*
- *$V = H - \frac{1}{3}z^3$ belongs to $\mathfrak{Y}_{1,1}$*

has Nevanlinna characteristic $O(r^2)$.

Proof The first two conditions are equivalent. We suppose that x and y satisfy some non-trivial algebraic equation $K(z, x, y) = 0$ and write

$$K(z, x, y) = \prod_{\nu=1}^{n} (y - K_\nu(z, x)).$$

Then $y(z) = K_\nu(z, x(z))$ locally holds for at least one ν, and so by analytic continuation, which transforms K_ν into some K_μ. We may assume that (x, y) has infinitely many poles p with residues $(1, -1)$. From Exercise 6.8 and $\mathbf{h}(p) = O(|p|^3)$ it then follows that $x(z) + y(z) = O(|p|)$ ($p \to \infty$ with $\mathrm{res}_p\, x = 1$) holds on $\Delta_\delta(p)$, hence also $K_\nu(z, x) = -x + O(|z|)$ holds as $x \to \infty$, uniformly with respect to z. Re-scaling along any such sequence (p_n) yields limit functions $\mathfrak{x}$ and $\mathfrak{y}$ satisfying $\mathfrak{y}(\mathfrak{z}) = -\mathfrak{x}(\mathfrak{z}) + \mathfrak{a}$ and also $\frac{1}{3}(\mathfrak{x}^3 + (\mathfrak{a} - \mathfrak{x})^3) + \mathfrak{x}(\mathfrak{a} - \mathfrak{x}) = (\mathfrak{a} - 1)(\mathfrak{x}^2 - \mathfrak{a}\mathfrak{x}) + \frac{1}{3}\mathfrak{a}^3 = \frac{1}{3}\mathfrak{c}$, which must be trivial: $\mathfrak{a} = \mathfrak{c} = 1$. This also holds if we consider poles with residues $(\omega, -\bar{\omega})$, hence we are in the reducible case and the set of poles $\mathfrak{P}$ has string structure. Then the corresponding algebraic differential equation $P(z, x, x') = 0$ has genus zero, x has only finitely many strings of poles, and $n(r, x) = O(r^2)$ holds. This is also true if $V \in \mathfrak{Y}_{1,1}$ since $n(r, x) = n(r, V) = O(r^2)$.

Exercise 6.27 Let (x, y) be any Weber–Hermite solution. Prove that H (hence also $V = H - \frac{1}{3}z^3$) satisfies some first-order algebraic differential equation.

(**Hint.** Use $H = \frac{1}{3}(x^3 + y^3) + zxy + \alpha y + \beta x$, $H' = xy$, and $K(z, x, y) = 0$ to derive $P(z, H, H') = 0$ by purely algebraic methods.)

Exercise 6.28 Set $(\tilde{x}, \tilde{y}) = \mathsf{B}_1(x, y)$ (see (6.6) for the definition of the Bäcklund transformation B_1) with $x + y - z \not\equiv 0$, $\tilde{\alpha} = \beta - 1$ and $\tilde{\beta} = \alpha + 1$, and prove (with self-explanatory notation)

$$\tilde{H}(z) - H(z) = \tilde{x}(z) - x(z) = y(z) - \tilde{y}(z).$$

Exercise 6.29 (Continued) Prove by induction on $\mathsf{ord}(x, y)$ that for every Weber–Hermite solution the function $V(z) = H(z) - \frac{1}{3}z^3 = H(z, x(z), y(z)) + \frac{1}{3}z^3$ belongs to $\mathfrak{Y}_{1,1}$. (**Hint.** $H = x + \alpha z + \frac{1}{3}z^3$ holds if $y = z - x$ and $\beta = \alpha + 1$, that is, if $\mathsf{ord}(x, y) = 0$. Use Exercise 6.28 for the step from H to $\tilde{H} = \mathsf{B}_1 H$, noting that the Hamiltonian for $(\hat{x}, \hat{y}) = (\omega x, \bar{\omega} y)$ is

$$\hat{H}(z, \hat{x}, \hat{y}) = H(z, x, y) + \beta(\omega - 1)x + \alpha(\bar{\omega} - 1)y.)$$

6.4.4 *Value Distribution of the Painlevé Transcendents*

To make a long story short, we will now define and discuss the corresponding cluster sets and transfer the properties of the solutions to the re-scaled differential equations to our main object, the Painlevé transcendents. Yosida Re-scaling yields limit functions $\mathfrak{w} = \lim_{h_n \to \infty} w_{h_n}$ satisfying

$$\mathfrak{w}'' = 1 + 6\mathfrak{w}^2,\ \mathfrak{w}'' = \mathfrak{w} + 2\mathfrak{w}^3,\ \text{and } 2\mathfrak{w}\mathfrak{w}'' = \mathfrak{w}'^2 + 3\mathfrak{w}^4 + 8\mathfrak{w}^3 + 4\mathfrak{w}^2,$$

respectively. In the first two cases it follows that $\mathfrak{w}'^2 = 4\mathfrak{w}^3 + 2\mathfrak{w} - 2\mathfrak{c}$ resp. $\mathfrak{w}'^2 = \mathfrak{w}^4 + \mathfrak{w}^2 - \mathfrak{c}$ holds, with constants of integration $\mathfrak{c} = \lim_{h_n \to \infty} h_n^{-\frac{3}{2}} W(h_n)$ resp. $\mathfrak{c} = \lim_{h_n \to \infty} h_n^{-2} W(h_n)$; this is true if the infimum of $|h_n|^{\frac{1}{4}} \operatorname{dist}(h_n, \mathfrak{P})$ resp. $|h_n|^{\frac{1}{2}} \operatorname{dist}(h_n, \mathfrak{P})$ is positive.

Exercise 6.30 For any fourth transcendent and limit function $\mathfrak{w} = \lim_{h_n \to \infty} w_{h_n}$ prove that $\mathfrak{w}''' = (6\mathfrak{w}^2 + 12\mathfrak{w} + 8)\mathfrak{w}'$ and $\mathfrak{w}'' = 2\mathfrak{w}^3 + 6\mathfrak{w}^2 + 8\mathfrak{w} - 2\mathfrak{c}$ holds with $\mathfrak{c} = \lim_{h_n \to \infty} W(h_n) h_n^{-3}$ ($\inf_n |h_n| \operatorname{dist}(h_n, \mathfrak{P}) > 0$). Deduce

$$\mathfrak{w}'^2 = \mathfrak{w}^4 + 4\mathfrak{w}^3 + 4\mathfrak{w}^2 - 4\mathfrak{c}\mathfrak{w}$$

without a new constant of integration!

The constants $\mathfrak{c}$ constitute the respective cluster sets $\mathfrak{C} = \mathfrak{C}(w)$. Just as in case of System (IV), the cluster sets are closed, bounded, and connected, and contain the limits as $p_n \to \infty$ of $14 p_n^{-\frac{3}{2}} \mathbf{h}(p_n)$, $\frac{7}{36} - 10\epsilon_n p_n^{-2} \mathbf{h}(p_n)$, and $2 p_n^{-3} \mathbf{h}(p_n)$, respectively, for every appropriate sequence of poles (p_n). The statements of Exercises 6.17–6.21 *mutatis mutandis* remain valid. In any case the limit functions satisfy some simple

Table 6.1 Exceptional parameters, trigonometric solutions, and periods

(i)	$\mathfrak{c} = \pm i\sqrt{\frac{2}{27}}$,	$\mathfrak{w} = \mp\frac{i}{\sqrt{6}}(2 + 3\tan^2(\sqrt[4]{\frac{3}{8}}(1 \mp i)\mathfrak{z}))$,	$\vartheta = (1 \pm i)\frac{\pi}{\sqrt[4]{6}}$,
(ii)	$\mathfrak{c} = 0$,	$\mathfrak{w} = \pm(\sinh\mathfrak{z})^{-1}$,	$\vartheta = 2\pi i$,
	$\mathfrak{c} = -\frac{1}{4}$,	$\mathfrak{w} = \pm\frac{1}{\sqrt{2}}\tan(\frac{\mathfrak{z}}{\sqrt{2}})$,	$\vartheta = \pi\sqrt{2}$,
(iv)	$\mathfrak{c} = 0$,	$\mathfrak{w} = \pm(\sinh 2\mathfrak{z})^{-1}$,	$\vartheta = \pi i$,
	$\mathfrak{c} = -\frac{8}{27}$,	$\mathfrak{w} = 8(9\tan^2(\frac{\mathfrak{z}}{\sqrt{3}}) - 3)^{-1}$,	$\vartheta = \pi\sqrt{3}$.

Briot–Bouquet equation

$$\mathfrak{w}'^2 = P(\mathfrak{w}; \mathfrak{c})$$

(the analogue to the algebraic curve (6.28)) depending on the parameter $\mathfrak{c}$. The discriminant of P with respect to $\mathfrak{w}$ is $27\mathfrak{c}^2 + 2$, $\mathfrak{c}(4\mathfrak{c} + 1)$, and $\mathfrak{c}^3(27\mathfrak{c} + 8)$, respectively, with zeros $\mathfrak{c} = \sqrt{-\frac{2}{27}}$, $\mathfrak{c} = 0, -\frac{1}{4}$ and $\mathfrak{c} = 0, -\frac{8}{27}$. The non-constant solutions are

- trigonometric functions of degree two if $\mathfrak{c}$ is exceptional (Table 6.1).
- elliptic functions of elliptic order two otherwise; the periods form a lattice $\mathscr{L}_\mathfrak{c}$ which is spanned by $(\vartheta, \vartheta') = (\vartheta_\mathfrak{c}, \vartheta'_\mathfrak{c})$ with $|\vartheta| \le |\vartheta'| \le |\vartheta \pm \vartheta'|$ and $\vartheta'_\mathfrak{c} \to \infty$ if $\mathfrak{c}$ approaches some exceptional value.

Not every constant solution is also a limit function. Constant limit functions only occur in the exceptional cases (i) $\mathfrak{w} = \sqrt{-\frac{1}{6}}$, (ii) $\mathfrak{w} = 0$ if $\mathfrak{c} = 0$ and $\mathfrak{w} = \sqrt{-\frac{1}{2}}$ if $\mathfrak{c} = -\frac{1}{4}$, (iv) $\mathfrak{w} = 0$ if $\mathfrak{c} = 0$ and $\mathfrak{w} = -\frac{2}{3}$ if $\mathfrak{c} = -\frac{2}{27}$. Note that the constant limit functions are Picard values of the corresponding trigonometric solutions. The analogue to Theorem 6.4 is

Theorem 6.8 *Suppose that the cluster set of some first, second, and fourth transcendent contains some* $\tilde{\mathfrak{c}} \ne \sqrt{-\frac{2}{27}}$, $\tilde{\mathfrak{c}} \ne 0, -\frac{1}{4}$, *and* $\tilde{\mathfrak{c}} \ne 0, -\frac{8}{27}$, *respectively. Then*

$$T(r, w) \asymp r^{\frac{5}{2}}, \quad T(r, w) \asymp r^3, \quad \text{and} \quad T(r, w) \asymp r^4, \tag{6.32}$$

respectively, holds, at least on some sequence $r = r_k \to \infty$.

Again we can reap the benefits of our work.

Theorem 6.9 *First Painlevé transcendents have the following properties:*

- $w \in \mathfrak{Y}_{\frac{1}{2},\frac{1}{4}}$ *and* $w^{(k)} \in \mathfrak{Y}_{\frac{1}{2}+\frac{k}{4},\frac{1}{4}}$;
- $T(r, w) = O(r^{\frac{5}{2}})$;
- $r^{\frac{5}{2}} / \log r = O(T(r, w))$;
- $m(r, w) \sim \frac{1}{2}\log r$ *and* $m(r, 1/w) = O(1)$;
- $w \asymp |z|^{\frac{1}{2}}$ *holds outside* $\mathfrak{P}_\delta \cup \mathfrak{Q}_\delta$, *where* $\mathfrak{Q}$ *is the set of non-zero zeros of* w.

Theorem 6.10 *Second Painlevé transcendents have the following properties:*

- $w \in \mathfrak{Y}_{\frac{1}{2},\frac{1}{2}}$ *and* $w^{(k)} \in \mathfrak{Y}_{\frac{1}{2}+\frac{k}{2},\frac{1}{2}}$;
- $T(r,w) = O(r^3)$;
- $r^{\frac{3}{2}} = O(T(r,w))$;
- $m(r,w) \le (\frac{1}{2} + o(1)) \log r$;
- $w = O(|z|^{\frac{1}{2}})$ *outside* $\mathfrak{P}_\delta$.

In both cases the fourth assertion follows from Theorem 4.17, hence only the third assertion in Theorems 6.9 and 6.10 requires a proof. In the first case we refer to Shimomura [155], and to the proof of Theorem 6.5 in the second.

Remark 6.10 The upper and lower estimates for $T(r,w)$ were proved by Shimomura [153, 155] in both cases. The lower estimate $\varrho \ge \frac{5}{2}$ for first transcendents is due to Mues and Redheffer [121]. Hinkkanen and Laine [91] derived lower estimates for the transcendental solutions to Painlevé (II) and (IV). There are also several published attempts by unnamed authors to obtain upper bounds. They failed for different reasons, one reason being that the ordinary spherical derivative does not qualify as a tool. The approach via the Yosida property (in several steps) can be traced back to the papers [176, 180, 182, 184, 186].

Exercise 6.31 The solutions to Painlevé's equation

$$2ww'' = w'^2 + 8\alpha w^3 - 2zw^2 - 1 \quad (\alpha \ne 0) \qquad \text{XXXIV.}$$

are given by $2\alpha w = y' + y^2 + \frac{1}{2}z$ with $y'' = -(2\alpha + \frac{1}{2}) + zy + 2y^3$, see [94], p. 340. Prove that $w \in \mathfrak{Y}_{1,\frac{1}{2}}$ and $T(r,w) = 2N_-(r,y) + O(\log r) = O(r^3)$, where $N_-(r,y)$ 'counts' the poles of y with residue -1.
Prove also $w''' + 2zw' + w = 0$ and $\nu(r) \sim \sqrt{2}r^{\frac{3}{2}}$ (central index) if $\alpha = 0$.

6.4.5 Airy Solutions

Airy solutions to Painlevé (II) are obtained by repeated application of the Bäcklund transformations (6.16) to the solutions to the Airy equations (6.15). They occur for parameters $\alpha \in \frac{1}{2} + \mathbb{Z}$. The analogue to Theorem 6.7 is

Theorem 6.11 *For every second transcendent each of the following conditions implies* $T(r,w) = O(r^{\frac{3}{2}})$:

- *w is an Airy solution.*
- $V = W + \frac{1}{4}z^2 \in \mathfrak{Y}_{\frac{1}{2},\frac{1}{2}}$.
- *w solves some first-order algebraic differential equation.*
- *V solves some first-order algebraic differential equation.*
- *w and V are algebraically dependent over* $\mathbb{C}(z)$.

Proof Airy solutions $w = R(z, w_0)$ (cf. Exercise 6.10) have Nevanlinna characteristic $T(r, w) = O(r^{\frac{3}{2}})$. This also follows from $V \in \mathfrak{Y}_{\frac{1}{2},\frac{1}{2}}$, which in combination with $\text{res}_p = -1$ yields $n(r, w) = n(r, V) = O(r^{\frac{3}{2}})$ and $T(r, w) = O(r^{\frac{3}{2}})$. Now assume that w also solves $P(z, w, w') = 0$, where

$$P(z, x, y) = \prod_{\nu=1}^{n} (y - G_\nu(z, x))$$

is irreducible. Then $w' = G_\nu(z, w)$ locally holds for at least one ν. On the discs $\Delta_\delta(p)$ with $\text{res}_p\, w = 1$, $w' + w^2 = O(|p|)$ holds as $p \to \infty$ (cf. Exercise 6.3), hence $G_\nu(z, x) = -x^2 + O(|z|)$ as $x \to \infty$, uniformly with respect to z. In the same way we obtain $w' = G_\mu(z, w)$ and $G_\mu(z, x) = x^2 + O(|z|)$ for some other branch if poles with residue -1 are considered. Re-scaling along any sequence of poles yields $\mathfrak{w}'_\nu = -\mathfrak{w}_\nu^2 + \mathfrak{c}_\nu$ and $\mathfrak{w}'_\mu = \mathfrak{w}_\mu^2 + \mathfrak{c}_\mu$, hence $\mathfrak{w}'^2_\nu = \mathfrak{w}_\nu^4 - 2\mathfrak{c}_\nu \mathfrak{w}_\nu^2 + \mathfrak{c}_\nu^2$ and $\mathfrak{w}'^2_\mu = \mathfrak{w}_\mu^4 + 2\mathfrak{c}_\mu \mathfrak{w}_\mu^2 + \mathfrak{c}_\mu^2$, respectively. Since, however, $\mathfrak{w}'^2 = \mathfrak{w}^4 + \mathfrak{w}^2 - \mathfrak{c}$ always holds this implies $-\mathfrak{c}_\nu = \frac{1}{2} = \mathfrak{c}_\mu$, hence $\mathfrak{c}_\nu^2 = \mathfrak{c}_\mu^2 = \frac{1}{4}$ and $\mathfrak{w}'^2 = (\mathfrak{w}^2 + \frac{1}{2})^2$ in any case. In particular, the set of poles $\mathfrak{P}$ has string structure, hence the algebraic curve $P(z, x, y) = 0$ has genus zero and $\mathfrak{P}$ consists of finitely many strings (p_k) with counting function $n(r, (p_k)) \sim \frac{\sqrt{2}}{3\pi} r^{\frac{3}{2}}$. Again $T(r, w) = O(r^{\frac{3}{2}})$ holds. The same is true if V satisfies some first-order algebraic differential equation or if V and w are algebraically dependent, since then w also satisfies some first-order equation. ☕

6.4.6 The Painlevé Hierarchies

The first Painlevé hierarchy is a sequence $_{2n}\text{PI}$ of algebraic differential equations of order $2n$; $_2\text{PI}$ coincides with Painlevé's first equation (I) $w'' = z + 6w^2$, and $_4\text{PI}$ is given by

$$w^{(4)} = 20ww'' + 10w'^2 - 40w^3 - 8aw - \tfrac{8}{3}z. \qquad {}_4\text{PI}$$

Shimomura [151, 154] proved that the solutions to $_4\text{PI}$ are meromorphic and have order of growth at least $\frac{7}{3}$. This is just a special case of a more general result, which says that the meromorphic solutions to $_{2n}\text{PI}$, if any, have order of growth at least $\frac{2n+3}{n+1}$. The Yosida test for equation $_4\text{PI}$ yields $4\beta = \max\{\alpha + 2\beta, 2\alpha, 0, 1 - \alpha\}$; the smallest value $\beta = \frac{1}{6}$ is obtained if $\alpha = \frac{1}{3}$. If $w \in \mathfrak{Y}_{\frac{1}{3},\frac{1}{6}}$ is true, this implies $T(r, w) = O(r^{2+2\beta}) = O(r^{\frac{7}{3}})$, in accordance with and completing Shimomura's result.

Exercise 6.32 Realise the Yosida test for

$$w^{(6)} = 42w''^2 + 56w'w''' + 28ww^{(4)} - 280w(w'^2 + ww'' - w^3) + z \qquad {}_6\text{PI}$$

and

$$w^{(4)} = 10w^2w'' + 10ww'^2 - 6w^5 + zw - \lambda \qquad {}_4\mathrm{PII}$$

(though it is not known whether or not the solutions are meromorphic).

Solution $\alpha = \frac{1}{4}$, $\beta = \frac{1}{8}$ and $\alpha = \beta = \frac{1}{4}$. For meromorphic solutions this suggests $T(r, w) = O(r^{\frac{9}{4}})$ and $T(r, w) = O(r^{\frac{5}{2}})$, respectively.

6.5 Asymptotic Expansions

The solutions to Riccati and implicit first-order differential equations of genus zero have asymptotic expansions on regularly distributed Stokes sectors, while the value distribution takes place on arbitrarily small sectors around the Stokes rays which separate the Stokes sectors. For families of systems and higher-order equations pole-free sectors and asymptotic expansions merely occur in exceptional cases, and only for particular solutions. As a rule, the plane is 'filled' with poles, but the exception proves the rule.

6.5.1 *Pole-Free Sectors and Asymptotic Expansions*

a) System (IV) Let (x, y) be any transcendental solution to System (IV) with 'pole-free' sector S in the sense that x and y have only finitely many poles on every closed sub-sector of S. Yosida Re-scaling with h_n restricted to any closed sub-sector of S leads to limit functions $(\mathfrak{x}, \mathfrak{y})$ without poles, hence to constants. The possible constants are $(0, 0)$ and $(-\omega, -\bar{\omega})$, again with $\omega^3 = 1$, hence we have

$$x(z) = o(|z|) \text{ and } y(z) = o(|z|) \quad \text{resp.}$$
$$x(z) = -\omega z + o(|z|) \text{ and } y(z) = -\bar{\omega} z + o(|z|)$$

as $z \to \infty$ on S. In each case we will prove that, as in the case of first-order equations, this leads to asymptotic expansions on specific sectors

$$\Sigma^1_\nu : |\arg z - (2\nu + 1)\tfrac{\pi}{4}| < \tfrac{\pi}{4} \quad \text{and} \quad \Sigma^2_\nu : |\arg z - \nu\tfrac{\pi}{2}| < \tfrac{\pi}{4},$$

also called Stokes sectors for System (IV) of the first and second kind. They are separated from each other by the Stokes rays

$$\sigma^1_\nu : \arg z = \nu\tfrac{\pi}{2} \quad \text{and} \quad \sigma^2_\nu : \arg z = (2\nu + 1)\tfrac{\pi}{4},$$

respectively. By Theorem 1.10, the size of pole-free sectors is determined by the differential equation (by its 'rank'), while their position is free. This is not the case in the context of Riccati equations, System (IV), and many other differential equations and systems. The reason is that pole-free sectors are 'bordered' by strings of poles, and these strings determine their own asymptotic direction. It is not hard to show that the potential asymptotic expansions are

$$\begin{aligned} x(z) &= -\omega z + \tfrac{1}{3}(\alpha + 2\bar\omega\beta + \omega)z^{-1} + O(|z|^{-3})\\ y(z) &= -\bar\omega z + \tfrac{1}{3}(\beta + 2\omega\alpha - \bar\omega)z^{-1} + O(|z|^{-3})\\ H(z) - \tfrac{1}{3}z^3 &= -(\omega\alpha + \bar\omega\beta)z + \tfrac{1}{3}(\omega\alpha^2 + \bar\omega\beta^2 + \alpha\beta - 1)z^{-1} + O(|z|^{-3}) \end{aligned} \tag{6.33}$$

and

$$\begin{aligned} x(z) &= -\alpha z^{-1} - (\alpha + \beta^2)z^{-3} + O(|z|^{-5})\\ y(z) &= -\beta z^{-1} + (\beta - \alpha^2)z^{-3} + O(|z|^{-5})\\ H(z) &= -\alpha\beta z^{-1} - \tfrac{1}{3}(\alpha^3 + \beta^3)z^{-3} + O(|z|^{-5}). \end{aligned} \tag{6.34}$$

We will now show how pole-free sectors and asymptotic expansions are determined by the regular behaviour of the solutions on non-Stokes rays $\arg z = \hat\theta$.

Theorem 6.12 *Suppose that $x(z) = -\omega z + o(|z|)$ and $y(z) = -\bar\omega z + o(|z|)$ resp. $x(z) = o(|z|)$ and $y(z) = o(|z|)$ holds as $z \to \infty$ on some non-Stokes ray $\hat\sigma : \arg z = \hat\theta$. Then x, y, and H have asymptotic expansions on some sector $\mu\frac{\pi}{2} < \arg z < \nu\frac{\pi}{2}$ resp. $(2\mu + 1)\frac{\pi}{4} < \arg z < (2\nu + 1)\frac{\pi}{4}$ that contains the ray $\hat\sigma$.*

Proof The main assertion concerns the size and position of pole-free sectors. We will exemplarily consider the second case. Re-scaling along any sequence (h_n) on $\hat\sigma$ yields limit functions $\mathfrak{x}$ and $\mathfrak{y}$ satisfying $\mathfrak{x}' = -\mathfrak{y}^2 - \mathfrak{x}$, $\mathfrak{y}' = \mathfrak{x}^2 + \mathfrak{y}$ with $\mathfrak{x}(0) = \mathfrak{y}(0) = 0$, hence $\mathfrak{x} = \mathfrak{y} \equiv 0$ and $\mathfrak{x}^3 + \mathfrak{y}^3 + 3\mathfrak{x}\mathfrak{y} = 0$. Thus there are pole-free discs $D_r : |z - re^{i\hat\theta}| < r^{-1}\kappa(r)$ $(r > r_0)$ such that $\kappa(r) \to \infty$ as $r \to \infty$. We will show that the presumably small domain $\bigcup_{r>r_0} D_r$ is part of a large pole-free sector. To this end we choose $r_0 > 0$ sufficiently large and define the sequence (r_n) inductively by $r_{n+1} = r_n + 4r_n^{-1}$ (the reason for choosing the factor 4 will soon become clear). By θ_n we denote the largest number such that

$$A_n = \{z : r_n \le |z| \le r_{n+1},\ \hat\theta \le \arg z < \theta_n\}$$

contains no pole, and note that $r_n\theta_n \to \infty$ at least like $\kappa(r_n) \to \infty$.

Re-scaling along any sequence $h_n \to \infty$ in $A = \bigcup_{n\ge0} A_n$ yields limit functions $\mathfrak{x}$ and $\mathfrak{y}$ that satisfy $\mathfrak{x}^3 + \mathfrak{y}^3 + 3\mathfrak{x}\mathfrak{y} = 3\mathfrak{c}$, and since A is pole-free, $\mathfrak{x}$ and $\mathfrak{y}$ cannot be elliptic functions (which have poles in 'four directions') even if $h_n \in \partial A$, hence $\mathfrak{c} = \lim_{h_n\to\infty} h_n^{-3}H(h_n) = 0$ or $\mathfrak{c} = 1$. These limits form the *cluster set* $\mathfrak{C}_A(x, y) \subset \{0, 1\}$ of $z^{-3}H(z)$ as $z \to \infty$ on A. Since A is locally connected at infinity, the cluster set is

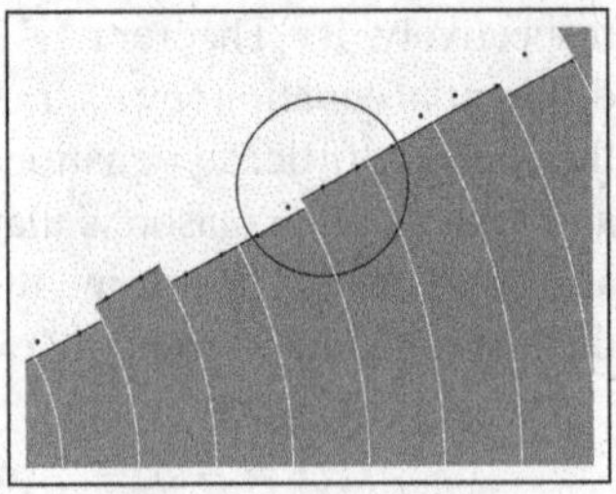

Fig. 6.1 The regions A_n. The circle $|z-p_n| = 5|p_n|^{-1}$ about $p_n \in \partial A_n$ encloses five poles

connected, and from $z^{-3}H(z) \to 0$ as $z = re^{i\hat{\theta}} \to \infty$ we obtain $\mathfrak{C}_A(x, y) = \{0\}$. We may assume that there exists some pole $\tilde{p}_0$ on ∂A_0, hence $\arg \tilde{p}_0 = \theta_0$. The same is true at least for some subsequence $\tilde{p}_{n_k} \in \partial A_{n_k}$ with $\arg p_{n_k} = \theta_{n_k}$. Re-scaling along the sequence $(\tilde{p}_{n_k})$ then yields limit functions $\mathfrak{x}$ and $\mathfrak{y}$ that satisfy $\mathfrak{x}^3 + \mathfrak{y}^3 + 3\mathfrak{x}\mathfrak{y} = 0$, hence the algebraic curve (6.28) has genus zero, and $\mathfrak{x}$ and $\mathfrak{y}$ have poles $\mathfrak{z}_\nu = \nu\vartheta$, $\vartheta = \frac{2\pi i}{3}$. It follows from Hurwitz' Theorem that to each pole p_{n_k} there exist five poles $z_{k,j} = \tilde{p}_{n_k} + j(\vartheta + o(1))\tilde{p}_{n_k}^{-1}$ $(-2 \le j \le 2)$ on $|z - \tilde{p}_n| < 5|\tilde{p}_n|^{-1}$, and no others (the reason for choosing the constants '4' and '5' was $4 < 2|\vartheta| < 5 < 3|\vartheta|$). Since $z_{k,-2}$ and $z_{k,2}$ do not belong to the annulus $r_{n_k} \le |z| \le r_{n_k+1}$, it follows that A_{n_k-1} and A_{n_k+1}, hence *each* A_n $(n \ge n_0)$ contains some pole $\tilde{p}_n$ on its boundary, and $(\tilde{p}_n)$ is a subsequence of some $(\pm\vartheta, 1)$-string (p_m). Strings of this kind satisfy $\lim_{m\to\infty} \arg p_m = (2\nu + 1)\frac{\pi}{4}$ for some integer ν, hence $\lim_{n\to\infty} \theta_n = (2\nu + 1)\frac{\pi}{4}$ holds, see Fig. 6.1.
The same argument applies to $\tilde{A}_n = \{z : r_n \le |z| \le r_{n+1},\ \theta_n < \arg z \le \hat{\theta}\}$, this showing that in the second case the natural pole-free sectors are given by Σ : $(2\mu + 1)\frac{\pi}{4} < \arg z < (2\nu + 1)\frac{\pi}{4}$. In the first case the same proof yields pole-free sectors $\Sigma : \mu\frac{\pi}{2} < \arg z < \nu\frac{\pi}{2}$; here the strings of poles have type $(\frac{\pi}{\sqrt{3}}, 1)$.

Also in the first case the principal terms $-\omega z$ and $-\bar{\omega} z$ are known. The argument used in the proof of Theorem 5.7 applies immediately, yielding asymptotic expansions for x and y. In the second case, however, only $x(z) = o(|z|)$ and $y(z) = o(|z|)$ is known. To determine the true principal terms we consider any closed sub-sector S of Σ and set $\max\{|x(z)|, |y(z)|\} = \epsilon|z|$. Since

$$|x(z)| + |y(z)| = o(|z|) \quad (z \to \infty \text{ on } S), \tag{6.35}$$

we may assume $\epsilon = \epsilon(z) < \frac{1}{2}$. From System (IV) and $|x'(z)| + |y'(z)| \to 0$ as $z \to \infty$ on every proper sub-sector $S' \subset S$ (this following from (6.35) and Cauchy's coefficient estimates), hence $|x'(z)| + |y'(z)| < 1$, say, we obtain

$$|zx(z)| = |x'(z) + y(z)^2 + \alpha| < \epsilon^2|z|^2 + |\alpha| + 1 \quad \text{and} \quad |zy(z)| < \epsilon^2|z|^2 + |\beta| + 1.$$

This yields $(\epsilon - \epsilon^2)|z|^2 < K$ and $\epsilon < 2K|z|^{-2}$, hence $|x(z)| + |y(z)| = O(|z|^{-1})$ and $|x'(z)| + |y'(z)| = O(|z|^{-2})$ as $z \to \infty$ on S, again by Cauchy's Theorem. From

System (IV) it then follows that $zx(z)+\alpha = O(|z|^{-2})$ and $zy(z)+\beta = O(|z|^{-2})$, hence the principal terms are $-\alpha z^{-1}$ and $-\beta z^{-1}$. From here the proof follows the lines of the proof of Theorem 5.7.

b) Painlevé (I), (II), and (IV) For first, second, and fourth transcendents the Stokes sectors and rays Σ_ν and σ_ν are as follows.

$$
\begin{array}{lll}
\text{(i)} & \Sigma_\nu : |\arg z - 2\nu\pi/5| < \pi/5 & \sigma_\nu : \arg z = (2\nu+1)\pi/5\\
\text{(ii)}_a & \Sigma_\nu : |\arg z - 2\nu\frac{\pi}{3}| < \frac{\pi}{3} & \sigma_\nu : \arg z = (2\nu+1)\frac{\pi}{3}\\
\text{(ii)}_b & \Sigma_\nu : |\arg z - (2\nu+1)\frac{\pi}{3}| < \frac{\pi}{3} & \sigma_\nu : \arg z = (2\nu+2)\frac{\pi}{3}\\
\text{(iv)}_a & \Sigma_\nu : |\arg z - (2\nu+1)\frac{\pi}{4}| < \frac{\pi}{4} & \sigma_\nu : \arg z = (\nu+1)\frac{\pi}{2}\\
\text{(iv)}_{bc} & \Sigma_\nu : |\arg z - \nu\frac{\pi}{2}| < \frac{\pi}{4} & \sigma_\nu : \arg z = (2\nu+1)\frac{\pi}{4}.
\end{array}
$$

Exercise 6.33 Confirm the following possible expansions on pole-free sectors:

$$
\begin{array}{ll}
\text{(i)} & w(z) = (-\frac{z}{6})^{1/2} - \frac{1}{48}z^{-2} + O(|z|^{-9/2})\\
\text{(ii)}_a & w(z) = \qquad -\alpha z^{-1} + 2\alpha(\alpha^2-1)z^{-4} + O(|z|^{-7})\\
\text{(ii)}_b & w(z) = (-\frac{z}{2})^{1/2} + \frac{1}{2}\alpha z^{-1} + O(|z|^{-5/2})\\
\text{(iv)}_a & w(z) = -\frac{2}{3}z + \alpha z^{-1} - \frac{1}{4}(3\alpha^2 - 9\gamma^2 + 1)z^{-3} + O(|z|^{-5})\\
\text{(iv)}_b & w(z) = -2z - \alpha z^{-1} + \frac{1}{4}(3\alpha^2 - \gamma^2 + 1)z^{-3} + O(|z|^{-5})\\
\text{(iv)}_c & w(z) = \qquad \gamma z^{-1} - \frac{1}{2}(2\gamma^2 - \alpha\gamma)z^{-3} + O(|z|^{-5}).
\end{array}
\tag{6.36}
$$

$$
\begin{array}{ll}
\text{(i)} & W(z) = -6(-\frac{z}{6})^{3/2} + \frac{1}{48}z^{-1} + O(|z|^{-7/2})\\
\text{(ii)}_a & W(z) = -\alpha^2 z^{-1} + \alpha^2(\alpha^2-1)z^{-4} + O(|z|^{-7}) \quad (\alpha \neq 0)\\
\text{(ii)}_b & W(z) = -\frac{1}{4}z^2 + \alpha\sqrt{2}(-z)^{1/2} + \frac{1}{8}(1+4\alpha^2)z^{-1} + O(|z|^{-5/2})\\
\text{(iv)}_a & W(z) = -\frac{8}{27}z^3 + \frac{2}{3}\alpha z - \frac{1}{6}(3\alpha^2 + 9\gamma^2 - 1)z^{-1} + O(|z|^{-3})\\
\text{(iv)}_b & W(z) = 2\alpha z + \frac{1}{2}(\alpha^2 - \gamma^2 + 1)z^{-1} + O(|z|^{-3})\\
\text{(iv)}_c & W(z) = 2\gamma z + (\gamma^2 - \alpha\gamma)z^{-1} + O(|z|^{-3})
\end{array}
\tag{6.37}
$$

(both branches of $(-z)^{1/2}$ and $\gamma = (-\beta/2)^{1/2}$). Note that the expansions (ii)$_a$ and (iv)$_c$ are only significant if $\alpha \neq 0$ and $\beta \neq 0$, respectively.

Exercise 6.34 (Continued) If $\beta = 0$ in (iv)$_c$ but $w \not\equiv 0$ set $y = w'/w$ to obtain $y' = P(z) - \frac{1}{2}y^2$ with $P(z) = 2z^2 - 2\alpha + \frac{3}{2}zw(z) + 2w(z)^2 \sim 2z^2 - 2\alpha$. Prove that y has an asymptotic expansion $y \sim \pm 2z + \cdots$ on S. Since $w(z) = e^{\int y(z)\,dz} \to 0$ on S requires $\operatorname{Re}(\pm z^2) < 0$ for some sign, this yields

$$
\begin{array}{lll}
\text{(iv)}_c & w'/w \sim -2z + (\alpha-1)z^{-1} + \frac{1}{4}(\alpha^2 - 4\alpha + 3)z^{-3} + \cdots & (S \subset \Sigma_0 \cup \Sigma_2)\\
\text{(iv)}_c & w'/w \sim 2z - (\alpha+1)z^{-1} - \frac{1}{4}(\alpha^2 + 4\alpha + 3)z^{-3} + \cdots & (S \subset \Sigma_1 \cup \Sigma_3),
\end{array}
$$

and $W = O(|z|^M)e^{-|\operatorname{Re} z^2|}$ for some $M > 0$.

If $\alpha = 0$ in (ii)$_a$ but $w \not\equiv 0$, prove that

$$\text{(ii)}_a \quad w'/w \sim \pm z^{1/2} - \tfrac{1}{4}z^{-1} \mp \tfrac{5}{32}z^{-5/2} + \cdots$$

and $W = O(|z|^M)e^{-\frac{2}{3}|\mathrm{Re}\, z^{3/2}|}$ holds on pole-free sectors with $\pm\mathrm{Re}\, z^{3/2} < 0$.

The analogue to Theorem 6.12, with exactly the same proof, is

Theorem 6.13 *Let w be any Painlevé transcendent having an asymptotics expansion* (6.36) *on some single non-Stokes ray $\arg z = \hat{\theta}$. Then the Stokes sector Σ which contains that ray is pole-free for w, and the asymptotics expansion holds on Σ.*

6.5.2 *Truncated Solutions*

Solutions to System (IV) or any Painlevé equation having no poles on some sector about any Stokes ray σ are called *truncated* (along σ).

a) System (IV) The following exercises show that truncated solutions exist with prescribed asymptotics.

Exercise 6.35 Theorem 1.10 does not immediately apply to System (IV). Suppose $\alpha\beta \neq 0$ and set $t = z^2$, $x(z) = t^{-1/2}u(t)^2$, and $y(z) = t^{-1/2}v(t)^2$ to obtain

$$\dot{u} = -\frac{\alpha + u^2}{4u} + \frac{u^2 - v^4}{4tu}, \quad \dot{v} = \frac{\beta + v^2}{4v} + \frac{u^4 + v^2}{4tv},$$

where $\dot{}$ denotes $\frac{d}{dt}$. Prove that given any half-plane $\mathbb{H}$, Theorem 1.10 now applies to the corresponding system for $\xi = u - \sqrt{-\alpha}$ and $\eta = v - \sqrt{-\beta}$ on $\mathbb{H}$. Thus to every quadrant $Q = \sqrt{\mathbb{H}}$ there exists some solution with the asymptotics (6.34). If $\mathbb{H}$ is chosen such that Q contains some Stokes ray $\arg z = (2\nu + 1)\frac{\pi}{4}$, the asymptotics for x, y, and H at least holds on $|\arg z - (2\nu + 1)\frac{\pi}{4}| < \frac{\pi}{2}$.

Exercise 6.36 (Continued) If $\alpha = 0$ and $\beta \neq 0$ resp. $\alpha \neq 0$ and $\beta = 0$ show that the transformation $t = z^2$ and $x(z) = t^{-3/2}u(t)^2$, $y(z) = t^{-1/2}v(t)^2$ resp. $x(z) = t^{-1/2}u(t)^2$, $y(z) = t^{-3/2}v(t)^2$ works in the same manner.

Exercise 6.37 Prove the analogue for solutions with expansions (6.33) without restriction on α and β. (**Hint.** Set $t = z^2$, $x(z) = t^{1/2}u(t)^2$, and $y(z) = t^{1/2}v(t)^2$.)

Exercise 6.38 Prove that in any case solutions with pole-free *half-plane* are uniquely determined by their asymptotics. (**Hint.** See the proof of Theorem 5.8.)

b) Painlevé Transcendents The existence of truncated transcendents is again based on Theorem 1.10. For a proof the Painlevé equations have to be transformed into appropriate systems.

Exercise 6.39 Painlevé's first equation $w'' = z + 6w^2$ is transformed by $z = t^{4/5}$, $w(z) = t^{2/5}v(t)$ into $\ddot{v} = \frac{96}{25}(6v^2 + 1) + \frac{4v}{25t^2} - \frac{\dot{v}}{t} = f(t, v, \dot{v})$. Prove that

Theorem 1.10 applies to the system $\dot{x} = y$, $\dot{y} = f(t, x \pm i/\sqrt{6}, y)$ of rank one, hence given any half-plane S, there exists a solution with prescribed asymptotics on S. Choosing $S^{\frac{4}{5}}$ to cover the Stokes rays $\sigma_{\nu-1}$ and σ_ν, the asymptotics extends to the sector $\Sigma_{\nu-1} \cup \sigma_{\nu-1} \cup \Sigma_\nu \cup \sigma_\nu \cup \Sigma_{\nu+1}$ with central angle $\frac{6}{5}\pi$, and determines the doubly truncated solution w uniquely.
(**Hint.** To prove uniqueness set $u = w_1 - w_2$ and note that u tends to zero and satisfies $u'' = Q(z)u$ with $Q(z) = 6(w_1(z) + w_2(z)) \sim \sqrt{-24z} + \cdots$ on S.) ☕

Exercise 6.40 Adapt the argument of the previous exercise to Painlevé's second equation $w'' = \alpha + zw + 2w^3$ with $\alpha \neq 0$.

Solution (ii)$_a$ $z = t^{2/3}$, $w(z) = t^{-2/3}v(t)$, $\ddot{v} = \frac{4}{9}(v+\alpha) + \frac{8}{9t^2}(v^3 - v) + \frac{\dot{v}}{t}$;

(ii)$_b$ $z = t^{2/3}$, $w(z) = t^{1/3}v(t)$, $\ddot{v} = \frac{4}{9}(v + 2v^3) + \frac{4\alpha}{9t} + \frac{v}{9t^2} - \frac{\dot{v}}{t}$. ☕

Remark 6.11 Equation (IV) could be dealt with in the same way. It is, however, more convenient to refer to Exercises 6.35–6.38, which show that given any Stokes ray σ_ν there exists some fourth transcendent that is truncated along σ_ν; w is uniquely determined by its asymptotics (iv)$_a$, (iv)$_b$, and (iv)$_c$, respectively, which may be prescribed. ☕

Transcendents that are truncated along some Stokes ray $\arg z = \hat{\theta}_\nu$ have an asymptotic expansion on $|\arg z - \hat{\theta}_\nu| < \epsilon$, which by Theorem 6.13 extends to $\Sigma_\nu \cup \sigma_\nu \cup \Sigma_{\nu+1}$. In particular, w has the same asymptotic expansion on adjacent Stokes sectors Σ_ν and $\Sigma_{\nu+1}$.[4] The converse is also true.

Theorem 6.14 *Let w be any Painlevé transcendent that has the same asymptotic expansion* (6.36) *on adjacent Stokes sectors Σ_ν and $\Sigma_{\nu+1}$. Then w is truncated along σ_ν and the asymptotic expansion holds on $\Sigma_\nu \cup \sigma_\nu \cup \Sigma_{\nu+1}$.*

Proof In any case $F(z) = e^{-\int W(z)\,dz}$ is an entire function of finite order with simple zeros at the poles of w. Since the idea of proof is the same in all cases we will consider exemplarily (iv)$_b$ with $W(z) = 2\alpha z + \lambda z^{-1} + O(|z|^{-3})$ on Σ_ν and $\Sigma_{\nu+1}$ ($\lambda = \frac{1}{2}(\alpha^2 - \gamma^2 + 1)$). Then

$$H(z) = F(z)e^{\alpha z^2} z^\lambda = \begin{cases} c_\nu + o(1) \text{ as } z \to \infty \text{ on } \Sigma_\nu \\ c_{\nu+1} + o(1) \text{ as } z \to \infty \text{ on } \Sigma_{\nu+1} \end{cases}$$

holds. We set $f(\zeta) = H(z)$ with $z = e^{i(2\nu+1)\frac{\pi}{4}}\sqrt{\zeta}$ and $\sqrt{\zeta} > 0$ on $\zeta > 0$. Since f has finite order ($\log^+\log^+|f(\zeta)| = O(\log|\zeta|)$) on $|\arg\zeta| < \frac{\pi}{2}$, say, and limits $c_\pm$ as $\zeta = re^{\pm i\delta} \to \infty$ for every $0 < \delta < \pi$, the Phragmén–Lindelöf Principle yields

[4]In cases (i) and (ii) 'the same expansion' means that the chosen square-root $\sqrt{-z}$ is continuous across σ_ν.

$c_- = c_+ = c$ and $f(\zeta) \to c$ on $|\arg\zeta| \le \delta$. Thus the half-plane $|\arg z - (2\nu+1)\frac{\pi}{4}| < \frac{\pi}{2}$ is pole-free for w, and w has the asymptotic expansion (6.36)(iv)$_b$ there.

c) Special Transcendents We conclude this section with two examples and a potential example of particular interest.

Example 6.6 Equation $w'' = zw + 2w^3$ has a unique solution, named after *Hastings and McLeod* [76], that is positive and decreasing on the real line. Moreover, it satisfies $w(x) \sim 0$ as $x \to +\infty$ and $w(x) \sim \sqrt{-x/2}$ as $x \to -\infty$. From Theorem 6.13 it follows that the asymptotic expansions $w \sim 0$ and $w(z) \sim \sqrt{-z/2} + \cdots$ hold on $|\arg z| < \frac{\pi}{3}$ and $|\arg z - \pi| < \frac{\pi}{3}$, respectively. The poles are asymptotically restricted to the sectors $|\arg z \mp \frac{\pi}{2}| < \frac{\pi}{6}$. Writing $w'' = (z + 2w(z)^2)w$ and noting that $w(z) = o(|z|^{-n})$ holds on $|\arg z| < \frac{\pi}{3}$ for every $n \in \mathbb{N}$, this yields $w(z) \sim k\mathsf{Ai}(z)$ for some real constant k, actually $k = 1$; Ai denotes the Airy function, a special solution to $y'' = zy$. Even more precisely, $w'/w \sim \pm z^{1/2} - \frac{1}{4}z^{-1} \mp \frac{5}{32}z^{-5/2} + \cdots$ holds as $z \to \infty$ on $|\arg z| < \frac{\pi}{3}$.

Example 6.7 For $\beta = 0$ and α real it is conjectured that there exists a unique real solution to Painlevé's equation (IV)—the so-called *Clarkson–McLeod solution* [25, 95]—satisfying $w(x) \sim 0$ as $x \to +\infty$ and $w(x) \sim -2x + \cdots$ as $x \to -\infty$. If it exists, and existence is supported by numerical experiments and by analogy with the Hastings–McLeod solution, the corresponding asymptotic expansions $w \sim 0$ (even $w'/w \sim -2z + \cdots$ by Exercise 6.34) and $w \sim -2z + \cdots$ hold on $|\arg z| < \frac{\pi}{4}$ and $|\arg z - \pi| < \frac{\pi}{4}$, respectively.

Example 6.8 According to Boutroux [16], Painlevé's equation (I) has five *triply truncated* solutions (also called *tritronquée*); for a recent existence proof, see Joshi and Kitaev [98]. Since (I) is invariant under the transformation $w \mapsto a^2 w(az)$ with $a^5 = 1$, it suffices to prove the existence of a triply truncated solution w_0 having the asymptotics $w_0(z) \sim -\sqrt{-z/6}$ with $\mathrm{Re}\sqrt{-z} > 0$ on $|\arg z - \pi| < \frac{4}{5}\pi$; this solution is truncated along σ_1, σ_2, and σ_3, with poles asymptotically restricted to $|\arg z| < \pi/5$. By Exercise 6.39 there exist uniquely determined first transcendents w_1 and w_2 with asymptotics $w_{1,2} = -\sqrt{-z/6} + O(|z|^{-2})$ and $\mathrm{Re}\sqrt{-z} > 0$ on $\frac{\pi}{5} < \arg z < \frac{7}{5}\pi$ and on $\frac{3}{5}\pi < \arg z < \frac{9}{5}\pi$, respectively. Then $u = w_1 - w_2$ satisfies $u'' = 6(w_1(z) + w_2(z))u = -(\sqrt{-24z} + O(|z|^{-2}))u$ on $\frac{3}{5}\pi < \arg z < \frac{7}{5}\pi$. We set $z = -6t^{4/5}$ and $u(z) = t^{-1/10}v(t)$ on the right half-plane $\mathrm{Re}\, t > 0$ ($t^{4/5} > 0$, $\sqrt{-z/6} = t^{2/5} > 0$ and $t^{-1/10} > 0$ if $t > 0$) to obtain

$$25\ddot{v} + \left[6412 + O(|t|^{-2})\right]v = 0. \tag{6.38}$$

Now every non-trivial solution to (6.38) tends to infinity exponentially as $t \to \infty$ at least on one of the sectors $\delta < \arg t < \frac{\pi}{2} - \delta$ and $-\frac{\pi}{2} + \delta < \arg t < -\delta$. This proves $v \equiv 0$ and $u \equiv 0$, and $w_0 = w_1 = w_2$ is truncated along σ_1, σ_2, and σ_3. The pole-free sector is 'bordered' by $(\frac{(1\pm i)\pi}{\sqrt[4]{6}}, \frac{1}{4})$-strings of poles (Fig. 6.2).

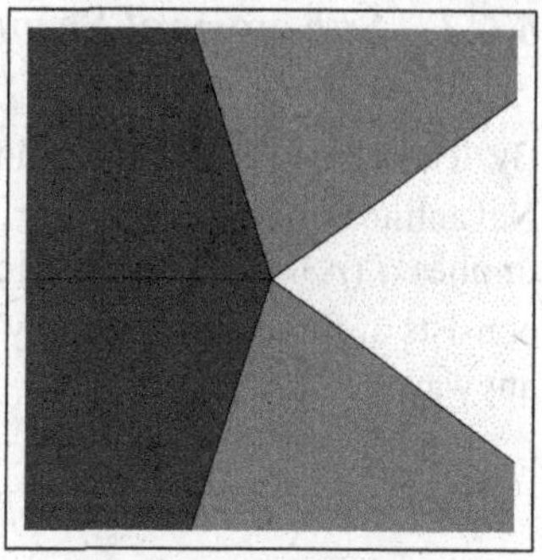

Fig. 6.2 The solutions w_1 and w_2 are truncated along $\arg z = \frac{3}{5}\pi$, $\arg z = \pi$ and $\arg z = \frac{7}{5}\pi$, $\arg z = \pi$, respectively, and agree on $|\arg z - \pi| < 2\pi/5$ *(dark grey)*. The sector $|\arg z - \pi| < 4\pi/5$ *(light and dark grey)* is pole-free for $w_0 = w_1 = w_2$ with asymptotics $w_0(z) \sim -\sqrt{-z/6}$

Remark 6.12 We note that $\overline{w_0(\bar{z})}$ is also a solution that is truncated along σ_1, σ_2, and σ_3, and hence coincides with w_0 by uniqueness. In other words, the tritronquée solution is real on the real axis. We also note that any solution that is truncated along σ_1 and σ_2 resp. σ_2 and σ_3 coincides with w_0 by Exercise 6.39 if it has the same asymptotics as w_0 on $\Sigma_1 \cup \sigma_1 \cup \Sigma_2$ resp. $\Sigma_2 \cup \sigma_2 \cup \Sigma_3$. In contrast to Exercise 6.39, the triply truncated solution w_0 is not only unique, but also determines its asymptotics. In other words, there is no solution $\tilde{w}_0$ that is truncated along σ_1, σ_2, and σ_3 and is asymptotic to $+\sqrt{-z/6}$ with $\operatorname{Re}\sqrt{-z} > 0$. Otherwise the solution $v(z) = e^{4\pi i/5}\tilde{w}_0(e^{2\pi i/5}z)$ would be truncated along the Stokes rays σ_0, σ_1, and σ_2, and two of the solutions w_0, $\tilde{w}_0$, and v would have the same asymptotic expansions on $\pi/5 < \arg z < 7\pi/5$, which by Exercise 6.39 is impossible.

6.6 Sub-normal Solutions

As a rule, given any linear differential equation with polynomial coefficients, all non-trivial solutions have maximal order of growth $\varrho_{\max}$. Only under the rarest of circumstances will it happen that transcendental solutions of order less than $\varrho_{\max}$ occur. These solutions are called *sub-normal.* The transcendental solutions to any Riccati equation with polynomial coefficients are normal (not sub-normal). In the present section we will discuss the question of how to define the term 'sub-normal' for nonlinear higher-order equations and systems, and how to identify the sub-normal solutions. Our goal is to solve this problem exemplarily for Painlevé's second equation and the Hamiltonian system (IV), hence also for Painlevé (IV). By the way we will settle an old question concerning the deficiency $\delta(0, w)$ for second and fourth Painlevé transcendents with parameter $\alpha = 0$ and $\beta = 0$, respectively.

6.6.1 Sub-normal Second Transcendents

By Theorem 6.8, second transcendents with cluster set $\mathfrak{C}(w) \not\subset \{0, -\frac{1}{4}\}$ have Nevanlinna characteristic $T(r, w) \asymp r^3$ as $r = r_n \to \infty$, while $\mathfrak{C}(w) \subset \{0, -\frac{1}{4}\}$ implies $T(r, w) = o(r^3)$. Moreover, the set of poles $\mathfrak{P}$ has string-structure ($\mathfrak{P}$ consists of finitely or infinitely many strings of poles). Since $r^{\frac{3}{2}} = O(T(r, w))$ holds anyway it is quite natural to call any second transcendent w *sub-normal* if

$$T(r, w) = O(r^{\frac{3}{2}}) \quad (r \to \infty)$$

holds. Recall that $V = W + \frac{1}{4}z^2 \in \mathfrak{Y}_{\frac{1}{2},\frac{1}{2}}$ is sufficient for w to be sub-normal. The next exercise will show that $V \in \mathfrak{Y}_{\frac{1}{2},\frac{1}{2}}$ is necessary for w to be an Airy solution.

Exercise 6.41 It follows from $w' = \pm(\frac{1}{2}z + w^2) = \pm\frac{d}{dz}(\frac{z^2}{4} + W) = \pm V'$ that $V = \pm w + \text{const}$, hence $V \in \mathfrak{Y}_{\frac{1}{2},\frac{1}{2}}$. To prove that this is true for *every* Airy solution assume $V \in \mathfrak{Y}_{\frac{1}{2},\frac{1}{2}}$ for *some* Airy solution with $\alpha \neq \frac{1}{2}$, and set $\tilde{w} = -w + \frac{\alpha-1/2}{w'-w^2-z/2} = -w + \frac{\alpha-1/2}{w'-V'}$, hence $w = -\tilde{w} - \frac{(\alpha-1)+1/2}{\tilde{w}'+\tilde{V}'}$. Prove that $\tilde{V} = V - w - \tilde{w} + \text{const}$ and deduce $\tilde{V}(z) = O(|z|^{\frac{1}{2}})$ outside $\tilde{\mathfrak{P}}_\delta$, hence $\tilde{V} \in \mathfrak{Y}_{\frac{1}{2},\frac{1}{2}}$ since the discs $\Delta_\delta(\tilde{p})$ are mutually disjoint and $\text{res}_{\tilde{p}}\, \tilde{V} = -1$; the notation $\tilde{V}$ and $\tilde{\mathfrak{P}}$ is self-explanatory.

Airy solutions are sub-normal. The goal of this section is to prove the converse.

Theorem 6.15 *For second transcendents the following statements are equivalent.*

- *w is sub-normal;*
- *w is an Airy solution;*
- *w satisfies some first-order algebraic differential equation;*
- *w and W (first integral) are algebraically dependent;*
- $V = W + \frac{1}{4}z^2 \in \mathfrak{Y}_{\frac{1}{2},\frac{1}{2}}$;
- *V satisfies some first-order algebraic differential equation.*

Sub-normal solutions occur if and only if $\alpha \in \frac{1}{2} + \mathbb{Z}$. They have Nevanlinna characteristic $T(r, w) \sim -2\alpha\Delta(w)\frac{\sqrt{2}}{3\pi} r^{\frac{3}{2}}$ with $\Delta(w) \in \{-3, -1, 1, 3\}$.

Proof We have to show that sub-normal solutions are Airy solutions; everything else has already been proved or discussed in the exercises. Since $\mathfrak{C}(w) \subset \{0, -\frac{1}{4}\}$, we have to consider two cases as follows.

a) $\mathfrak{C}(w) = \{0\}$. For every sequence $h_n \to \infty$ the re-scaling method yields limit functions $\mathfrak{w} = \lim_{h_n\to\infty} w_{h_n}$ satisfying $\mathfrak{w}'' = \mathfrak{w} + 2\mathfrak{w}^3$ and $\mathfrak{w}'^2 = \mathfrak{w}^4 + \mathfrak{w}^2$. The limit functions are $\mathfrak{w} \equiv 0$ and $\mathfrak{w} = \pm 1/\sinh(\mathfrak{z} + \mathfrak{z}_0)$. The zeros of $\sinh \mathfrak{z}$ form a πi-periodic sequence, and this leads as usual to the conclusion that the poles of w are arranged in $(\pm\pi i, \frac{1}{2})$-strings (p_k) with alternating residues ± 1 and counting function $n(r,p_k) \sim \frac{2}{3\pi}r^{\frac{3}{2}}$. Each string is asymptotic to some ray $\arg z = (2\nu+1)\frac{\pi}{3}$, and from $T(r,w) = O(r^{\frac{3}{2}})$ it follows that there are only finitely many strings. On the open sectors $\Sigma_\nu : (2\nu-1)\frac{\pi}{3} < \arg z < (2\nu+1)\frac{\pi}{3}$, $w = o(|z|^{\frac{1}{2}})$ holds (re-scaling with (h_k) restricted to any closed sub-sector yields the limit function $\mathfrak{w} \equiv 0$). From $w^3 = o(|zw|)$ and $w'' = o(|z|^{-3/2})$ on every closed sub-sector of Σ_ν it then follows that $zw + \alpha = o(|zw|) + o(|z|^{-3/2})$, hence $w = -\alpha/z + o(|z|^{-1})$ holds. The usual technique then shows that w and $W = w^4 + zw^2 + 2\alpha w - w'^2$ have asymptotic expansions (6.36) (ii)$_a$ and (6.37) (ii)$_a$, one and the same on each sector Σ_ν. Now

$$F(z) = e^{-\int W(z)\,dz}$$

is an entire function[5] of finite order which satisfies $F(z) = C_\nu z^{\alpha^2}(1+o(1))$ on every sector $(2\nu-1)\frac{\pi}{3} + \epsilon < \arg z < (2\nu+1)\frac{\pi}{3} - \epsilon$. Applying the Phragmén–Lindelöf Principle to $F(z)z^{-\alpha^2}$ on the sectors $|\arg z - (2\nu+1)\frac{\pi}{3}| < \epsilon$ then shows that F is a polynomial (of degree α^2) and $w^2 = -(F'/F)'$ is a rational function in contrast to our general assumption that w is transcendental.

b) $\mathfrak{C}(w) = \{-\frac{1}{4}\}$. The re-scaling method yields limit functions again satisfying $\mathfrak{w}'' = \mathfrak{w} + 2\mathfrak{w}^3$, but now also $\mathfrak{w}'^2 = (\mathfrak{w}^2 + \frac{1}{2})^2$ with constant and non-constant limit functions $\mathfrak{w} = \pm i/\sqrt{2}$ and $\mathfrak{w} = \pm\frac{1}{\sqrt{2}}\tan\frac{3}{\sqrt{2}}$, respectively. Thus the poles of w form $(\pm\frac{\pi}{\sqrt{2}}, \frac{1}{2})$-strings with constant residues and counting function $n(r,p_k) \sim \frac{\sqrt{2}}{3\pi}r^{\frac{3}{2}}$, each being asymptotic to some ray $\arg z = 0 \bmod 2\pi/3$. Again the number of strings is finite and the constant limit functions lead to asymptotic expansions (6.36) (ii)$_b$ and (6.37) (ii)$_b$ on $\Sigma_\nu : 2\nu\frac{\pi}{3} < \arg z < 2(\nu+1)\frac{\pi}{3}$. Let $\Delta(w)$ denote the difference between the number of strings of poles with residues 1 and -1, respectively, hence

$$\frac{1}{2\pi i}\int_{\Gamma_r} w(z)\,dz = \Delta(w)\frac{\sqrt{2}}{3\pi}r^{\frac{3}{2}} + o(r^{\frac{3}{2}})$$

holds by the Residue Theorem. The contribution of the sectors $|\arg z - 2\nu\frac{\pi}{3}| < \epsilon$ and $2\nu\frac{\pi}{3} + \epsilon < \arg z < 2(\nu+1)\frac{\pi}{3} - \epsilon$ is $O(\epsilon r^{\frac{3}{2}})$ and $\pm(-1)^\nu\frac{\sqrt{2}}{3\pi}r^{\frac{3}{2}} + O(\epsilon r^{\frac{3}{2}})$, respectively, this following from the asymptotics on Σ_ν, $w = O(|z|^{\frac{1}{2}})$ on Γ_r, and the fact that the length of the part of Γ_r in $|\arg z - 2\nu\frac{\pi}{3}| < \epsilon$ is $O(\epsilon r)$; the sign $\pm$ depends on the branch of $\sqrt{-z/2}$. This proves that $\Delta(w) \in \{-3,-1,1,3\}$. We may assume that $\Delta(w)$ is negative (consider $(-w,-\alpha)$ in place of (w,α) otherwise).

Exercise 6.42 Prove that the asymptotic expansions of w and its Bäcklund transform $\tilde{w} = -w + \frac{\alpha-1/2}{w'-w^2-z/2}$ $(\alpha \neq 1/2)$ have the same principal term by showing that $w' - w^2 - z/2 = O(|z|^{-\frac{1}{2}})$. In particular, $\Delta(\tilde{w}) = \Delta(w)$ holds.

[5] Also called a τ-function, see Okamoto [131].

From $\tilde V = V - w - \tilde w + \text{const}$ (cf. Exercise 6.41) it then follows that

$$\begin{aligned} n(r,w) - n(r,\tilde w) &= \frac{1}{2\pi i}\int_{\Gamma_r} (\tilde V(z) - V(z))\,dz = -\frac{1}{2\pi i}\int_{\Gamma_r} (\tilde w(z) + w(z))\,dz \\ &= -(\Delta(\tilde w) + \Delta(w))\frac{\sqrt{2}r^{\frac32}}{3\pi} + o(r^{\frac32}) = -2\Delta(w)\frac{\sqrt{2}r^{\frac32}}{3\pi} + o(r^{\frac32}). \end{aligned}$$

In each step $w \mapsto \tilde w$, $\alpha \mapsto \tilde\alpha = \alpha - 1$ the number of strings decreases by $-2\Delta(w) \geq 2$, and after a finite number of steps we end up with $\alpha = \frac12$ and

$$w'' = \tfrac12 + zw + 2w^2. \tag{6.39}$$

We note that $\alpha \in \frac12 + \mathbb{N}_0$ is necessary for w to be sub-normal with $\Delta(w) < 0$, and we have to prove that sub-normal solutions to (6.39) also satisfy $w' = \frac12 z + w^2$. Assuming the contrary, define a meromorphic function $u \not\equiv 0$ locally by

$$w' - w^2 - \tfrac12 z = -u^2.$$

Exercise 6.43 From (6.39) deduce $(z + 2w^2 - 2w')w = -2uu'$, $w = -u'/u$, and $u'' = -\frac12 zu + u^3$. To prove that u is meromorphic, transform the differential equation for u into $v'' = zv + 2v^3$.

Solution $v(z) = \kappa u(\sqrt{2}\kappa z)$, $\kappa^3 = -\frac12\sqrt{2}$.

By the first part of the proof, the function v in Exercise 6.43 is not sub-normal in contrast to $T(r,v) = T(\sqrt[3]{2}r, u) = O(T(\sqrt[3]{2}r, w)) = O(r^{\frac32})$. Thus $w' - w^2 - \frac12 z = -u^2 \equiv 0$ holds, and Theorem 6.15 is completely proved.

Exercise 6.44 Theorem 6.15 also provides a new proof of **Theorem 21.1** in [60]. *Second transcendents satisfying some first-order algebraic differential equation are Airy solutions.* Prove also that first transcendents do *not* solve any first-order algebraic differential equation. (**Hint.** Assuming $P(z,w,w') = 0$, prove that $\mathfrak{w} = \lim_{h_n\to\infty} w_{h_n}$ satisfies $\prod_{\nu=1}^n(\mathfrak{w}' - 2\epsilon_\nu \mathfrak{w}^{\frac32} + \mathfrak{c}_\nu) = 0$ with $\epsilon_\nu^2 = 1$, in contrast to $\mathfrak{w}'^2 = 4\mathfrak{w}^3 + 2\mathfrak{w} - 2\mathfrak{c}$.)

Exercise 6.45 The solutions to $y' = \frac12 z + y^2$ generically have three strings of poles, one along each Stokes ray $\arg z = 2\nu\frac{\pi}{3}$, with residue -1 ($\Delta(y) = -3$). Prove that the sub-normal solutions to $w'' = k + \frac12 + zw + 2w^3$ ($k \in \mathbb{N}$) generically have k strings of poles with residue 1 and $k+1$ strings with residue -1 along $\arg z = 2\nu\frac{\pi}{3}$; 'generically' means 'up to three exceptional solutions', in which case the strings of poles are asymptotic to just one Stokes ray ($\Delta(w) = -1$). In any case, $w = R_k(z,y)$ holds with $\deg_y R_k = 2k+1$, and w solves some first-order equation of degree $2k+1$ with respect to w'. Prove also that any two strings of poles $\sigma = (p_j)$ and $\tilde\sigma = (\tilde p_j)$ absent themselves from each other: $\lim_{j\to\infty} |p_j|^{\frac12}\operatorname{dist}(p_j, \tilde\sigma) = \infty$.

Remark 6.13 The special Bäcklund transformation $w \leftrightarrow v$, cf. [60] Thm. 19.2, maps Airy solutions w to $v = 0$ (Airy solutions come from out of nowhere). It is worth emphasising the role of the first integral V and the Yosida class $\mathfrak{Y}_{\frac{1}{2},\frac{1}{2}}$. While $w \in \mathfrak{Y}_{\frac{1}{2},\frac{1}{2}}$ holds for *every* second transcendent, $V \in \mathfrak{Y}_{\frac{1}{2},\frac{1}{2}}$ is true *if and only if* w is sub-normal. In any case V certainly belongs to some Yosida class $\mathfrak{Y}_{\lambda,\mu}$ with $\frac{1}{2} \le \lambda = \mu \le 2$ ($\mu = \lambda$ is necessary since V has simple poles with residue -1), this implying $n(r, w) = O(r^{\lambda+1})$. However, re-scaling in $\mathfrak{Y}_{\lambda,\lambda}$ with $\lambda > \frac{1}{2}$ leads to limit functions either $\mathfrak{w} = \pm 1/\mathfrak{z}$ and $\mathfrak{V} = -1/\mathfrak{z}$ or else $\mathfrak{w} = 0$ and $\mathfrak{V} =$ const, hence is completely useless.

6.6.2 *The Deficiency of Zero of Second Transcendents*

For any second transcendent, standard arguments from Nevanlinna theory yield $m\big(r, \frac{1}{w-c}\big) = O(\log r)$ and $\delta(c, w) = 0$ if $|\alpha| + |c| > 0$; in case of $\alpha = c = 0$ the best known estimate is $\delta(0, w) \le \frac{1}{2}$, see [60]. We are now able to settle this case completely.

Theorem 6.16 *Second transcendents have deficiency* $\delta(0, w) = 0$.

Proof Of course, the statement is relevant only if $\alpha = 0$. We use the special transformation of the preceding proof (with w and v interchanged) to switch from $w'' = zw + 2w^3$ to $v'' = \frac{1}{2} + zv + 2v^3$. The poles and zeros of w coincide with the poles of v with residue 1 and -1, or *vice versa*. For a proof of $n(r, 1/w) = n(r, w) + O(r^{\frac{3}{2}})$ and $N(r, 1/w) = N(r, w) + O(r^{\frac{3}{2}})$ we refer to Exercise 6.46 below. Thus $N(r, 1/w) = T(r, w) + o(T(r, w))$ holds, at least on some sequence $r = r_n \to \infty$, since w is not sub-normal. This proves $m(r_n, 1/w) = o(T(r_n, w))$ and $\delta(0, w) = 0$.

Exercise 6.46 Let v be any second transcendent and denote by $n_\epsilon(r, v)$ the number of poles on $|z| \le r$ with residue ϵ. Prove that $|n_+(r, v) - n_-(r, v)| = O(r^{\frac{3}{2}})$ is always true, and $N_-(r, v) \le 2N_+(r, v)$ holds except when v satisfies $v' = \frac{1}{2}z + v^2$. (**Hint.** Use $n_+(r, v) - n_-(r, v) = \frac{1}{2\pi i}\int_{\Gamma_r} v(z)\, dz$, where Γ_r is a simple closed curve like in Exercise 4.17. Consider $\phi = v' - \frac{1}{2}z - v^2$ at poles with residue 1 and residue -1, respectively, and apply the First Main Theorem whenever possible.)

Exercise 6.47 Though our focus is on *transcendental* solutions (and not on rational solutions and other *Special Functions* results) we note that the method of proof of Theorem 6.15 also applies to rational solutions. Necessary for their occurrence is that $\alpha^2 \in \mathbb{N}_0$, hence also $(\alpha + k)^2 \in \mathbb{N}_0$ for every $k \in \mathbb{Z}$. Prove that rational solutions exist if and only if $\alpha \in \mathbb{Z}$, exactly one for each α. They are obtained by repeated application of Bäcklund transformations to the trivial solution to $w'' = zw + 2w^3$.

Exercise 6.48 Let ϕ be the unique rational solution to $w'' = \alpha + zw + 2w^3$ ($\alpha \in \mathbb{N}$), w any transcendental solution, and B the Bäcklund transformation (6.16) which transforms α into $\alpha - 1$. Then $\tilde{\phi} = B[\phi]$ is the unique rational solution to $\tilde{w}'' = \alpha - 1 + z\tilde{w} + 2\tilde{w}^3$, and $\tilde{w} = B[w]$ is a transcendental solution. Prove

$$m\Big(r, \frac{\tilde{w} - \tilde{\phi}}{w - \phi}\Big) = O(\log r),$$

and deduce $\delta(\phi, w) = 0$ from Theorem 6.16 by induction on α.
(**Hint.** Prove $\frac{\tilde{w}-\tilde{\phi}}{w-\phi} = P\big(z, w, \tilde{w}, \frac{w'-\phi'}{w-\phi}\big)$ and $\frac{w-\phi}{\tilde{w}-\tilde{\phi}} = \tilde{P}\big(z, w, \tilde{w}, \frac{\tilde{w}'-\tilde{\phi}'}{\tilde{w}-\tilde{\phi}}\big)$, where P and $\tilde{P}$ are rational in z and polynomials in the other variables.

6.6.3 *Sub-normal Solutions to System (IV) and Painlevé (IV)*

Sub-normal solutions to System (IV) and Painlevé (IV) are defined by the growth condition $T(r,x)+T(r,y) = O(r^2)$ and $T(r,w) = O(r^2)$, respectively. The analogue to Theorem 6.15 and the main result of this section will be Theorem 6.17 below. To be prepared for the proof we will start with some exercises. Here (x, y) is any solution to System (IV) and $(\tilde{x}, \tilde{y}) = \mathsf{B}_1(x, y)$ (see (6.6)) denotes its Bäcklund transform.

Exercise 6.49 Prove that $x \sim -\tau z + \cdots$ and $y \sim -\bar{\tau} z + \cdots$ ($\tau^3 = 1$) on some sector S implies $\tilde{x} \sim -\bar{\tau} z + \cdots$ and $\tilde{y} \sim -\tau z + \cdots$ on S.

Exercise 6.50 Prove that the poles of (x, y) with residues $(e^{\pm 2\pi i/3}, -e^{\mp 2\pi i/3})$ are also poles of $(\tilde{x}, \tilde{y})$ with the same residues, while $(\tilde{x}, \tilde{y})$ is regular at poles of (x, y) with residues $(1, -1)$. It is known that $v = x + y - z$ vanishes at poles of (x, y) with residues $(1, -1)$. Prove that the zeros of v different from those coincide with the poles of $(\tilde{x}, \tilde{y})$ with residues $(1, -1)$.

Exercise 6.51 Prove that solutions having no poles with residues $(\omega, -\bar{\omega})$ for some ω with $\omega^3 = 1$ are Weber–Hermite solutions of order $\mathsf{ord}(x, y) \le 1$.

Theorem 6.17 *For every transcendental solution* (x, y) *to System* (IV) *the following statements are equivalent.*

- *(x, y) is sub-normal;*
- *(x, y) is a Weber–Hermite solution;*
- *x and y are algebraically dependent over $\mathbb{C}(z)$;*
- *x and y separately satisfy first-order algebraic differential equations;*
- $V = H - \frac{1}{3}z^3 \in \mathfrak{Y}_{1,1}$;
- *V satisfies some first-order algebraic differential equation.*

Proof Weber–Hermite solutions are sub-normal, and the subsequent statements in Theorem 6.17 hold. Conversely, each of the above conditions implies that (x, y) is sub-normal. Again we just have to prove that sub-normal solutions are also Weber–Hermite solutions. Sub-normal solutions have cluster set $\mathfrak{C}(x, y) \subset \{0, 1\}$, and the set of poles $\mathfrak{P}$ consists of finitely many strings. Although all examples suggest that $\mathfrak{C}(x, y) = \{1\}$, we also have to discuss the case
a) $\mathfrak{C}(x, y) = \{0\}$. The strings of poles (if any) are distributed along the rays $\arg z = (2\nu + 1)\frac{\pi}{4}$, and x, y, and particularly $H \sim -\alpha\beta/z + \cdots$ have asymptotic expansions on the sectors between these rays. Again $F(z) = e^{\int H(z)\,dz}$ is an entire function of finite order, and the Phragmén–Lindelöf Principle applied to $F(z)z^{\alpha\beta}$ on the small sectors $|\arg z - (2\nu + 1)\frac{\pi}{4}| < \epsilon$ shows that F is a polynomial (of degree $-\alpha\beta$), and $H = F'/F$, x and y are rational functions.
b) $\mathfrak{C}(x, y) = \{1\}$. The idea of the proof is simple, while the technical part itself is involved. Third roots of unity ω, ρ, and τ will occur in different contexts: $\omega = e^{2\pi i/3}$ is definitely fixed, ρ and $-\bar\rho$ represent residues, and $-\tau z$ and $-\bar\tau z$ are principal terms in asymptotic expansions. The poles of (x, y) are arranged in $(\pm\frac{2\pi}{\sqrt{3}}, 1)$-strings which now are asymptotic to the rays $\arg z = \nu\frac{\pi}{2}$. We denote by $\mathfrak{P}^\rho$ the set of non-zero poles with $\mathrm{res}_p(x, y) = (\rho, -\bar\rho)$. On each sector $\Sigma_\nu : (\nu - 1)\frac{\pi}{2} < \arg z < \nu\frac{\pi}{2}$, $x \sim -\tau_\nu z + \lambda_\nu z^{-1} + \cdots$ and $y \sim -\bar\tau_\nu z + \kappa_\nu z^{-1} + \cdots$ $(\tau_\nu^3 = 1)$ have asymptotic expansions, cf. (6.33). This will be expressed by the symbol $\begin{array}{c|c} \tau_2 & \tau_1 \\ \hline \tau_3 & \tau_4 \end{array}$. In combination with the asymptotic expansions, the Residue Theorem yields

$$\sum_{\rho^3=1} \rho n(r, \mathfrak{P}^\rho) = \frac{1}{2\pi i}\int_{\Gamma_r} x(z)\,dz = -\frac{r^2}{2\pi i}\sum_{\nu=1}^{4}(-1)^\nu\tau_\nu + o(r^2) \qquad (6.40)$$

(for the construction of Γ_r, see Exercise 4.17). Let $s_x(\rho)$ denote the number of strings of x with residue ρ. Taking the real- and imaginary part we obtain

$$\begin{aligned} s_x(1) - \tfrac{1}{2}(s_x(\omega) + s_x(\bar\omega)) &= -\tfrac{2}{\sqrt{3}}\sum_{\nu=1}^{4}(-1)^\nu \mathrm{Im}\,\tau_\nu \\ s_x(\omega) - s_x(\bar\omega) &= \tfrac{4}{3}\sum_{\nu=1}^{4}(-1)^\nu \mathrm{Re}\,\tau_\nu. \end{aligned} \qquad (6.41)$$

To proceed we need two technical results.

Lemma 6.6 *Suppose* $s_x(1) > \frac{1}{2}(s_x(\omega) + s_x(\bar\omega))$ *with* $\omega = e^{2\pi i/3}$ *and* $\beta = \alpha + 1$ *holds. Then* $x + y - z$ *vanishes identically.*

Proof Since $\alpha - \beta + 1 = 0$, the function $v = x + y - z$ vanishes at least twice at poles of (x, y) with residues $(1, -1)$. If v does not vanish identically, it follows from $m(r, v) = O(\log r)$ and Nevanlinna's First Main Theorem that

$$2N(r, \mathfrak{P}^1) \le N(r, 1/v) \le T(r, v) + O(1) = N(r, \mathfrak{P}^\omega) + N(r, \mathfrak{P}^{\bar\omega}) + O(\log r),$$

holds, hence also $2s_x(1) \le s_x(\omega) + s_x(\bar\omega)$, against our hypothesis.

Lemma 6.7 *For any solution with symbol* $\dfrac{\tau_2|\tau_1}{\tau_3|\tau_4}$, $\sum_{\nu=1}^{4}(-1)^\nu \operatorname{Im}\tau_\nu$ *is non-zero.*

Proof The sum vanishes if and only if $\tau_1 = \tau_2$ and $\tau_3 = \tau_4$ (or $\tau_2 = \tau_3$ and $\tau_4 = \tau_1$). Again we consider some entire function associated with (x, y), this time

$$F(z) = e^{-\frac{z^4}{12}} e^{\int H(z)\,dz}.$$

From $H(z) \sim z^3/3 + 2az + a_1/z + \cdots$ and $H(z) \sim z^3/3 + 2bz + b_1/z + \cdots$ on the upper and lower half-plane ($2a = -\tau_1\alpha - \bar\tau_1\beta$ and $2b = -\tau_3\alpha - \bar\tau_3\beta$) it follows that

$$F(z) = Ae^{az^2} z^{a_1}(1 + o(1)) \quad \text{and} \quad F(z) = Be^{bz^2} z^{b_1}(1 + o(1))$$

holds as $z \to \infty$, uniformly on $\epsilon \le \arg z \le \pi - \epsilon$ and $-\pi + \epsilon \le \arg z \le -\epsilon$, respectively. In particular, the upper and lower half-plane are 'pole-free'. Since F has finite order, Corollary 1.1 applies to $f(z) = F(\sqrt{z})$ ($\sqrt{z} > 0$ if $z > 0$) on $|\arg z| < \frac{\pi}{2}$, say, and yields $\operatorname{Re} a = \operatorname{Re} b$ and $\operatorname{Im} a \le \operatorname{Im} b$. The same argument, however, also applies to $(F(-z), b, a, b_1, a_1)$ in place of $(F(z), a, b, a_1, b_1)$. The second part of Corollary 1.1 then yields $a = b$, $a_1 = b_1$, $A = B$, and $F(z) = Ae^{az^2} z^{a_1}(1 + o(1))$ as $z \to \infty$ on $|\arg z| \le \epsilon$ and also on $|\arg z - \pi| \le \epsilon$, hence F has only finitely many zeros along the real axis. Thus $H = F'/F + z^3/3$ has only finitely many poles and is a rational function in contrast to our general assumption. This proves the lemma.

We claim that to any sub-normal solution satisfying $\bar\rho x + \rho y - z \not\equiv 0$ for every third root of unity there exists some sub-normal solution $(\tilde{x}, \tilde{y})$ that has fewer strings of poles than (x, y), and this will prove Theorem 6.17.[6] To reach our claim we note that by (6.41) and Lemma 6.7 there exists a ρ_1 such that $\frac{1}{2}(s_x(\rho_2) + s_x(\rho_3)) < s_x(\rho_1)$ holds. Set

$$(\hat{x}, \hat{y}) = (\bar\rho_1 x, \rho_1 y) \quad \text{and} \quad (\tilde{x}, \tilde{y}) = \mathsf{B}_1(\hat{x}, \hat{y});$$

the latter is well-defined since $s_{\hat{x}}(1) > \frac{1}{2}(s_{\hat{x}}(\omega) + s_{\hat{x}}(\bar\omega))$ and $\hat{x} + \hat{y} - z \not\equiv 0$. From Exercise 6.28, which says $\tilde{H} - \hat{H} = \tilde{x} - \hat{x}$, and $n(r, x) = n(r, \hat{x})$ we obtain

$$n(r, \tilde{x}) - n(r, x) = \frac{1}{2\pi i}\int_{\Gamma_r} (\tilde{H}(z) - \hat{H}(z))\,dz = \frac{1}{2\pi i}\int_{\Gamma_r} (\tilde{x}(z) - \hat{x}(z))\,dz.$$

[6] Note that $x + y - z \equiv 0$ already yields $x' = -\alpha - z^2 + zx - x^2$.

Order one:

$w \sim -\bar\omega z + \cdots$
◇ ◇ ◇ ◇ ◇ ◇
$w = -z + o(|z|)$
● ● ● ● ● ●
$w \sim -\omega z + \cdots$

$\begin{array}{c|c} \omega & \bar\omega \\ \hline \bar\omega & \omega \end{array}$

Order two:

$w \sim -\omega z + \cdots$
○ ○ ○ ○ ○ ○
$w = -\bar\omega z + o(|z|)$
◇ ◇ ◇ ◇ ◇ ◇
$w = -z + o(|z|)$
● ● ● ● ● ●
$w = -\omega z + o(|z|)$
○ ○ ○ ○ ○ ○
$w \sim -\bar\omega z + \cdots$

$\begin{array}{c|c} \bar\omega & \omega \\ \hline \omega & \bar\omega \end{array}$

Order three:

$w \sim -\bar\omega z + \cdots$
◇ ◇ ◇ ◇ ◇ ◇
$w = -z + o(|z|)$
● ● ● ● ● ●
$w = -\omega z + o(|z|)$
○ ○ ○ ○ ○ ○
$w = -\bar\omega z + o(|z|)$
◇ ◇ ◇ ◇ ◇ ◇
$w = -z + o(|z|)$
● ● ● ● ● ●
$w \sim -\omega z + \cdots$

$\begin{array}{c|c} \omega & \bar\omega \\ \hline \bar\omega & \omega \end{array}$

Fig. 6.3 Asymptotics and distribution of poles ○ ● ◇ with residues $(1,-1)$, $(\omega,-\bar\omega)$, and $(\bar\omega,-\omega)$ ($\omega = e^{2\pi i/3}$) along the positive real axis of generic sub-normal solutions of order one, two, and three (*from left to right*). How to continue?

Since $(\hat x,\hat y)$ and $(\tilde x,\tilde y)$ have symbols $\begin{array}{c|c}\tau_2&\tau_1\\ \hline \tau_3&\tau_4\end{array}$ and $\begin{array}{c|c}\bar\tau_2&\bar\tau_1\\ \hline \bar\tau_3&\bar\tau_4\end{array}$, $\tilde x - \hat x \sim -(\bar\tau_\nu - \tau_\nu)z + \cdots = 2i(\operatorname{Im}\tau_\nu)z + \cdots$ holds on Σ_ν. This implies

$$n(r,\tilde x) - n(r,x) = -(2s_{\hat x}(1) - s_{\hat x}(\omega) - s_{\hat x}(\bar\omega))\frac{\sqrt{3}r^2}{4\pi} + o(r^2),$$

proving our claim and also Theorem 6.17. ☕

Analysis of the Proof Starting with any sub-normal solution (x,y) we obtain after finitely many steps some sub-normal solution (x_0,y_0) such that $x_0 + y_0 - z \equiv 0$, $\alpha_0 - \beta_0 + 1 = 0$, $x_0' = -\alpha_0 - z^2 + zx_0 - x_0^2$, and $y_0' = \beta_0 + z^2 - zy_0 + y_0^2$ hold (cf. Exercise 6.3). Each step requires some trivial Bäcklund transformation $\mathsf{M}_\rho : (x,y) \mapsto (\rho x, \bar\rho y) = (\hat x,\hat y)$ followed by $\mathsf{B}_1 : (\hat x,\hat y) \mapsto (\tilde x,\tilde y)$; this may be abbreviated by writing

$$(x,y) \xrightarrow{\mathsf{M}_\rho} (\hat x,\hat y) \xrightarrow{\mathsf{B}_1} (\tilde x,\tilde y). \tag{6.42}$$

The very last step only requires some M_ρ to achieve $x_0 + y_0 - z \equiv 0$. Of course the sequence (6.42) may be reversed (Fig. 6.3).

Exercise 6.52 Prove that generic sub-normal solutions have

- $k+1$ strings with residues $(\omega_1,-\bar\omega_1)$, and k strings with residues $(\omega_2,-\bar\omega_2)$ and $(\omega_3,-\bar\omega_3)$, respectively in each Stokes direction if $\mathsf{ord}(x,y) = 2k$;

– k strings with residues $(\omega_1, -\bar\omega_1)$ and $(\omega_2, -\bar\omega_2)$, respectively, and $k-1$ strings with residues $(\omega_3, -\bar\omega_3)$ (not necessarily in the same order as in the first case) in each Stokes direction if $\mathsf{ord}(x, y) = 2k - 1$

Exercise 6.53 Prove that generic sub-normal solutions with

– $\mathsf{ord}(x, y) = 0$ and $s_x(1) = 4$ have symbol $\begin{array}{c|c} \bar\omega & \omega \\ \hline \omega & \bar\omega \end{array}$ and $s_x(\omega) = s_x(\bar\omega) = 0$;

– $\mathsf{ord}(x, y) = 1$ and $s_x(1) = 0$ have symbol $\begin{array}{c|c} \omega & \bar\omega \\ \hline \bar\omega & \omega \end{array}$ and $s_x(\omega) = s_x(\bar\omega) = 4$;

$(\omega = e^{2\pi i/3})$. For exceptional (non-generic) solutions one ω resp. $\bar\omega$ has to be replaced with $\bar\omega$ resp. ω, and the number 4 by 2.

Generic symbols have 'diagonal structure', which simplifies matters significantly. Obviously, the method of proof provides the 'steepest descent', that is, the number of non-trivial steps equals the order $\mathsf{ord}(x, y)$, which now may be assigned to any sub-normal solution.

For the convenience of the reader we will state the analogue to Theorem 6.17 for fourth transcendents. In this form it was proved by Classen [26]. Here Weber–Hermite solutions are obtained from the solutions to the Weber–Hermite equations $w' = \sqrt{-2\beta} \pm (2zw + w^2)$ by repeated application of the Bäcklund transformations $w \mapsto \frac{w' - \sqrt{-2\beta} - 2zw - w^2}{2w}$ and $w \mapsto -\frac{w' + \sqrt{-2\beta} + 2zw + w^2}{2w}$ (for both branches of $\sqrt{-2\beta}$) and the elementary transformations $w \mapsto -iw(iz)$ and $w \mapsto \overline{w(\bar z)}$.

Theorem 6.18 *For every fourth transcendent the following statements are equivalent.*

– *w is sub-normal;*
– *w is a Weber–Hermite solution;*
– *w satisfies some first-order algebraic differential equation;*
– *$W \in \mathfrak{Y}_{1,1}$ (first integral);*
– *w and W are algebraically dependent;*
– *W satisfies some first-order algebraic differential equation.*

6.6.4 The Deficiency of Zero of Fourth Transcendents

Just like for second transcendents with parameter $\alpha = 0$, the Wittich–Mokhon'ko method does not apply to estimate $m(r, 1/w)$ or even to compute the deficiency $\delta(0, w)$ for fourth transcendents with parameter $\beta = 0$. The same problem occurs for non-trivial solutions to the differential equation

$$2vv'' = v'^2 - v^4 - 4zv^3 - (3z^2 + 4\alpha + 2)v^2 \qquad (\mathrm{I}\Lambda')$$

which is derived from our System (IV) with $\alpha-\beta+1=0$ via $v=x+y-z$ (note that the parameters α and β have different meanings in Painlevé (IV) and System (IV) resp. ($\mathsf{I}\Lambda'$)). Since $v=x+y-z$ has simple poles at the poles of (x,y) with residues $(\omega,-\bar\omega)$ and $(\bar\omega,-\omega)$ $(\omega=e^{2\pi i/3})$, respectively, and vanishes at least twice at poles with residues $(1,-1)$,

$$n(r,v)=\tfrac{2}{3}n(r,\mathfrak{P})+O(r^2)=2n(r,\mathfrak{P}^1)+O(r^2)\le n(r,1/v)+O(r^2)$$

and $T(r,v) = N(r,v)+O(\log r) \le N(r,1/v)+O(r^2)$ holds. Here we used $n(r,\mathfrak{P}^\rho)=\frac{1}{3}N(r,\mathfrak{P})+O(r^2)$ for every third root of unity, which is true for every solution and follows from the general form of (6.40), namely $\sum_{\rho^3=1}\rho n(r,\mathfrak{P}^\rho)=O(r^2)$. If v is not sub-normal, $r_k^2=o(T(r_k,v))$ holds on some sequence $r_k\to\infty$, and so $\delta(0,v)=0$. This was the first part of the proof of the following, rather unexpected

Theorem 6.19 *The deficiency $\delta(0,v)$ of any non-trivial solution to ($\mathsf{I}\Lambda'$) is zero, except when v is sub-normal of even order $\mathsf{ord}(v)=2k$. Then $\delta(0,v)=\frac{1}{2k+1}$ holds.*

Proof We may restrict ourselves to sub-normal solutions $v=x+y-z\not\equiv 0$ with parameters satisfying $\alpha-\beta+1=0$. Then v has poles at poles of (x,y) with residues $(\omega,-\bar\omega)\ne(1,-1)$, and double zeros at poles with residues $(1,-1)$, while from

$$v'=(x(z)-y(z))v\quad (v\not\equiv 0)$$

and the uniqueness part of Cauchy's Theorem it follows that v is non-zero otherwise. This implies (note that $s_x(\rho)$ denotes the number of strings with residue ρ)

$$N(r,1/v)\sim\frac{2s_x(1)}{s_x(\omega)+s_x(\bar\omega)}T(r,v)\quad\text{and}\quad \delta(0,v)=\frac{s_x(\omega)+s_x(\bar\omega)-2s_x(1)}{s_x(\omega)+s_x(\bar\omega)}$$

$(\omega=e^{2\pi i/3})$, and requires $2s_x(1)\le s_x(\omega)+s_x(\bar\omega)$, hence

- $s_x(1)=\mu k$ and $s_x(\omega)+s_x(\bar\omega)=\mu(2k+1)$ if $\mathsf{ord}(v)=2k$ is even, and
- $s_x(\omega)=s_x(\bar\omega)=\mu k$ and $s_x(1)=\mu(k-1)$ if $\mathsf{ord}(v)=2k-1$ is odd;

generically the factor μ equals 4, and equals 2 only in special (truncated) cases. In any case, this yields $\delta(0,v)=\frac{1}{2k+1}$ and $\delta(0,v)=\frac{1}{k}$, respectively. We need to check in which case $\alpha-\beta+1=0$ can hold, and will restrict ourselves to the generic case.

a) $\mathsf{ord}(v)=2k\ge 2$. We may assume $s_x(\omega)=4(k+1)$ and $s_x(1)=s_x(\bar\omega)=4k$. Then x has the symbol $\begin{array}{c|c}1&\bar\omega\\ \hline \bar\omega&1\end{array}$, and the step

$$\begin{array}{c|c}1&\bar\omega\\ \hline \bar\omega&1\end{array}\xrightarrow{\mathsf{M}_\omega}\begin{array}{c|c}\bar\omega&\omega\\ \hline \omega&\bar\omega\end{array}\xrightarrow{\mathsf{B}_1}\begin{array}{c|c}\omega&\bar\omega\\ \hline \bar\omega&\omega\end{array}\xrightarrow{\mathsf{M}_\omega}\begin{array}{c|c}1&\omega\\ \hline \omega&1\end{array}\xrightarrow{\mathsf{B}_1}\begin{array}{c|c}1&\bar\omega\\ \hline \bar\omega&1\end{array}$$

$$(\alpha, \beta) \mapsto (\bar{\omega}\alpha, \omega\beta) \mapsto (\omega\beta - 1, \bar{\omega}\alpha + 1) \mapsto (\beta - \bar{\omega}, \alpha + \omega) \mapsto (\alpha + \omega - 1, \beta - \bar{\omega} + 1)$$

has to be applied k times to obtain the parameters $\alpha + k(\omega - 1)$ and $\beta - k(\bar{\omega} - 1)$. In the final step $\begin{array}{c|c} 1 & \bar{\omega} \\ \hline \bar{\omega} & 1 \end{array} \xrightarrow{\mathsf{M}_\omega} \begin{array}{c|c} \bar{\omega} & \omega \\ \hline \omega & \bar{\omega} \end{array}$ we get the parameters $\alpha_0 = \bar{\omega}\alpha + k(1 - \bar{\omega})$ and $\beta_0 = \omega\beta - k(1 - \omega)$; they have to satisfy $\alpha_0 - \beta_0 + 1 = 0$. In terms of α and β this means $\bar{\omega}\alpha - \omega\beta + 3k + 1 = 0$, which in combination with $\alpha - \beta + 1 = 0$ yields $\alpha = \bar{\omega} - ik\sqrt{3}$ and $\beta = -\omega - ik\sqrt{3}$. Since the procedure may be reversed, the sub-normal solutions of even order $\mathsf{ord}(v) = 2k$ with corresponding parameters $\alpha = \bar{\omega} - ik\sqrt{3}$ and $\beta = -\omega - ik\sqrt{3}$ have deficiency $\delta(0, v) = \frac{1}{2k+1}$.
b) $\mathsf{ord}(v) = 2k - 1 \geq 1$. Then v has a predecessor $\tilde{v}$ of even order $\mathsf{ord}(\tilde{v}) = 2k$ and parameters $\tilde{\alpha} = \omega\beta + \omega$ and $\tilde{\beta} = \bar{\omega}\alpha - \bar{\omega}$. After $2k$ steps we arrive at some solution of order 0 with parameters $\alpha_0 = \bar{\omega}\tilde{\alpha} + k(1 - \bar{\omega}) = \beta + 1 + k(1 - \bar{\omega})$ and $\beta_0 = \omega\tilde{\beta} - k(1 - \omega) = \alpha - 1 - k(1 - \omega)$. Again $\alpha_0 - \beta_0 + 1 = 0$, thus $\beta - \alpha + 3k + 3 = 0$ is necessary, which, however, is incompatible with $\alpha - \beta + 1 = 0$.

Remark 6.14 The parameters $(\alpha, \beta) = (\bar{\omega} - ik\sqrt{3}, -\omega - ik\sqrt{3})$ in System (IV) and (IΛ′) correspond to $(\alpha, \beta) = (2k + 1, 0)$ in Painlevé (IV).

Example 6.9 ([60], p. 127) Sub-normal solutions to

$$2ww'' = w'^2 + 3w^4 + 8zw^3 + 4(z^2 - 3)w^2 \quad (\alpha = 3,\ \beta = 0)$$

also satisfy $w'^3 + w(w + 2z)w'^2 - w(w(w + 2z)^2 - 16w + 32)w' + P_0(z, w) = 0$ with $P_0(z, w) = -w^2(w^4 + 6zw^3 + (12z^2 - 16)w^2 + 8z(z^2 - 6)w - 32(z^2 - 1))$ and deficiency $\delta(0, w) = \frac{1}{3}$.

References

1. L. Ahlfors, Beiträge zur Theorie der meromorphen Funktionen, in *VII Congrés des Mathematiciens Scandinavia* (Oslo, 1929), pp. 84–88
2. L. Ahlfors, *Complex Analysis* (McGraw-Hill, New York, 1979)
3. S. Bank, On zero-free regions for solutions of *n*th order linear differential equations. Comment. Math. Univ. St. Pauli **36**, 199–213 (1987)
4. S. Bank, A note on the zeros of solutions of $w'' + P(z)w = 0$, where P is a polynomial. Appl. Anal. **25**, 29–41 (1988)
5. S. Bank, A note on the location of complex zeros of solutions of linear differential equations. Complex Variables **12**, 159–167 (1989)
6. S. Bank, R. Kaufman, On meromorphic solutions of first order differential equations. Comment. Math. Helv. **51**, 289–299 (1976)
7. S. Bank, R. Kaufman, On the order of growth of meromorphic solutions of first-order differential equations. Math. Ann. **241**, 57–67 (1979)
8. S. Bank, R. Kaufman, On Briot–Bouquet differential equations and a question of Einar Hille. Math. Z. **177**, 549–559 (1981)
9. S. Bank, G. Frank, I. Laine, Über die Nullstellen von Lösungen linearer Differentialgleichungen. Math. Z. **183**, 355–364 (1983)
10. A. Beardon, T.W. Ng, Parametrizations of algebraic curves. Ann. Acad. Sci. Fenn. **31**, 541–554 (2006)
11. W. Bergweiler, On a theorem of Gol'dberg concerning meromorphic solutions of algebraic differential equations. Complex Variables **37**, 93–96 (1998)
12. W. Bergweiler, Rescaling principles in function theory, in *Proceedings of the International Conference on Analysis and Its Applications*, 14 p. (2000)
13. W. Bergweiler, Bloch's principle. Comput. Meth. Funct. Theory (CMFT) **6**, 77–108 (2006)
14. W. Bergweiler, A. Eremenko, On the singularities of the inverse to a meromorphic function of finite order. Rev. Matem. Iberoam. **11**, 355–373 (1995)
15. L. Bieberbach, *Theorie der gewöhnlichen Differentialgleichungen* (Springer, New York, 1965)
16. P. Boutroux, Recherches sur les transcendentes de M. Painlevé et l'étude asymptotique des équations différentielles du seconde ordre. Ann. École Norm. Supér. **30**, 255–375 (1913); **31**, 99–159 (1914)
17. D.A. Brannan, W.K. Hayman, Research problems in complex analysis. Bull. Lond. Math. Soc. **21**, 1–35 (1989)
18. F. Brüggemann, On the zeros of fundamental systems of linear differential equations with polynomial coefficients. Complex Variables **15**, 159–166 (1990)

N. Steinmetz, *Nevanlinna Theory, Normal Families, and Algebraic Differential Equations*, Universitext, DOI 10.1007/978-3-319-59800-0

19. F. Brüggemann, On solutions of linear differential equations with real zeros; proof of a conjecture of Hellerstein and Rossi. Proc. Am. Math. Soc. **113**, 371–379 (1991)
20. F. Brüggemann, Proof of a conjecture of Frank and Langley concerning zeros of meromorphic functions and linear differential polynomials. Analysis **12**, 5–30 (1992)
21. H. Cartan, Un nouveau théorème d'unicité relatif aux fonctions méromorphes. C. R. Acad. Sci. Paris **188**, 301–303 (1929)
22. H. Cartan, Sur les zéros des combinaisons linéaires de p fonctions holomorphes données. Math. Cluj **7**, 5–31 (1933)
23. H. Chen, Y. Gu, An improvement of Marty's criterion and its applications. Sci. China Ser. A **36**, 674–681 (1993)
24. C.T. Chuang, Une généralisation d'une inégalité de Nevanlinna. Sci. Sinica **13**, 887–895 (1964)
25. P. Clarkson, J. McLeod, Integral equations and connection formulae for the Painlevé equations, in *Painlevé Transcendents, Their Asymptotics and Physical Applications*, ed. by P. Winternitz, D. Levi (Springer, New York, 1992), pp. 1–31
26. C. Classen, Subnormale Lösungen der vierten Painlevéschen Differentialgleichung, Ph.D. thesis, TU Dortmund (2015)
27. J. Clunie, The derivative of a meromorphic function. Proc. Am. Math. Soc. **7**, 227–229 (1956)
28. J. Clunie, On integral and meromorphic functions. J. Lond. Math. Soc. **37**, 17–27 (1962)
29. J. Clunie, The composition of entire and meromorphic functions, in *Mathematical Essays Dedicated to A.J. Macintyre* (Springer, New York, 1970), pp. 75–92
30. J. Clunie, W.K. Hayman, The spherical derivative of integral and meromorphic functions. Comment. Math. Helv. **40**, 117–148 (1966)
31. T.P. Czubiak, G.G. Gundersen, Meromorphic functions that share pairs of values. Complex Variables **34**, 35–46 (1997)
32. W. Doeringer, Exceptional values of differential polynomials. Pac. J. Math. **98**, 55–62 (1982)
33. A. Edrei, W.H.J. Fuchs, S. Hellerstein, Radial distribution of the values of a meromorphic function. Pac. J. Math. **11**, 135–151 (1961)
34. A. Eremenko, Meromorphic solutions of algebraic differential equations. Russ. Math. Surv. **37**, 61–95 (1982)
35. A. Eremenko, Meromorphic solutions of first-order algebraic differential equations. Funct. Anal. Appl. **18**, 246–248 (1984)
36. A. Eremenko, Normal holomorphic curves from parabolic regions to projective spaces. arXiv:0710.1281v1 (2007)
37. A. Eremenko, Lectures on Nevanlinna Theory (2012, preprint)
38. A. Eremenko, On the second main theorem of Cartan. Ann. Acad. Sci. Fenn. **39**, 859–871 (2014). Correction to the paper "On the second main theorem of Cartan". Ann. Acad. Sci. Fenn. **40**, 503–506 (2015)
39. A. Eremenko, A. Gabrielov, Singular pertubation of polynomial potentials with application to PT-symmetric families. Mosc. Math. J. **11**, 473–503 (2011)
40. A. Eremenko, S. Merenkov, Nevanlinna functions with real zeros. Ill. J. Math. **49**, 1093–1110 (2005)
41. A. Eremenko, M. Sodin, Iteration of rational functions and the distribution of the values of the Poincaré function. J. Sov. Math. **58**, 504–509 (1992)
42. A. Eremenko, L.W. Liao, T.W. Ng, Meromorphic solutions of higher order Briot–Bouquet differential equations. Math. Proc. Camb. Phil. Soc. **146**, 197–206 (2009)
43. S.J. Favorov, Sunyer-i-Balaguer's almost elliptic functions and Yosida's normal functions. J. d'Anal. Math. **104**, 307–340 (2008)
44. A. Fokas, A. Its, A. Kapaev, V. Novokshënov, *Painlevé Transcendents: The Riemann–Hilbert Approach*. Mathematical Surveys and Monographs, vol. 128 (American Mathematical Society, Providence, RI, 2006)
45. G. Frank, Picardsche Ausnahmewerte bei Lösungen linearer Differentialgleichungen. Manuscripta Math. **2**, 181–190 (1970)

46. G. Frank, Über eine Vermutung von Hayman über Nullstellen meromorpher Funktionen. Math. Z. **149**, 29–36 (1976)
47. G. Frank, S. Hellerstein, On the meromorphic solutions of nonhomogeneous linear differential equations with polynomial coefficients. Proc. Lond. Math. Soc. **53**, 407–428 (1986)
48. G. Frank, G. Weissenborn, Rational deficient functions of meromorphic functions. Bull. Lond. Math. Soc. **18**, 29–33 (1986)
49. G. Frank, G. Weissenborn, On the zeros of linear differential polynomials of meromorphic functions. Complex Variables **12**, 77–81 (1989)
50. G. Frank, H. Wittich, Zur Theorie linearer Differentialgleichungen im Komplexen. Math. Z. **130**, 363–370 (1973)
51. M. Frei, Über die Lösungen linearer Differentialgleichungen mit ganzen Funktionen als Koeffizienten. Comment. Math. Helvet. **35**, 201–222 (1961)
52. F. Gackstatter, I. Laine, Zur Theorie der gewöhnlichen Differentialgleichungen im Komplexen. Ann. Polon. Math. **38**, 259–287 (1980)
53. V.I. Gavrilov, The behavior of a meromorphic function in the neighbourhood of an essentially singular point. Am. Math. Soc. Transl. **71**, 181–201 (1968)
54. V.I. Gavrilov, On classes of meromorphic functions which are characterised by the spherical derivative. Math. USSR Izv. **2**, 687–694 (1968)
55. V.I. Gavrilov, On functions of Yosida's class (A). Proc. Jpn. Acad. **46**, 1–2 (1970)
56. A.A. Gol'dberg, On single-valued solutions of first order differential equations (Russian). Ukr. Math. Zh. **8**, 254–261 (1956)
57. A.A. Gol'dberg, I.V. Ostrovskii, *Value Distribution of Meromorphic Functions*. Translations of Mathematical Monographs, vol. 236 (Springer, Berlin, 2008)
58. W.W. Golubew, *Vorlesungen über Differentialgleichungen im Komplexen* [German transl.] (Dt. Verlag d. Wiss. Berlin, 1958)
59. J. Grahl, Sh. Nevo, Spherical derivatives and normal families. J. d'Anal. Math. **117**, 119–128 (2012)
60. V. Gromak, I. Laine, S. Shimomura, *Painlevé Differential Equations in the Complex Plane*. De Gruyter Studies in Mathematical, vol. 28 (Walter de Gruyter, New York, 2002)
61. F. Gross, On the equation $f^n + g^n = 1$. Bull. Am. Math. Soc. **72**, 86–88 (1966). Erratum ibid., p. 576
62. F. Gross, On the equation $f^n + g^n = 1$, II. Bull. Am. Math. Soc. **74**, 647–648 (1968)
63. F. Gross, C.F. Osgood, On the functional equation $f^n + g^n = h^n$ and a new approach to a certain class of more general functional equations. Indian J. Math. **23**, 17–39 (1981)
64. X.-Y. Gu, A criterion for normality of families of meromorphic functions (Chinese). Sci. Sin. Special Issue on Math. **1**, 267–274 (1979)
65. G.G. Gundersen, Meromorphic functions that share three or four values. J. Lond. Math. Soc. **20**, 457–466 (1979)
66. G.G. Gundersen, Meromorphic functions that share four values. Trans. Am. Math. Soc. **277**, 545–567 (1983); Correction to "Meromorphic functions that share four values." Trans. Am. Math. Soc. **304**, 847–850 (1987)
67. G. Gundersen, On the real zeros of solutions of $f'' + A(z)f = 0$, where A is entire. Ann. Acad. Sci. Fenn. **11**, 275–294 (1986)
68. G.G. Gundersen, Meromorphic functions that share three values IM and a fourth value CM. Complex Variables **20**, 99–106 (1992)
69. G. Gundersen, Meromorphic solutions of $f^6 + g^6 + h^6 = 1$. Analysis (München) **18**, 285–290 (1998)
70. G. Gundersen, Solutions of $f'' + P(z)f = 0$ that have almost all real zeros. Ann. Acad. Sci. Fenn. **26**, 483–488 (2001)
71. G. Gundersen, Meromorphic solutions of $f^5 + g^5 + h^5 = 1$. Complex Variables **43**, 293–298 (2001)
72. G. Gundersen, Meromorphic functions that share five pairs of values. Complex Variables Elliptic Equ. **56**, 93–99 (2011)

73. G. Gundersen, W.K. Hayman, The strength of Cartan's version of Nevanlinna theory. Bull. Lond. Math. Soc. **36**, 433–454 (2004)
74. G. Gundersen, E. Steinbart, A generalization of the Airy integral for $f'' - z^n f = 0$. Trans. Am. Math. Soc. **337**, 737–755 (1993)
75. G. Gundersen, N. Steinmetz, K. Tohge, Meromorphic functions that share four or five pairs of values. Preprint (2016)
76. S. Hastings, J. McLeod, A boundary value problem associated with the second Painlevé transcendent and the Korteweg–de Vries equation. Arch. Ration. Mech. Anal. **73**, 31–51 (1980)
77. W.K. Hayman, Picard values of meromorphic functions and their derivative. Ann. Math. **70**, 9–42 (1959)
78. W.K. Hayman, *Meromorphic Functions* (Oxford University Press, Oxford, 1964)
79. W.K. Hayman, The local growth of power series: a survey of the Wiman–Valiron method. Can. Math. Bull. **17**, 317–358 (1974)
80. W.K. Hayman, Waring's Problem für analytische Funktionen. Bayer. Akad. Wiss. Math.-Natur. Kl. Sitzungsber. **1984**, 1–13 (1985)
81. S. Hellerstein, J. Rossi, Zeros of meromorphic solutions of second-order differential equations. Math. Z. **192**, 603–612 (1986)
82. S. Hellerstein, J. Rossi, On the distribution of zeros of solutions of second-order differential equations. Complex Variables Theory Appl. **13**, 99–109 (1989)
83. G. Hennekemper, W. Hennekemper, Picardsche Ausnahmewerte von Ableitungen gewisser meromorpher Funktionen. Complex Variables **5**, 87–93 (1985)
84. E. Hille, *Lectures on Ordinary Differential Equations* (Addison-Wesley, Reading, MA, 1969)
85. E. Hille, *Ordinary Differential Equations in the Complex Domain* (Wiley, New York, 1976)
86. E. Hille, Some remarks on Briot–Bouquet differential equations II. J. Math. Anal. Appl. **65**, 572–585 (1978)
87. E. Hille, Second-order Briot–Bouquet differential equations. Acta Sci. Math. (Szeged) **40**, 63–72 (1978)
88. A. Hinkkanen, I. Laine, Solutions of the first and second Painlevé equations are meromorphic. J. d'Anal. Math. **79**, 345–377 (1999)
89. A. Hinkkanen, I. Laine, Solutions of a modified third Painlevé equation are meromorphic. J. d'Anal. Math. **85**, 323–337 (2001)
90. A. Hinkkanen, I. Laine, The meromorphic nature of the sixth Painlevé transcendents. J. d'Anal. Math. **94**, 319–342 (2004)
91. A. Hinkkanen, I. Laine, Growth results for Painlevé transcendents. Math. Proc. Camb. Phil. Soc. **137**, 645–655 (2004)
92. P.C. Hu, P. Li, C.C. Yang, *Unicity of Meromorphic Mappings* (Kluwer Academic Publishers, Dordrecht/Boston/London, 2003)
93. B. Huang, On the unicity of meromorphic functions that share four values. Indian J. Pure Appl. Math. **35**, 359–372 (2004)
94. E.L. Ince, *Ordinary Differential Equations* (Dover Publications, New York, 1956)
95. A. Its, A. Kapaev, Connection formulae for the fourth Painlevé transcendent; Clarkson–McLeod solution. J. Phys. A: Math. Gen. **31**, 4073–4113 (1998)
96. G. Jank, L. Volkmann, *Meromorphe Funktionen und Differentialgleichungen* (Birkhäuser, Basel, 1985)
97. Y. Jiang, B. Huang, A note on the value distribution of $f^l(f^{(k)})^n$. arXiv:1405.3742.v1 [math.CV] (2014)
98. N. Joshi, A. Kitaev, On Boutroux's tritronqée solutions of the first Painlevé equation. Stud. Appl. Math. **107**, 253–291 (2001)
99. T. Kecker, A cubic polynomial Hamiltonian system with meromorphic solutions, in *Computational Methods and Function Theory (CMFT)*, vol. 16 (Springer, Berlin, 2016), pp. 307–317
100. T. Kecker, Polynomial Hamiltonian systems with movable algebraic singularities. J. d'Anal. Math. **129**, 197–218 (2016)

101. S. Krantz, *Function Theory of Several Complex Variables* (AMS Chelsea Publishing, Providence, RI, 1992)
102. I. Laine, *Nevanlinna Theory and Complex Differential Equations*. De Gruyter Studies in Mathematics, vol. 15 (De Gruyter, Boston, 1993)
103. J.K. Langley, G. Shian, On the zeros of certain linear differential polynomials. J. Math. Anal. Appl. **153**, 159–178 (1990)
104. J.K Langley, Proof of a conjecture of Hayman concerning f and f''. J. Lond. Math. Soc. **48**, 500–514 (1993)
105. J.K. Langley, On the zeros of the second derivative. Proc. R. Soc. Edinb. **127**, 359–368 (1997)
106. J.K. Langley, An inequality of Frank, Steinmetz and Weissenborn. Kodai Math. J. **34**, 383–389 (2011)
107. P. Lappan, A criterion for a meromorphic functions to be normal. Comment. Math. Helv. **49**, 492–495 (1974)
108. O. Lehto, K. Virtanen, Boundary behaviour and normal meromorphic functions. Acta Math. **97**, 47–65 (1957)
109. A. Lohwater, Ch. Pommerenke, On normal meromorphic functions. Ann. Acad. Sci. Fenn. Ser. A I **550**, 12 p. (1973)
110. B.J. Lewin [B.Ya. Levin], *Nullstellenverteilung Ganzer Funktionen* [German transl.] (Akademie Verlag, Berlin, 1962)
111. B.Ya. Levin, *Lectures on Entire Functions*. Translations of Mathematical Monographs, vol. 150 (American Mathematical Society, Providence, RI, 1996)
112. S.A. Makhmutov, Distribution of values of meromorphic functions of class $\mathscr{W}_p$. Sov. Math. Dokl. **28**, 758–762 (1983)
113. J. Malmquist, Sur les fonctions à un nombre fini de branches satisfaisant à une équation différentielle du premier ordre. Acta Math. **36**, 297–343 (1913)
114. J. Malmquist, Sur les fonctions à un nombre fini de branches satisfaisant à une equation différentielle du premier ordre. Acta Math. **42**, 59–79 (1920)
115. A.Z. Mokhon'ko, The Nevanlinna characteristics of certain meromorphic functions (Russian). Teor. Funkcii. Funkc. Anal. Prilozen **14**, 83–87 (1971)
116. A.Z. Mokhon'ko, V.D. Mokhon'ko, Estimates for the Nevanlinna characteristics of some classes of meromorphic functions and their applications to differential equations. Sib. Math. J. **15**, 921–934 (1974)
117. E. Mues, Über eine Defekt- und Verzweigungsrelation für die Ableitung meromorpher Funktionen. Manuscripta Math. **5**, 275–297 (1971)
118. E. Mues, Zur Faktorisierung elliptischer Funktionen. Math. Z. **120**, 157–164 (1971)
119. E. Mues, Über ein Problem von Hayman. Math. Z. **164**, 239–259 (1979)
120. E. Mues, Meromorphic functions sharing four values. Complex Variables **12**, 169–179 (1989)
121. E. Mues, R. Redheffer, On the growth of logarithmic derivatives. J. Lond. Math. Soc. **8**, 412–425 (1974)
122. E. Mues, N. Steinmetz, The theorem of Tumura–Clunie for meromorphic functions. J. Lond. Math. Soc. **23**, 113–122 (1981)
123. T. Muir, *A Treatise on the Theory of Determinants* (Dover, New York, 1960)
124. R. Nevanlinna, Zur Theorie der meromorphen Funktionen. Acta. Math. **46**, 1–99 (1925)
125. R. Nevanlinna, Einige Eindeutigkeitssätze in der Theorie der meromorphen Funktionen. Acta. Math. **48**, 367–391 (1926)
126. R. Nevanlinna, *Le théorème de Picard–Borel et la théorie des fonctions méromorphes* (Gauthier-Villars, Paris, 1929)
127. R. Nevanlinna, Über Riemannsche Flächen mit endlich vielen Windungspunkten. Acta. Math. **58**, 295–273 (1932)
128. R. Nevanlinna, *Eindeutige Analytische Funktionen* (Springer, Berlin, 1936)
129. V. Ngoan, I.V. Ostrovskii, The logarithmic derivative of a meromorphic function (Russian). Akad. Nauk. Armjan. SSR Dokl. **41**, 742–745 (1965)
130. K. Noshiro, Contributions to the theory of meromorphic functions in the unit-circle. J. Fac. Sci. Hokkaido Univ. **7**, 149–159 (1938)

131. K. Okamoto, On the τ-function of the Painlevé equations. Physica D **2**, 525–535 (1981)
132. K. Okamoto, K. Takano, The proof of the Painlevé property by Masuo Hukuhara. Funkcial. Ekvac. **44**, 201–217 (2001)
133. C.F. Osgood, Sometimes effective Thue–Siegel–Roth–Schmidt–Nevanlinna bounds, or better. J. Number Theory **21**, 347–389 (1985)
134. P. Painlevé, *Lecons sur la théorie analytique des équations différentielles, profesées à Stockholm* (Paris, 1897)
135. P. Painlevé, Mémoire sur les équations différentielles dont l'intégrale générale est uniforme. Bull. Soc. Math. Fr. **28**, 201–261 (1900)
136. X. Pang, Bloch's principle and normal criterion. Sci. China Ser. A **32**, 782–791 (1989)
137. X. Pang, On normal criterion of meromorphic functions. Sci. China Ser. A **33**, 521–527 (1990)
138. X. Pang, Y. Ye, On the zeros of a differential polynomial and normal families. J. Math. Anal. Appl. **205**, 32–42 (1997)
139. X. Pang, L. Zalcman, On theorems of Hayman and Clunie. N. Z. J. Math. **28**, 71–75 (1999)
140. G. Pólya, G. Szegö, *Aufgaben und Lehrsätze aus der Analysis I, II* (Springer, Berlin, 1970/1971)
141. Ch. Pommerenke, Estimates for normal meromorphic functions. Ann. Acad. Sci. Fenn. Ser. A I **476**, 10 p. (1970)
142. M. Reinders, Eindeutigkeitssätze für meromorphe Funktionen, die vier Werte teilen. Mitt. Math. Sem. Giessen **200**, 15–38 (1991)
143. M. Reinders, A new example of meromorphic functions sharing four values and a uniqueness theorem. Complex Variables **18**, 213–221 (1992)
144. M. Reinders, A new characterisation of Gundersen's example of two meromorphic functions sharing four values. Results Math. **24**, 174–179 (1993)
145. A. Ros, The Gauss map of minimal surfaces, in *Differential Geometry, Valencia 2001* (World Scientific Publishing Co., River Edge, NJ, 2002), pp. 235–252
146. L.A. Rubel, *Entire and Meromorphic Functions*. Springer Universitext (Springer, New York, 1996)
147. E. Rudolph, Über meromorphe Funktionen, die vier Werte teilen, Diploma Thesis, Karlsruhe (1988)
148. J. Schiff, *Normal Families*. Springer Universitext (Springer, New York, 1993)
149. H. Selberg, Über die Wertverteilung der algebroiden Funktionen. Math. Z. **31**, 709–728 (1930)
150. T. Shimizu, On the theory of meromorphic functions. Jpn. J. Math. **6**, 119–171 (1929)
151. S. Shimomura, Painlevé property of a degenerate Garnier system of (9/2)-type and of a certain fourth order non-linear ordinary differential equation. Ann. Scuola Norm. Sup. Pisa Cl. Sci. **XXIX**, 1–17 (2000)
152. S. Shimomura, Proofs of the Painlevé property for all Painlevé equations. Jpn. J. Math. **29**, 159–180 (2003)
153. S. Shimomura, Growth of the first, the second and the fourth Painlevé transcendents. Math. Proc. Camb. Phil. Soc. **134**, 259–269 (2003)
154. S. Shimomura, Poles and α-points of meromorphic solutions of the first Painlevé hierarchy. Publ. RIMS Kyoto Univ. **40**, 471–485 (2004)
155. S. Shimomura, Lower estimates for the growth of the fourth and the second Painlevé transcendents. Proc. Edinb. Math. Soc. **47**, 231–249 (2004)
156. K. Shin, New polynomials P for which $f'' + P(z)f = 0$ has a solution with almost all real zeros. Ann. Acad. Sci. Fenn. **27**, 491–498 (2002)
157. G.D. Song, J.M. Chang, Meromorphic functions sharing four values. Southeast Asian Bull. Math. **26**, 629–635 (2002)
158. L. Sons, Deficiencies of monomials. Math. Z. **111**, 53–68 (1969)
159. R. Spigler, The linear differential equation whose solutions are the products of solutions of two given differential equations. J. Math. Anal. Appl. **98**, 130–147 (1984)
160. N. Steinmetz, Zur Theorie der binomischen Differentialgleichungen. Math. Ann. **244**, 263–274 (1979)

161. N. Steinmetz, Ein Malmqistscher Satz für algebraische Differentialgleichungen erster Ordnung. J. Reine Angew. Math. **316**, 44–53 (1980)
162. N. Steinmetz, Über die Nullstellen von Differentialpolynomen. Math. Z. **176**, 255–264 (1981)
163. N. Steinmetz, Über eine Klasse von Painlevéschen Differentialgleichungen. Arch. Math. **41**, 261–266 (1983)
164. N. Steinmetz, Eine Verallgemeinerung des zweiten Nevanlinnaschen Hauptsatzes. J. Reine Angew. Math. **368**, 134–141 (1986)
165. N. Steinmetz, Ein Malmquistscher Satz für algebraische Differentialgleichungen zweiter Ordnung. Results Math. **10**, 152–166 (1986)
166. N. Steinmetz, On the zeros of $(f^{(p)} + a_{p-1}f^{(p-1)} + \cdots + a_0 f)f$. Analysis **7**, 375–389 (1987)
167. N. Steinmetz, Meromorphe Lösungen der Differentialgleichung $Q(z,w)\frac{d^2w}{dz^2} = P(z,w)\left(\frac{dw}{dz}\right)^2$. Complex Variables **10**, 31–41 (1988)
168. N. Steinmetz, A uniqueness theorem for three meromorphic functions. Ann. Acad. Sci. Fenn. **13**, 93–110 (1988)
169. N. Steinmetz, On the zeros of a certain Wronskian. Bull. Lond. Math. Soc. **20**, 525–531 (1988)
170. N. Steinmetz, Meromorphic solutions of second order algebraic differential equations. Complex Variables **13**, 75–83 (1989)
171. N. Steinmetz, Exceptional values of solutions of linear differential equations. Math. Z. **201**, 317–326 (1989)
172. N. Steinmetz, Linear differential equations with exceptional fundamental sets. Analysis **11**, 119–128 (1991)
173. N. Steinmetz, Linear differential equations with exceptional fundamental sets II. Proc. Am. Math. Soc. **117**, 355–358 (1993)
174. N. Steinmetz, *Iteration of Rational Functions. Complex Analytic Dynamical Systems*. De Gruyter Studies in Mathematics, vol. 16 (Walter de Gruyter, Berlin, 1993)
175. N. Steinmetz, On Painlevé's equations I, II and IV. J. d'Anal. Math. **82**, 363–377 (2000)
176. N. Steinmetz, Value distribution of the Painlevé transcendents. Isr. J. Math. **128**, 29–52 (2002)
177. N. Steinmetz, Zalcman functions and rational dynamics. N. Z. J. Math. **32**, 1–14 (2003)
178. N. Steinmetz, Normal families and linear differential equations. J. d'Anal. Math. **117**, 129–132 (2012)
179. N. Steinmetz, The Yosida class is universal. J. d'Anal. Math. **117**, 347–364 (2012)
180. N. Steinmetz, Sub-normal solutions to Painlevé's second differential equation. Bull. Lond. Math. Soc. **45**, 225–235 (2013)
181. N. Steinmetz, Reminiscence of an open problem. Remarks on Nevanlinna's four-points theorem. South East Asian Bull. Math **36**, 399–417 (2012)
182. N. Steinmetz, Complex Riccati differential equations revisited. Ann. Acad. Sci. Fenn. **39**, 503–511 (2014)
183. N. Steinmetz, Remark on meromorphic functions sharing five pairs. Analysis **36**, 169–179 (2016)
184. N. Steinmetz, An old new class of meromorphic functions. Preprint (2014), to appear in J. d'Anal. Math.
185. N. Steinmetz, First order algebraic differential equations of genus zero. Bull. Lond. Math. Soc. **49**, 391–404 (2017). doi:10.1112/blms.12035
186. N. Steinmetz, A unified approach to the Painlevé transcendents. Ann. Acad. Sci. Fenn. **42**, 17–49 (2017)
187. W. Sternberg, Über die asymptotische Integration von Differentialgleichungen. Math. Ann. **81**, 119–186 (1920)
188. E.C. Titchmarsh, *Eigenfunction Expansions Associated with Second-Order Differential Equations, Part I*, 2nd edn. (Oxford University Press, London, 1962)
189. M. Tsuji, On the order of a meromorphic function. Tôhoku Math. J. **3**, 282–284 (1951)
190. H. Ueda, Some estimates for meromorphic functions sharing four values. Kodai Math. J. **17**, 329–340 (1994)

191. G. Valiron, Sur le théorème de M. Picard. Enseignment **28**, 55–59 (1929)
192. G. Valiron, Sur la dérivée des fonctions algebroides. Bull. Soc. Math. Fr. **59**, 17–39 (1931)
193. G. Valiron, *Lectures on the General Theory of Integral Functions* (Chelsea Publishing, New York, 1949)
194. S.P. Wang, On meromorphic functions that share four values. J. Math. Anal. Appl. **173**, 359–369 (1993)
195. Y. Wang, On Mues conjecture and Picard values. Sci. China Ser. A **36**, 28–35 (1993)
196. J.P. Wang, Meromorphic functions sharing four values. Indian J. Pure Appl. Math. **32**, 37–46 (2001)
197. W. Wasow, *Asymptotic Expansions for Ordinary Differential Equations* (Wiley, New York, 1965)
198. G. Weissenborn, The theorem of Tumura and Clunie. Bull. Lond. Math. Soc. **18**, 371–373 (1986)
199. J.M. Whittaker, The order of the derivative of a meromorphic function. Proc. Lond. Math. Soc. **40**, 255–272 (1936)
200. A. Wiman, Über den Zusammenhang zwischen dem Maximalbetrage einer analytischen Funktion und dem größten Gliede der zugehörigen Taylorschen Reihe. Acta Math. **37**, 305–326 (1914)
201. H. Wittich, Eindeutige Lösungen der Differentialgleichungen $w'' = P(z, w)$. Math. Ann. **125**, 355–365 (1953)
202. H. Wittich, *Neuere Untersuchungen über eindeutige Analytische Funktionen* (Springer, Berlin, 1968)
203. K. Yamanoi, The second main theorem for small functions and related problems. Acta Math. **192**, 225–294 (2004)
204. K. Yamanoi, Defect relation for rational functions as targets. Forum Math. **17**, 169–189 (2005)
205. K. Yamanoi, Zeros of higher derivatives of meromorphic functions in the complex plane. Proc. Lond. Math. Soc. **106**, 703–780 (2013)
206. S. Yamashita, On K. Yosida's class (A) of meromorphic functions. Proc. Jpn. Acad. **50**, 347–378 (1974)
207. N. Yanagihara, Meromorphic solutions of some difference equations. Funkc. Ekvac. **23**, 309–326 (1980)
208. K. Yosida, A generalisation of a Malmquist's theorem. Jpn J. Math. **9**, 253–256 (1932)
209. K. Yosida, On algebroid solutions of ordinary differential equations. Jpn. J. Math. **10**, 253–256 (1933)
210. K. Yosida, On a class of meromorphic functions. Proc. Phys. Math. Soc. Jpn. **16**, 227–235 (1934)
211. K. Yosida, A note on Malmquist's theorem on first order algebraic differential equations. Proc. Jpn. Acad. **53**, 120–123 (1977)
212. L. Zalcman, A heuristic principle in function theory. Am. Math. Monthly **82**, 813–817 (1975)
213. L. Zalcman, Normal families: new perspectives. Bull. Am. Math. Soc. **35**, 215–230 (1998)

Index

N. Steinmetz, *Nevanlinna Theory, Normal Families, and Algebraic Differential Equations*, Universitext, DOI 10.1007/978-3-319-59800-0

Printed in the United States
By Bookmasters